Bioresources and Bioprocessing

Bioresources and Bioprocessing

Editor: Angela Santos

R CALLISTO REFERENCE

www.callistoreference.com

Callisto Reference,
118-35 Queens Blvd., Suite 400,
Forest Hills, NY 11375, USA

Visit us on the World Wide Web at:
www.callistoreference.com

ISBN: 978-1-64116-154-1 (Hardback)

Cataloging-in-Publication Data

Bioresources and bioprocessing / edited by Angela Santos.
    p. cm.
Includes bibliographical references and index.
ISBN 978-1-64116-154-1
1. Biotechnology. 2. Biochemical engineering. 3. Biodiversity. I. Santos, Angela.
TP248.2 .B56 2019
660.6--dc23

Table of Contents

Preface

This book has been an outcome of determined endeavour from a group of educationists in the field. The primary objective was to involve a broad spectrum of professionals from diverse cultural background involved in the field for developing new researches. The book not only targets students but also scholars pursuing higher research for further enhancement of the theoretical and practical applications of the subject.

Bioresource engineering deals with modifying biological and agricultural feedstock to create usable products. It covers all aspects of biomass, bioremediation, biological waste treatment, bioenergy, etc. Various biochemical conversion technologies such as aerobic methods, anaerobic digestion, enzymatic method and composting are also within the scope of this field. The products developed from bioresource engineering include fibers, fuels, fertilizers, polymers, etc. Research in bioresource engineering explores animal manure management, wastewater irrigation and storm water management, besides others. The process that uses living cells and their components to develop products is called a bioprocess. The fundamental theories that are central to biological processes are the transport of energy and mass. Bioprocessing can be classified as cell bioprocessing, upstream bioprocessing and downstream bioprocessing. This book discusses the fundamentals as well as modern approaches of bioresource engineering and bioprocessing. It strives to provide a fair idea about these disciplines and to help develop a better understanding of the latest advances within these fields. Researchers and students in these fields will be assisted by this book.

It was an honour to edit such a profound book and also a challenging task to compile and examine all the relevant data for accuracy and originality. I wish to acknowledge the efforts of the contributors for submitting such brilliant and diverse chapters in the field and for endlessly working for the completion of the book. Last, but not the least; I thank my family for being a constant source of support in all my research endeavours.

Editor

Hydrolysis of carotenoid esters from *Tagetes erecta* by the action of lipases from *Yarrowia lipolytica*

Abraham Figueiras Abdala[1], Alfonso Pérez Gallardo[2], Lorenzo Guevara Olvera[3] and Eleazar Máximo Escamilla Silva[1*]

Abstract

The present study was conducted to evaluate the feasibility of enzymatic hydrolysis of carotenoid esters from *Tagetes erecta* using lipases from the yeast of *Yarrowia lipolytica*, with the aim of obtaining free lutein. The optimal concentrations of seven nutrients, considering the production of lipases relative to biomass ($Y_{p/x}$) as the response variable, were determined in flask fermentations. In addition, we studied the effect on hydrolysis of growing *Y. lipolytica* in the presence of the oleoresin of the marigold flower in flask and stirred tank. Furthermore, hydrolysis of the oleoresin using the lipases from this microorganism was compared with the hydrolysis using lipases from *Rhizopus oryzae*. Cultured in the presence of marigold oleoresin, *Y. lipolytica* showed an increase in free carotenoids of 12.41% in flask and 8.8% in stirred tank, representing a fourfold and a threefold increase compared to the initial value in the fermentation, respectively. When lipases from the supernatant from both microorganisms were used for only 14 h hydrolysis experiments, a slight increase was achieved compared to a blank. We concluded that carotenoid esters of the oleoresin could not be completely hydrolyzed in 14 h by these lipases, but that growing *Y. lipolytica* in the presence of marigold oleoresin gives until fourfold production of free carotenoids in 72 h fermentations.

Keywords: *Yarrowia lipolytica*, *Tagetes erecta*, Enzymatic hydrolysis, Carotenoid esters, Free lutein

Background

Marigold flower (*Tagetes erecta*) is a plant capable of synthesizing various carotenoids of which, once extracted, lutein esters make up around 72% of them (Lim 2014). Studies aiming to increase the extraction yield of lutein esters have encouraged researches about the effect of enzymatic pretreatments to degrade cell walls and membranes in marigold flower (Barzana et al. 2002; Benitez-Garcia et al. 2014; Delgado-Vargas and Paredes-López 1997; Navarrete-Bolanos et al. 2004); however, this native plant of Mexico is mainly used as an ornament. The type and proportion of carotenoids of the plant depend on the variety being mostly lutein and zeaxanthin in yellow flowers, while for white flowers it is mostly lutein, zeaxanthin, β-cryptoxanthin and β-carotene (Benitez-Garcia et al. 2014). Studies showed that hydrolysis of zeaxanthin esters to achieve their free forms enhances the bioavailability of this carotenoid (Chitchumroonchokchai and Failla 2006). Unesterified lutein (free lutein) is the most interesting carotenoid since it is in great demand in the food, pharmaceutical, and cosmetic industries; its commercial market is expected to grow up to US $ 308 million in 2018 (Lim 2014; Lin et al. 2015a). Thus, carotenoid esters are usually subjected to saponification, which consists in making the oleoresin react with concentrated alcohol solutions of potassium or sodium hydroxide. The disadvantages of saponification are the degradation and isomerization of carotenoids, as well as the power costs and the need to implement safety measures for handling corrosive chemicals and waste products. In addition, the food industry is always looking for more natural alternatives for obtaining their products. Some studies have aimed to replace saponification with

*Correspondence: eleazar@iqcelaya.itc.mx
[1] Departamento de Ingeniería Química, Instituto Tecnológico de Celaya, Av. Tecnológico y A.G. Cubas s/n, 38010 Celaya, Gto, Mexico
Full list of author information is available at the end of the article

the aid of lipases from different microorganisms, but few studies have used marigold oleoresin (Zorn et al. 2003). After pretreatments with bile salts and protease from *Streptomyces griseus*, mature human milk was treated with lipases from *Candida rugosa* in order to hydrolyze retinyl esters to obtain free β-carotene and retinol from milk; however, this method still required further chemical hydrolysis (Liu et al. 1998). Hydrolysis of esters of astaxanthin was achieved using the non-specific cholesterol esterase which has been demonstrated to hydrolyze vitamins (Howles and Hui 2001; Jacobs et al. 1982). Carboxyl ester lipase (cholesterol esterase) achieved high activity for papaya and loquat extracts but low activity in incubations with paprika and marigold oleoresins. A porcine pancreatic lipase and a lipase from *Candida rugosa* was also tested and showed some activity on xanthophyll extracts (Breithaupt et al. 2002). Alkali labile carotenoids were hydrolyzed with a pig liver esterase converting astaxanthin dipalmitate to the monopalmitate and free astaxanthin (Aakermann et al. 1996). Astaxanthin, from *Haematococcus pluvialis* algal cell extracts, was effectively hydrolyzed by 5 fungal lipases in Tween 80-emulsified systems; under optimal conditions of pH, temperature, reaction time, and lipase dosage, free astaxanthin recoveries of 63.2% were achieved (Zhao et al. 2011). A process for enzymatic hydrolysis of carotenoid esters and other esters with aims of human and animal consumption has been presented as a patent; this process consists in the following: (1) incubate the esters with ester- cleaving lipases, and (2) appropriately isolate the resulting free forms (Flachmann et al. 2005). *Yarrowia lipolytica* is a yeast that has the potential industrial application of producing α-ketonic, acetic, citric, isocitric, pyruvic, and succinic acid; furthermore, it produces extracellular enzymes such as proteases and lipases of great industrial interest (Fletcher 2006; Gajdos et al. 2015); recently, metabolically engineered *Y. lipolytica* has been applied in fermentations where omega-3 eicosapentaenoic acid, a fatty acid with a wide range of health benefits, is produced by carefully balancing expression levels of pathway enzymes and modifying fatty acid and lipid metabolism (Xie et al. 2015). This microorganism has the ability to use fatty acids as a carbon source but the metabolism of these hydrophobic compounds is not yet fully understood; nevertheless, there are various proposed mechanisms in the literature for the use of fatty acids and alcohols by *Y. lipolytica* (Fickers et al. 2005; Fukuda and Ohta 2013; Hirakawa et al. 2009). It is considered a "safe to use" organism either as final product or as Yarrowia-derived product (Groenewald et al. 2014). Because of the safety of its use and its industrial interest, this work aims to explore the use of lipases produced by this microorganism to assess the lutein esters hydrolysis

from an industrially obtained oleoresin into free lutein in flasks and 7 L stirred tank with 7 optimized nutrients. These results will also be compared with the use of lipases from *Rhizopus oryzae*.

Methods

All of the reagents were reagent grade obtained from Sigma Chemical Co. (St. Louis, MO) and solvents were ACS grade unless otherwise specified.

Biological material

The microorganism used for this study was the non-pathogenic yeast *Yarrowia lipolytica* strain CX39-74B (Strain number: ATCC 32339), which has a regulated dimorphism growth (Guevara-Olvera et al. 1993).

Propagation of the strain of *Yarrowia lipolytica*

The strain was propagated by taking a sample by swab from a colony and streaking on YPD agar (yeast extract, peptone, dextrose; 1, 1, 2%). It was incubated for 24 h at 30 °C. Afterwards, 10 petri dishes with the yeast culture were stored as reserve at 4 °C. These reserve petri dishes were reseeded each month to prevent the aging of the culture. The preculture of *Y. lipolytica* in sterile saline solution was carried out in 500 mL Erlenmeyer flasks with 200 mL of YPD broth for 24 h at 30 °C and 250 rpm.

Pigments

A sample of marigold oleoresin used in this work was donated by the company ALCOSA S.A. de C.V. (Celaya, Gto. Mexico); it had an average content of xanthophyll carotenoids (equivalent to all-trans-lutein) of 137.59 g/kg according to the official AOAC method 430.18, used by the company for quality control. It was stored under refrigeration at 4 °C in a sealed container until use.

Analytical methods
Cell growth

The growth of cells in the culture broths was measured by the dry weight technique. A known volume of broth (5 mL) was filtered through a cellulose membrane (Merck ®) with a pore size of 0.22 μm previously dried to constant weight. Subsequently, the membrane was placed in an oven at 90 °C until a constant weight was obtained, and the weight difference was expressed as grams of dry cell weight per liter. In the case of the samples of culture media with oil, a known volume of broth was centrifuged in 50 mL tubes at 2600 rpm for 10 min; afterwards, the supernatant was discarded and the pellet was resuspended in distilled water. The process was repeated two more times. The final pellet was resuspended and filtered through the cellulose membrane. Cell growth was calculated by the weight difference.

Enzyme activity

The lipolytic activity was measured by the increase in absorbance at 401 nm caused by the release of ρ-nitrophenol as a result of the hydrolysis of ρ-nitrophenol palmitate by the action of the lipases present in the culture medium, using a Lambda 2 Perkin Elmer spectrophotometer. To perform the reaction, the sample of culture broth was first filtered through a cellulose membrane with a pore diameter of 0.22 μm to remove the suspended solids. This filtrate was the crude extract of the extracellular lipase enzyme. We added 2.5 mL of solution A (0.1 mM phosphate buffer solution, pH 7.8, and 100 mg of arabic gum per 90 mL) to a reaction tube, as well 0.2 mL of solution B (40.5 mg added of ρ-nitrophenol palmitate in 10 mL of isopropyl alcohol). This was placed in a water bath at 37 °C for 5 min. Subsequently, we added 0.3 mL of the crude extract and vortexed for 20 s. The mixture was placed again into the water bath for 10 min; after that time, the reaction was stopped with 2.5 mL of ethanol. The liquid was filtered through a nylon membrane with a pore size of 0.22 μm and the absorbance was read at 401 nm. Comparing the result with a calibration curve allowed us to determine the amount of released ρ-nitrophenol. Lipase units are defined as the μmols of ρ-nitrophenol released per minute per ml of sample.

Saponification

In order to carry out the saponification of the carotenoids contained in the oleoresin, we first extracted the pigments from 50 mg of the oleoresin using 30 mL of HEAT solution (hexane, ethanol, acetone, toluene; 10/7/6/7; v/v/v/v) in a 100 mL volumetric flask. Subsequently, we added 2 mL of methanolic potassium hydroxide, stirred for 1 min and heated to 56 °C for 20 min. The neck of the flask was connected to a condenser to prevent evaporation and loss of solvent. After the heating time had passed, the flask was left to cool and kept in the dark for 1 h. We then added 30 mL of hexane, stirred the mix, and filled the flask to the calibration mark with a solution of Na_2SO_4 (10%). Before performing any analysis, we allowed the mix to stand in the dark for 1 h. To assess the effect of saponification on pigments, we performed a thin layer chromatography. For this, we used silica gel plates (60 F254; 5 × 10 cm) for analytical chromatography. After drawing the reference line at 0.5 cm from the bottom, we placed 20 μL of each sample solution in the plates. The concentration of the unsaponified oleoresin solution was 0.5 mg/mL, while the concentration of the saponified solution was 1 mg/mL. The sample was eluted with increasing proportions of a mixture of hexane and acetone (80/20, v/v) within a closed chamber. Before the stains migrated to the upper end, the plate was removed and dried with nitrogen flow.

Extraction, characterization, and quantification of pigments

For the sample characterization according to its degree of esterification and amount of pigments, we used the AOAC official method 970.64 with the modifications introduced by Fletcher (2006). The difference with the original method is that the saponification step is omitted and larger volumes of elution solvents are used; also, the final dilution is greater than in the original method. With this method we were able to separate the fractions, those which had their xanthophylls quantified and those which had their degree of esterification measured. Thus, the pigments were classified as di-esters (E), monohydroxy pigments (MHP), free pigments or dihydroxy pigments (DHP), and residual pigments or polyols (RP). The content of xanthophylls was expressed as equivalents of trans-luteins. To extract the pigments, we dissolved 50 mg of oleoresin and 5 mL of the culture broth sample in 30 mL of HEAT using a 100 mL volumetric flask, adding 2 mL of methanol. After stirring, we left this mix in the dark for 1 h before conducting any analysis. Subsequently, the pigments were separated by open column chromatography. The column was 12.5 mm i.d. X 30 cm, with Teflon stopcock; the column was 2 mm i.d. and 10 cm long. The stationary phase comprised a mixture of silica gel–diatomaceous earth (1/1, w/w) activated by drying. The activation was carried out by heating in an oven for 24 h at 90 °C. The mixture was then cooled and stored in a desiccator before use. The column was packed with 7 cm of stationary phase and 2 cm of anhydrous sodium sulfate. After the column was packed and connected to a Buchner flask, an aspirator was connected to one of the ends of the flask to initiate the separation; the amount of sample of the extraction solution was 5 mL. We used four eluents: for carotenes and esters, hexane/acetone (96/4, v/v); for MHP, hexane/acetone (90/10, v/v); for DHP, hexane/acetone (80/20, v/v); for RP, hexane/methanol/acetone (80/10/10, v/v/v). The eluents were added in the same order until the corresponding fraction was recovered, which was then immediately transferred to the next solvent. Once the desired fraction was obtained, it was diluted with a known volume (25, 50 or 100 mL) of water, and its absorbance was read at 474 nm using the elution solvent as the blank. After determining the amount of pigments in each band or fraction, we calculated their percentage share of the total. In order to obtain ester-free lutein, we saponified a known amount of oleoresin, followed by open column chromatography in which the third band (corresponding to the DHP) was recovered. Lutein was identified by its spectral properties and its chromatographic behavior. No other tests were carried out because we were working with marigold oleoresin, in which lutein represents about 80% of the pigment content.

Statistical method to investigate the factors effect involved in the production of lipases by *Y. lipolytica* on the composition of the culture medium

The effect of nutrients on the production of lipases was analyzed using a fractionated design 2^{7-4}. The factors and levels of this design are summarized in Table 1. The design also involves a central experiment whose purpose is to determine the average of the levels studied as shown in Table 2. The variables that were measured are biomass (X), expressed as dry cell weight (g/L), and enzyme production, expressed as lipolytic activity in units per liter (U/L). We used these data to determine the fermentation kinetics of all of the experiments by sampling every 8 h for 72 h. The response variable of each treatment was the specific production of lipases or the biomass yield $Y_{p/x}$ (U/g), which was obtained as the slope of the straight line fitted to the data of lipolytic activity (U/L) vs X (g/L), but only from the beginning to the end of the exponential growth phase.

The fermentations of the design were carried out in stirred flasks containing 200 mL of culture medium in duplicate. The temperature was 30 °C and the stirring speed was 200 rpm; Tween 20 (0.01%) was added to each flask to improve the emulsification of the oil. The pH was adjusted to 6.0 with sodium hydroxide at the start of the fermentation. The inoculum consisted of a preculture of *Y. lipolytica* in YEPD medium, of which we added a volume corresponding to 10% of the total fermentation volume. The cells of this culture were harvested in the period between the first third and the end of the exponential growth phase. We took samples every 8 h and we measured the enzymatic activity of lipases and the biomass. These measurements were performed in triplicate. The growth kinetics and the enzyme production of each data point were estimated from the average of the results of these three measurements for every treatment (data not shown). Data analysis was performed using JMP statistical software v.5.0.1.2 (SAS Institute 2015) to determine which factors influence the production of lipases and how they do it.

Lipase production in stirred tank

In order to carry out the fermentations in a more controlled environment, we used a 7 L stirred tank bioreactor (Applikon, The Nederlands). The fermenter conditions were as follows: 4 L working volume, aeration of 0.8 vvm, stirring speed of 250 rpm, temperature of 30 °C, and pH 6. We inoculated with the strain propagated in YEPD medium using a volume equal to 10% of the working volume. The cells were harvested in the period between the first quarter

Table 1 Factors and levels for the stirred flasks experiments

Factor	Compound	Levels	
		−	+
A	Olive oil	10 g/L	20 g/L
B	Yeast extract	3.0 g/L	5 g/L
C	KH_2PO_4	1.0 g/L	2 g/L
D	$MgSO_4 \cdot 7H_2O$	1.4 g/L	2 g/L
E	NaCl	1.0 g/L	2 g/L
F	$CaCl_2$	0.8 g/L	2 g/L
G	Trace elements solution[a]	2.0 mL	4 mL

[a] Each 100 mL contains: 50 mg of boric acid; 4 mg of $CuSO_4$; 10 mg of KI; 20 mg of $FeCl_3$; 20 mg of $MnCl_2$; 20 mg of NaMoO4; and 40 mg and $ZnSO_4$

Table 2 Treatments of the fractionated design with a central point

Factor		A	B	C	D	E	F	G
No.	Design coordinate	G/L						ml/L
1	−−++−−+	10	3	2	2	1	0.8	4
2	+++++++	20	5	2	2	2	2	4
3	−++−+−−	10	5	2	1.4	2	0.8	2
4	−+−+−+−	10	5	1	2	1	2	2
5	−−−−+++	10	3	1	1.4	2	2	4
6	+−−++−−	20	3	1	2	2	0.8	2
7	++−−−−+	20	5	1	1.4	1	0.8	4
8	0000000	15	4	1.5	1.7	1.5	1.4	3
9	+−+−−+−	20	3	2	1.4	1	2	2

+ and − represent high and low level, respectively, for factors ABCDEFG. 0 stands for a medium level which results from the average of the maximum and minimum value

and the end of the exponential growth phase. Sampling was initially done at 8, 12, and 18 h, finishing with 8-h sampling from 24 to 72 h, measuring lipase enzymatic activity and the biomass. Experiments were done in triplicate and expressed as the mean (SD) (standard deviation).

Fitting the cell growth kinetics of Y. lipolytica to a mathematical model

The data obtained on the kinetics of cell growth were fitted to the logistic equation of population growth described by Verhulst (Peleg et al. 2007) which, applied to microbial growth, takes the form of Eq. 1.

$$X(t) = \frac{X_0 e^{\mu t}}{1 - \left(\frac{X_0}{X\text{max}}\left(1 - e^{\mu t}\right)\right)} \quad (1)$$

This logistic model equation was entered in Microsoft Excel 2007 and, using the solver tool, we obtained the fitted parameters μ, X_0 and $X_{\max}$ that minimized the squared differences between experimental values and calculated values.

Effect of including the oleoresin in the culture medium of Y. lipolytica on the hydrolysis of carotenoid esters

The culture of Y. lipolytica was performed in the presence of 5 g of marigold oleoresin. Experiments were conducted in flask and stirred tank. The oleoresin was added to culture broth that had been previously emulsified. The emulsion composition was water/oleoresin/tween 20 (78/20/2, v/v/v). The yeast was cultured in 500 mL Erlenmeyer flasks using a working volume of 200 mL. In the preparation of the culture medium, we considered the volume and dilution that would increase the emulsion of marigold oleoresin. The medium composition was based on the results of the experimental design carried out in flasks. The culture conditions were temperature of 30 °C, stirring speed of 250 rpm, and pH 6 (adjusted with NaOH at the start of the fermentation). The inoculum volume was 10% of the working volume. Samples were taken every 12 h, assessing the amount of total carotenoid pigments and the fractions representing MHP, DHP, and RP. Three independent experiments were carried out; the assessments were made in triplicate and expressed as mean (SD). The culture conditions for experiments carried out in stirred tank were working volume of 4 L, aeration of 0.8 vvm, stirring speed of 250 rpm, temperature of 30 °C, and pH 6 adjusted at the beginning of fermentation. Samples were taken every 12 h, assessing the amount of total carotenoid pigments and the fractions representing E, MHP, DHP, and RP. Three independent experiments were carried out; the assessments were done in triplicate and expressed as mean (SD).

Comparison between the activity of the lipases from Yarrowia lipolytica and from commercial Rhizopus oryzae in the hydrolysis of carotenoids esters from the marigold flowers

Tests were made in flasks to compare the activity of lipases produced by Y. lipolytica in the emulsified oleoresin, and the activity of commercial lipases produced by R. oryzae (Fluka, BioChemika). Yarrowia lipolytica was cultured in a medium with a composition based on the results of the experimental design carried out in flasks. The supernatant was harvested during the phase of greater lipolytic activity (48 h); 50 mL of this was taken and filtered through cellulose membrane with a pore size of 0.22 µm. The filtrate was placed in a 250 mL Erlenmeyer flask, adding $CaCl_2$ to a concentration of 20 mM. Immediately, afterwards, we added approximately 50 mg of emulsified oleoresin in 6 g of triton X-100. The flask was placed under stirring at 150 rpm and 30 °C for 14 h. The hydrolysis test performed with lipases produced by R. oryzae was carried out as follows: 50 mL of phosphate buffer solution 0.1 M at pH 7.9 was added to a 250 mL Erlenmeyer flask, together with 5 mL of a solution containing the enzyme dissolved in distilled water at a concentration of 4 mg/mL. We then added about 50 mg of emulsified oleoresin in 6 g of triton X-100. The flask was stirred at 150 rpm and 30 °C for 14 h. The blank was prepared with 50 ml of distilled water and approximately 50 mg of emulsified oleoresin in 6 g of triton X-100 under the same experimental conditions. In both experiments, once the reaction time was complete, the pigments were extracted and quantified according to AOAC official method 970.64 with the modifications introduced by Fletcher (2006), which were described above. Three experiments were conducted. The assessments were done in triplicate and a variance analysis was done at the end to detect differences between treatments.

Results and discussion

Effect of the factors involved in the production of lipases by Y. lipolytica

In order to stablish the best culture medium, we performed a kinetic study aiming to determine the time to achieve the maximum value of the enzyme activity and the growth curve for this microorganism. After the model was fitted to the experimental data, we performed the regression formerly described to obtain the specific production of lipase ($Y_{p/x}$). Table 3 shows the model parameters and the $Y_{p/x}$ values for the 18 experiments, where we can see that the highest $Y_{p/x}$ was achieved in treatment 3.

Figure 1 shows the behavior of the culture of Y. lipolytica for treatment 3; it can be seen that the time needed

Table 3 Experimental and fitted data for the experimental design 2^{7-4}

Treatment	Design coordinate[d]	Lipolytic activity[a] U/L	Biomass[a] X_{max} (g/L)	μ[c] h^{-1}	Specific yield of lipases[b] $Y_{p/x}$ (U/g)
3	−+ +−+−−	1153.6	3.556	0.080	510.56
1	−−+ +−−+	944.5	4.972	0.122	254.24
2	+++ ++++	1029.06	4.589	0.123	301.39
2	+++ ++++	1229.4	4.179	0.124	375.68
3	−+ +−+−−	1273.4	3.428	0.086	507.64
4	−+−+−+−	1111.7	4.391	0.100	391.71
1	−−+ +−−+	1023.3	4.909	0.124	220.09
5	−−−−+++	945.4	4.181	0.126	287.19
6	+−−+ +−−	1018.0	4.819	0.149	253.20
7	++−−−−+	981.9	4.598	0.118	310.47
4	−+−+−+−	1121.2	4.345	0.097	338.53
5	−−−−+++	936.0	4.026	0.129	276.97
8	0	947.9	5.106	0.109	233.71
6	+−−+ +−−	1030.0	4.256	0.129	324.98
7	++−−−−+	1050.0	4.509	0.128	262.01
8	0	1100.4	4.521	0.094	313.86
9	+−+−−+−	805.1	5.425	0.122	174.16
9	+−+−−+−	888.3	4.594	0.157	284.31

[a] Measures at 40 h of culturing

[b] Computed through the linear regression of lipolytic activity (U/L) vs X (g/L)

[c] Computed through the logistic model

[d] + and − represent high and low level, respectively, for factors ABCDEFG. 0 stands for a medium level which results from the average of the maximum and minimum value

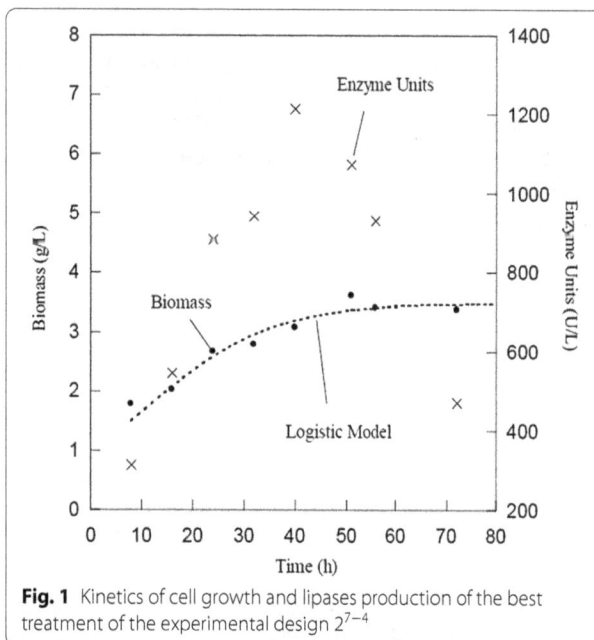

Fig. 1 Kinetics of cell growth and lipases production of the best treatment of the experimental design 2^{7-4}

for reaching maximum lipolytic activity (1213.5 U/L) is around 40 h. Similar lipolytic activity values were obtained by Pereira-Meirelles et al. (1997). It can also be seen that

the maximum production of the enzyme coincides with the end of the exponential growth phase; for *Y. lipolytica*, this tendency may be explained by the extracellular lipase association with the cell membrane at the beginning of the growth phase and the release of lipases in the culture media at the end of the growth phase as confirmed in the literature by Western blot analysis (Fickers et al. 2004). However, the lipolytic activity decreased with the onset of the stationary growth phase which is probably due to the Zn^{2+} inhibitory effect since this ion is present in the trace elements solution (Fickers et al. 2013), caused by proteolysis as demonstrated in previous literature by adding a serine protease inhibitor (Pereira-Meirelles et al. 1997) or caused by a combination of both effects. The adjusted value of the X_{max} fitted for the model was 3.49 g/L. The obtained μ (0.083 h^{-1}) indicated that growth was slow and that under these conditions the yeast would take longer to reach its maximum development. The specific production of lipase ($Y_{p/x}$) used as a response variable was 508 U/g. Using the statistical program JMP, we estimated $Y_{p/x}$ of 504 U/g resulting in a 0.8% of error between estimation and experimental value.

The analysis of variance for the 18 treatments of the design (Table 4) in combination with the effect of the factors shown in Table 5 allowed us to discern that olive oil, yeast extract, NaCl, and trace elements solution concentration produced

Table 4 Analysis of variance of the design

Source	Degrees of freedom	Sum of squares	Mean square	F
Model	7	85,682.80	12,240.4	5.69*
Error	10	21,526.57	2152.7	
Total	17	107,209.37		

* Significant differences ($p < 0.05$)

Table 6 One-way analysis of variance

Source	Degrees of freedom	Sum of squares	Mean square	F
Design	8	116,760.21	14595.0	7.36*
Error	9	17,846.22	1982.9	
Total	17	134,606.43		

* Significant differences ($p \leq 0.05$)

Table 5 Effect of the factors selected for the experimental design 2^{7-4} on lipase production in terms of biomass ($Y_{p/x}$)

Factor	Degrees of freedom	Sum of squares	F
Olive oil	1	15,645.63	7.39*
Yeast extract	1	53,274.41	25.16*
KH_2PO_4	1	2102.45	0.99
$MgSO_4$	1	1464.78	0.69
NaCl	1	22,626.92	10.69*
$CaCl_2$	1	2852.895	1.35
Solution of trace mineral nutrients	1	15,466.031	7.3*

* Significant differences ($p \leq 0.05$)

significant differences ($p < 0.05$). The study of the effects of the concentrations of the selected nutrients allowed us to propose that the best composition of the culture medium of *Y. lipolytica* (the one with the greatest specific activity) was 10 g/L of olive oil, 5 g/L of yeast extract, 2 g/L of KH_2PO_4, 1.4 g/L of $MgSO_4$, 2 g/L of NaCl, 0.8 g of $CaCl_2$, and 2 mL of the solution of trace mineral nutrients.

A one-way direction ANOVA was studied and results are summarized in Table 6. From this, we can conclude that at least one of the means is significantly different from others. Moreover, a Tukey test was used in order to find means that are significantly different from each other (Table 7); we could determine that the mean of treatment 3 was significantly different ($p < 0.05$) from the means of the other treatments.

Confirmatory tests for lipase production in stirred flask
The fermentations in stirred flask under the optimal medium composition (see Additional file 1: Figure S1; Table S1) confirmed that the highest lipolytic activity occurs around 40 h. The $Y_{p/x}$ was 481.2 U/g indicating that the model has a relative error of 4.8% for the confirmatory tests. The lipolytic activity obtained at 40 h was 1077 U/L, the maximum cell growth reached a value of 3.01 g/L, and the μ obtained was 0.127 h^{-1} was obtained.

Results of the production of lipases in stirred tank
Table 8 shows the results of cell growth and enzyme production in the culture of *Y. lipolytica* in stirred tank.

We obtained a maximum biomass of 8.9 g/L and μ of 0.236 h^{-1}. The growth rate, higher than that obtained in the flask fermentations, indicated that the culture evolved rapidly. The maximum lipolytic activity of 1598 U/L was obtained at 40 h. Similar lipolytic activities were achieved by other authors (Pereira-Meirelles et al. 1997); however, an analysis of Fig. 2 indicated that there were no large increases in enzyme units between 24 and 40 h. Under these conditions, the microorganism achieved greater cell concentration and enzyme production than in stirred flask fermentations. Furthermore, the lipolytic activity decreased as the stationary phase continued beyond 40 h. Yield productivity in terms of biomass ($Y_{p/x}$) was lower (113.8 U/g); thereby, the cell concentration increased. The increase in cell density and lipolytic activity compared to stirred flask experiments was attributed to the culture conditions in stirred tank, since we added stirring, aeration, and better temperature control. Alonso et al. (2005) pointed out the importance of oxygen in cell development; a higher content of oxygen accelerates the consumption of lipids, leading to increased biomass. In addition, stirring results in better dispersion of hydrophobic particles in the medium, enhancing their bioavailability. In contrast with our findings, Pereira-Meirelles et al. (2000) reported that the released lipase achieved its maximum concentration in the late stationary phase, and not in the beginning.

Characterization of marigold oleoresin
The results of thin layer chromatography showed that carotenoid esters were eluted faster than free carotenoids when silica gel was used as stationary phase. Table 9 summarizes the results of the fractionation by open column chromatography. This separation of carotenoid pigments, based on their degree of esterification, showed a predominance of di-esters, which represented 88.4% of the total. The monohydroxy and dihydroxy pigments represented 7.9 and 2.7%, respectively. The result of total xanthophylls (141.66 g/kg) was similar to the value of 137.59 g/kg reported by ALCOSA SA, differing by only 2.96%. It is noteworthy that, with their own extraction and identification protocols, Lin et al. (2015a) obtained samples containing free lutein equivalents of 0.44 g/kg of

Table 7 Tukey's test for the means of the treatments

Treatment	Design coordinate	Groups[a]		Mean $Y_{p/x}$ value
3	−++−+−−	A		509.10
4	−+−+−+−	A	B	365.12
2	+++++++	A	B	338.53
6	+−−++−−		B	289.09
7	++−−−−+		B	286.24
5	−−−−+++		B	281.88
8	0000000		B	273.78
1	−−+++−−+		B	237.16
9	+−+−−+−		B	229.23

[a] Treatments with the same groups are not significantly different among themselves

Table 8 Results of the kinetics of growth and lipase production in stirred tank

	X_0^a g/L	X_{max}^a g/L	Enzyme units (40 h) U/L	μ h^{-1}	$Y_{p/x}$ U/g
Average of three experiments	0.44	8.97	1598 (163.9)[b]	0.236	113.8[c]

[a] The data of X_0 and X_{max} were calculated from the numerical solution of the logistic model applied to the average of three experimental runs

[b] The detected amount of enzyme units is shown as mean (SD) with $n = 3$

[c] The specific yield, or specific productivity, in terms of biomass ($Y_{p/x}$) was calculated as described in the "Methods" section

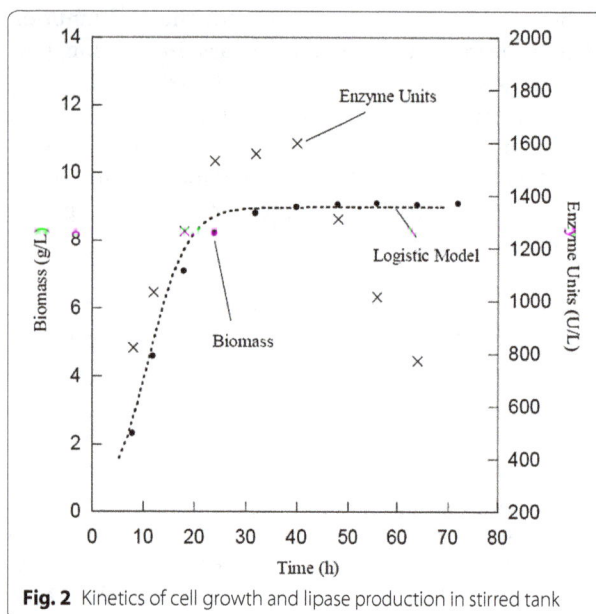

Fig. 2 Kinetics of cell growth and lipase production in stirred tank

crude extract and Lin et al. (2015b) estimated 20 g/kg dry material.

Influence of the growth of *Y. lipolytica* cultured in oleoresin on the hydrolysis of carotenoid esters

The hydrolysis of esters in stirred flask in the presence of oleoresin showed the behavior observed in Fig. 3a, b, i.e., a suddenly change in the content of esters and MHP during the first 12 h. This can be attributed to the aggregation of lipid particles—which at the beginning of the culture were emulsified and dispersed—since the enzyme cannot act properly if the substrate is not emulsified. Table 10 shows the average of three experiments on the behavior of the DHP during hydrolysis. It can be seen that, compared to the beginning of fermentation, there is an increase in the fractions of 4.62 and 12.41% after 48 and 72 h, respectively. This increase of 2.7–11.6 mg of DHP, representing a fourfold increase, is promoted by the aqueous solution, since DHP can disperse better in it than di-esters and MHP in the presence of stirring and emulsifying agents. The technical difficulties of using semi-solid fatty substances such as animal tallow for cultures of *Y. lipolytica* in stirred tank have been reported by Papanikolaou et al. (2002). These authors reported that the aggregation of lipid particles limits the proper performance of the cultures, and that higher stirring speeds are thus required to disperse these substrates in the medium.

When *Y. lipolytica* was cultured in stirred tank in the presence of marigold oleoresin, we obtained an average of 8.8% of DHP after 96 h of culture. Table 11 shows the results of these three experiments. Figure 4 shows the average of 3 experiments of the changes in the degree of esterification of the carotenoids due to the effect of the growth of *Y. lipolytica* in stirred tank. It can be seen that the real change was from 4.9 to 15 mg of DHP, i.e., a threefold increase. There are two main differences with the experiments in flask. The first is that the contents of DHP and MHP are higher at the start of the culture probably due to the effect of mechanical stirring and to the coalescence of lipid particles, which may cause a decrease in pigment content at 48 h. The second difference is that after this period, mechanical stirring probably promoted the dispersion of the particles in the culture medium, and thus the concentration of DHP and MHO remained similar between 60 and 96 h. Finally, we should remark that the concentrations of MHP and DHP reached similar levels to those of flask experiments.

Table 9 Results of the characterization of the marigold oleoresin by open column separation

Separated fraction	Xanthophylls[a,b] g/kg	% of total carotenoids
Esters	125.23 (0.22)	88.4 (0.11)
MHP	12.53 (0.15)	7.92 (0.07)
DHP + RP[c]	3.9 (0.37)	2.68 (0.02)
Total xanthophylls	141.66	100.00

[a] Data are expressed as mean (SD) with $n = 3$

[b] Xanthophylls were quantified as equivalent to trans-lutein

[c] DHP + RP = free lutein + residual pigments

Hydrolysis by lipases Y. lipolytica compared with commercial Rhizopus oryzae

Table 12 summarizes the results of the hydrolysis by *Y. lipolytica* and by *R. oryzae*. When the lipases secreted by *Y. lipolytica* were used for the hydrolysis of carotenoid esters, we obtained an average of 2.8% of free pigments out of the total amount of pigments; however, this represents a low increase in free carotenoids compared to a sample without lipase (blank). The grade of hydrolysis by *R. oryzae* was in accordance with previous literature (Zorn et al. 2003). The low yield was attributed to the

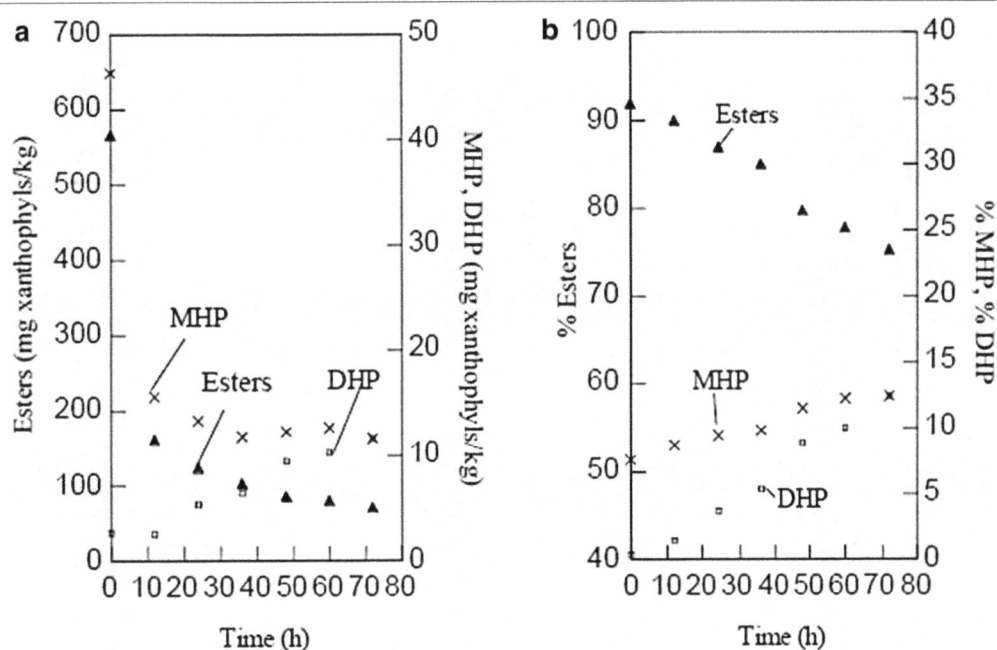

Fig. 3 Changes in the esterification levels of carotenoids during fermentation in stirred flask. **a** Expressed as mg/kg of xanthophylls; **b** expressed in *percentages* of the total pigments

Table 10 Increase of DHP due to the effect of culturing Y. lipolytica in stirred flask in a medium with marigold oleoresin, in different experiments

Experiment	% DHP[a]			% increase in free pigments[b]	
	Start (0 h)	Intermediate (48 h)	End (72 h)	Intermediate (48 h)	End (72 h)
1	0.34 (0.02)	4.3 (1.10)	16.04 (0.65)	3.96	15.7
2	0.48 (0.04)	3.6 (0.10)	10.75 (0.37)	3.12	10.27
3	0.5 (0.03)	7.3 (0.02)	11.76 (0.64)	6.8	11.26
Average	0.44	5.07	12.51	4.62	12.41

[a] Data are presented as mean (SD) with $n = 3$

[b] Calculated as the result of the difference between the mean percentages of DHP at the start and at the end of the fermentation

Table 11 Variation in the content of DHP due to the effect of growth of *Y. lipolytica* in a 7 L stirred tank in a culture medium with marigold oleoresin, in different experiments

Experiment	% DHP[a]				Increase in the % of free pigments[b]		
	Start (0 h)	Intermediate		End (96 h)	Intermediate		End (96 h)
		(48 h)	(72 h)		(48 h)	(72 h)	
1	3.8 (0.1)	7.5 (0.2)	10.7 (0.2)	12.6 (0.3)	3.7	6.9	8.8
2	3.7 (0.2)	8.4 (0.5)	10.2 (0.7)	11.7 (0.7)	4.7	6.5	8.0
3	4.1 (0.3)	7.2 (1.2)	10.5 (0.6)	13.7 (0.8)	3.1	6.2	9.6
Average	3.9	7.8	10.5	12.7	3.8	6.5	8.8

[a] Data are presented as mean (SD) with $n = 3$

[b] Calculated as the result of the difference between the mean percentages of DHP at the start and at the end of the fermentation

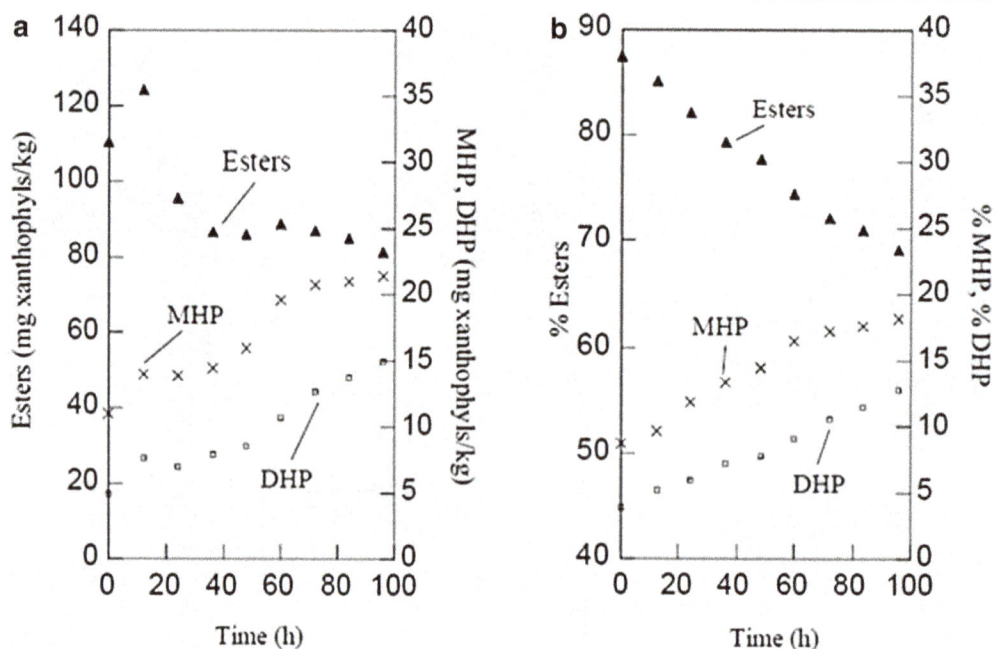

Fig. 4 Changes in the esterification of carotenoids during fermentation in stirred tank. **a** Expressed as mg/kg of xanthophylls and; **b** expressed in their *percentage* of the total pigments

fact the following facts: (1) the produced enzymes were not specific for the substrate used here, (2) the reaction was permitted to be proceeded for only 14 h, which was apparently not enough, and (3) reaction conditions of temperature, pH, aeration rate, and stirring speed have not been yet optimized. On this respect, it has been shown that other lipases show little or no activity in marigold extracts (Zorn et al. 2003). Even though it was minimal, the conversion rate was significantly different ($p < 0.05$) from the blank (1.8% of DHP) and the *R. ory-zae* (2.6% of DHP) according to the variance analysis presented in Table 13.

Conclusions

Given the great interest in natural free lutein for industrial purposes, it is highly important to find alternative methods for producing this xanthophyll such as the use of enzymes. The present work studied the enzymatic reaction of an industrially obtained oleoresin from marigold flower into free lutein to replace the chemical saponification using enzymes from *Y. lipolytica*. The yield of lipases by *Y. lipolytica* was, in descending order, mostly affected by the concentration of olive oil, yeast extract, sodium chloride, and trace mineral nutrients in the culture medium. Confirmatory experiments supported the

Table 12 Increase in free pigments (DHP) in the hydrolysis of carotenoid esters from the marigold flower

Source of lipases[b]	Test	Mg/kg[a]			% of total pigments[a]		
		Esters	MHP	DHP	Esters	MHP	DHP
A	1	75.87 (0.02)	7.84 (0.06)	2.36 (0.02)	88.2 (0.04)	9.1 (0.06)	2.7 (0.03)
	2	83.65 (1.44)	8.03 (0.45)	2.53 (0.13)	88.8 (0.56)	8.5 (0.56)	2.7 (0.11)
	3	75.38 (0.06)	6.96 (0.07)	2.80 (0.07)	80.0 (0.98)	7.4 (0.10)	3.0 (0.04)
	Avg	78.30	7.61	2.56	85.7	8.3	2.8
B	1	76.47 (1.93)	5.93 (0.50)	2.23 (0.10)	90.3 (0.80)	7.0 (0.65)	2.6 (0.15)
	2	70.57 (0.20)	8.26 (0.10)	1.95 (0.06)	87.4 (0.15)	10.2 (0.09)	2.4 (0.07)
	3	70.61 (0.14)	7.94 (0.29)	2.23 (0.03)	87.4 (0.51)	9.8 (0.38)	2.8 (0.02)
	Avg	72.55	7.38	2.14	88.4	9.0	2.6
C	1	72.43 (0.13)	5.99 (0.21)	1.44 (0.08)	90.7 (0.29)	7.5 (0.24)	1.8 (0.09)
	2	72.39 (0.69)	6.05 (0.08)	1.51 (0.09)	90.7 (1.25)	7.6 (0.11)	1.9 (0.10)
	3	71.48 (0.06)	6.62 (0.08)	1.42 (0.11)	89.5 (0.50)	8.2 (0.18)	1.8 (0.13)
	Avg	72.10	6.22	1.46	90.3	7.8	1.8

[a] Data are presented as mean (SD) with $n = 3$

[b] A—*Yarrowia lipolytica*; B—*Rhizopus oryzae*; C—Blank

Table 13 Analysis of variance of the use of the hydrolysis of carotenoid esters from the marigold flower

Source	Degrees of freedom	Sum of squares	Mean square	F
Source of lipases	2	4.787	2.39	115.34*
Error	24	0.498	0.021	
Total	26	5.285		

* Significant differences ($p \leq 0.05$)

statistical optimal values of the experimental design. With the addition of the oleoresin into the fermentation medium, the hydrolysis of carotenoid esters was achieved showing a strong increase in the free pigments fraction of the oleoresin (12.41%) in flask, and in stirred tank (8.8%) at 72 h, i.e., a fourfold and a threefold increase in mg of free carotenoids, respectively. Conducting 14 h hydrolysis tests with the enzymes from *Y. lipolytica* may not be enough time to transform most of the lutein esters; nevertheless, we obtained a poor conversion (2.8%) of the total oleoresin pigments to DHP against an experiment without lipases. Results were similar with lipases from *R. oryzae*. Therefore, carotenoid esters cannot be completely hydrolyzed with the lipases produced by *Y. lipolytica* in only 14 h, but culturing this microorganism in the presence of marigold oleoresin generates a high increase in the production of free carotenoids in 3-day fermentations. Further studies have to be carried out to optimize hydrolysis conditions and to better understand the system biology of this hydrolysis in order to evaluate the feasibility to achieve a better conversion.

Abbreviations

DHP: free pigments or dihydroxy pigments; E: di-esters; HEAT: solvent solution (hexane, ethanol, acetone, toluene; 10/7/6/7; v/v/v/v); MHP: monohydroxy pigments; RP: residual pigments or polyols; SD: standard deviation; YPD: agar (yeast extract, peptone, dextrose; 1, 1, 2%)

Nomenclature

t: fermentation time in h; μ: specific rate of growth in h^{-1}; X_0: initial biomass concentration in g/L; X_{max}: final biomass concentration in g/L; $Y_{p/x}$: production of lipases relative to biomass in lipase units U/L

Authors' contributions

AFA, APG, and EMES made substantial contributions to conception and design; APG developed the experimental acquisition of data; AFA and APG made substantial contributions to the statistical analysis, parameter estimation, and interpretation of data; EMES, AFA, and LGO were involved in drafting the manuscript, revising it critically for important intellectual content; EMES gave final approval of the version to be published. Each author participated sufficiently in the work to take public responsibility for appropriate portions of the content, and agreed to be accountable for all aspects of the work in ensuring that questions related to the accuracy or integrity of any part of the work are appropriately investigated and resolved. All authors read and approved the final manuscript.

Author details

[1] Departamento de Ingeniería Química, Instituto Tecnológico de Celaya, Av. Tecnológico y A.G. Cubas s/n, 38010 Celaya, Gto, Mexico. [2] Facultad de Química, Universidad Autónoma de Querétaro, Cerro de las Campanas s/n, 76010 Santiago de Querétaro, Querétaro de Arteaga, Mexico. [3] Departamento de ingeniería Bioquímica, Instituto Tecnológico de Celaya, Av. Tecnológico y A.G. Cubas s/n, 38010 Celaya, Gto, Mexico.

Competing interests

The authors declare that they have no competing interests.

Funding
This research was funded by Tecnológico Nacional de México, Grant: 5539.15-P.

References

Aakermann T, Hertzberg S, Liaaen-Jensen S (1996) Enzymatic hydrolysis of esters of alkali labile carotenols. Biocatal Biotransform 13(3):157–163

Alonso FOM, Oliveira EBL, Dellamora-Ortiz GM, Pereira-Meirelles FV (2005) Improvement of lipase production at different stirring speeds and oxygen levels. Braz J Chem Eng 22(1):9–18

Barzana E, Rubio D, Santamaria RI, Garcia-Correa O, Garcia F, Ridaura Sanz VE, Lopez-Munguia A (2002) Enzyme-mediated solvent extraction of carotenoids from marigold flower (*Tagetes erecta*). J Agric Food Chem 50(16):4491–4496

Benitez-Garcia I, Emilio Vanegas-Espinoza P, Melendez-Martinez AJ, Heredia FJ, Paredes-Lopez O, Angelica Del Villar-Martinez A (2014) Callus culture development of two varieties of *Tagetes erecta* and carotenoid production. Electron J Biotech 17(3):107–113

Breithaupt DE, Bamedi A, Wirt U (2002) Carotenol fatty acid esters: easy substrates for digestive enzymes? Comp Biochem Physiol B Biochem Mol Biol 132(4):721–728

Chitchumroonchokchai C, Failla ML (2006) Hydrolysis of zeaxanthin esters by carboxyl ester lipase during digestion facilitates micellarization and uptake of the xanthophyll by Caco-2 human intestinal cells. J Nutr 136(3):588–594

Delgado-Vargas F, Paredes-López O (1997) Effects of enzymatic treatments on carotenoid extraction from marigold flowers (*Tagetes erecta*). Food Chem 58(3):255–258

Fickers P, Nicaud JM, Gaillardin C, Destain J, Thonart P (2004) Carbon and nitrogen sources modulate lipase production in the yeast *Yarrowia lipolytica*. J Appl Microbiol 96(4):742–749

Fickers P, Benetti PH, Wache Y, Marty A, Mauersberger S, Smit MS, Nicaud JM (2005) Hydrophobic substrate utilisation by the yeast *Yarrowia lipolytica*, and its potential applications. FEMS Yeast Res 5(6–7):527–543

Fickers P, Sauveplane V, Nicaud J-M (2013) The lipases from *Y. lipolytica*: genetics, production, regulation, and biochemical characterization. In: Barth G (ed) *Yarrowia lipolytica*: biotechnological applications. Springer, Berlin, pp 99–119

Flachmann R, Sauer M, Schopfer C, Klebsattel M (2005) Method for hydrolysing carotenoids esters. Method for hydrolysing carotenoids esters. US patent US20050255541 A1, 17 Nov 2005

Fletcher DL (2006) A method for estimating the relative degree of saponification of xanthophyll sources and feedstuffs. Poult Sci 85(5):866–869

Fukuda R, Ohta A (2013) Utilization of hydrophobic substrate by *Yarrowia lipolytica*. In: Barth G (ed) *Yarrowia lipolytica*, microbiology monographs. Springer, Berlin, pp 111–119

Gajdos P, Nicaud J-M, Rossignol T, Certik M (2015) Single cell oil production on molasses by *Yarrowia lipolytica* strains overexpressinq DGA2 in multicopy. Appl Microbiol Biot 99(19):8065–8074

Groenewald M, Boekhout T, Neuveglise C, Gaillardin C, van Dijck PW, Wyss M (2014) *Yarrowia lipolytica*: safety assessment of an oleaginous yeast with a great industrial potential. Crit Rev Microbiol 40(3):187–206

Guevara-Olvera L, Calvo-Mendez C, Ruiz-Herrera J (1993) The role of polyamine metabolism in dimorphism of *Yarrowia lipolytica*. J Gen Microbiol 139(3):485–493

Hirakawa K, Kobayashi S, Inoue T, Endoh-Yamagami S, Fukuda R, Ohta A (2009) Yas3p, an Opi1 family transcription factor, regulates cytochrome P450 expression in response to *n*-alkanes in *Yarrowia lipolytica*. J Biol Chem 284(11):7126–7137

Howles PN, Hui DY (2001) Cholesterol Esterase. In: Mansbach CM, Tso P, Kuksis A (eds) Intestinal lipid metabolism. Springer, Boston, pp 119–134

Jacobs PB, LeBoeuf RD, McCommas SA, Tauber JD (1982) The cleavage of carotenoid esters by cholesterol esterase. Comp Biochem Physiol Part B Comp Biochem 72(1):157–160

Lim TK (2014) Tagetes erecta. Edible medicinal and non-medicinal plants. Springer, Berlin, pp 432–447

Lin J-H, Lee D-J, Chang J-S (2015a) Lutein in specific marigold flowers and microalgae. J Taiwan Inst Chem Eng 49:90–94

Lin J-H, Lee D-J, Chang J-S (2015b) Lutein production from biomass: marigold flowers versus microalgae. Bioresour Technol 184:421–428

Liu Y, Xu MJ, Canfield LM (1998) Enzymatic hydrolysis, extraction, and quantitation of retinol and major carotenoids in mature human milk. J Nutr Biochem 9(3):178–183

Navarrete-Bolanos JL, Jimenez-Islas H, Botello-Alvarez E, Rico-Martinez R, Paredes-Lopez O (2004) Improving xanthophyll extraction from marigold flower using cellulolytic enzymes. J Agric Food Chem 52(11):3394–3398

Papanikolaou S, Chevalot I, Komaitis M, Marc I, Aggelis G (2002) Single cell oil production by *Yarrowia lipolytica* growing on an industrial derivative of animal fat in batch cultures. Appl Microbiol Biotechnol 58(3):308–312

Peleg M, Corradini MG, Normand MD (2007) The logistic (Verhulst) model for sigmoid microbial growth curves revisited. Food Res Int 40(7):808–818

Pereira-Meirelles FV, Rocha-LeãO MHM, Anna GLS (1997) A stable lipase from *Candida lipolytica*. Appl Biochem Biotech 63(1):73

Pereira-Meirelles F, Rocha-Leão MM, Sant'Anna G Jr (2000) Lipase location in *Yarrowia lipolytica* cells. Biotechnol Lett 22(1):71–75

Xie D, Jackson E, Zhu Q (2015) Sustainable source of omega-3 eicosapentaenoic acid from metabolically engineered *Yarrowia lipolytica*: from fundamental research to commercial production. Appl Microbiol Biot 99(4):1599–1610

Zhao Y, Guan F, Wang G, Miao L, Ding J, Guan G, Li Y, Hui B (2011) Astaxanthin preparation by lipase-catalyzed hydrolysis of its esters from *Haematococcus pluvialis* Algal Extracts. J Food Sci 76(4):C643–C650

Zorn H, DE Breithaupt, Takenberg M, Schwack W, Berger RG (2003) Enzymatic hydrolysis of carotenoid esters of marigold flowers (*Tagetes erecta* L.) and red paprika (*Capsicum annuum* L.) by commercial lipases and *Pleurotus sapidus* extracellular lipase. Enzyme Microb Technol 32(5):623–628

Recent progress on deep eutectic solvents in biocatalysis

Pei Xu[1,2], Gao-Wei Zheng[3], Min-Hua Zong[1,2], Ning Li[1] and Wen-Yong Lou[1,2]*

Abstract

Deep eutectic solvents (DESs) are eutectic mixtures of salts and hydrogen bond donors with melting points low enough to be used as solvents. DESs have proved to be a good alternative to traditional organic solvents and ionic liquids (ILs) in many biocatalytic processes. Apart from the benign characteristics similar to those of ILs (e.g., low volatility, low inflammability and low melting point), DESs have their unique merits of easy preparation and low cost owing to their renewable and available raw materials. To better apply such solvents in green and sustainable chemistry, this review firstly describes some basic properties, mainly the toxicity and biodegradability of DESs. Secondly, it presents several valuable applications of DES as solvent/co-solvent in biocatalytic reactions, such as lipase-catalyzed transesterification and ester hydrolysis reactions. The roles, serving as extractive reagent for an enzymatic product and pretreatment solvent of enzymatic biomass hydrolysis, are also discussed. Further understanding how DESs affect biocatalytic reaction will facilitate the design of novel solvents and contribute to the discovery of new reactions in these solvents.

Keywords: Deep eutectic solvents, Biocatalysis, Catalysts, Biodegradability, Influence

Introduction

Biocatalysis, defined as reactions catalyzed by biocatalysts such as isolated enzymes and whole cells, has experienced significant progress in the fields of either biocatalysts or reaction medium during the past decades. One noteworthy example is the greener synthetic pathway developed by Savile for the redesigned synthesis of sitagliptin, an active ingredient used to treat diabetes (Savile et al. 2010). Generally, biocatalysis refrains many weaknesses, such as the lack of enantio-, chemo- and regioselectivity and utilization of metal catalysts in organic synthesis in the production of valuable compounds by the chemical method (Bommarius and Paye 2013).

Along with the rapid emergence of new biocatalyst technology, the shift to green chemistry puts high stress on the biocatalytic process. Among the Twelve Principles of Green Chemistry developed by Anastas and Warner (1998), one concept is "Safer Solvents and Auxiliaries". Solvents represent a permanent challenge for green and sustainable chemistry due to their vast majority of mass used in catalytic processes (Anastas and Eghbali 2010). Environmental benignity is the ideal pursuit of exploiting each generation of solvents.

Solvents for a biocatalysis reaction have experienced several generations of development. Water has been considered as the greenest solvent considering its quality and quantity. However, the high polarity of the water molecule hinders its application in all biocatalytic reactions, because of some substrates' insolubility in aqueous solution due to their high hydrophobicity. Traditional organic solvents (water miscible or water immiscible), in the form of co-solvents or second phase, can provide solutions for the above-described challenges to some degree. But inevitably, organic solvents face their own disadvantages such as high volatility, inflammability and activity inhibition to the biocatalyst. Ionic liquids (ILs) are the first enzyme-compatible untraditional media developed by the green and sustainable concept (given their low vapor pressure). Numerous reactions, e.g., hydrolytic and redox reactions as well as formation of C–C bond, have been successfully performed in such ILs-containing media

*Correspondence: wylou@scut.edu.cn
[1] Laboratory of Applied Biocatalysis, School of Food Sciences and Engineering, South China University of Technology, Guangzhou 510640, China
Full list of author information is available at the end of the article

(Potdar et al. 2015; Roosen et al. 2008; van Rantwijk and Sheldon 2007). Despite the excellent performance of ILs in biocatalysis, more doubts about their ungreenness and environmental influence have been gradually presented (Zhao et al. 2007; Thuy Pham et al. 2010; Petkovic et al. 2012).

Deep eutectic solvents (DESs), the recognized alternative of ILs, first came to the public vision in 2001 (Abbott et al. 2001). Since then, research on DESs faced a prosperous increase in many fields, such as extraction, materials synthesis, and biotransformation (Atilhan and Aparicio 2016; Carriazo et al. 2012; Dai et al. 2013b; Garcia et al. 2015; Gonzalez-Martinez et al. 2016; van Osch et al. 2015). A brief statistics of the published literature starting form 2004 is presented in Fig. 1. This review will provide an overview of the (un)greenness of DESs and recent applications in biocatalysis. Finally, how DESs affect biocatalytic reactions are discussed.

DESs and their properties
DESs
DESs are eutectic mixtures of salts and hydrogen bond donors (HBDs) with melting point low enough to be used as solvents. A eutectic liquid is generally formed by two solids at eutectic temperature, in which the process follows a thermal equilibrium. A deep eutectic system refers to a eutectic mixture with a melting point much lower than either of the individual components. A classic example is that ChCl/urea (1:2 mol ratio) mixture has a melting point of 12 °C (far lower than the melting point of the ChCl and urea, 302 and 133 °C, respectively), which makes it a liquid to be used as a solvent at room temperature. In addition, the complex hydrogen bonding network existing in DES also makes it a special solvent for different applications (Hammond et al. 2016).

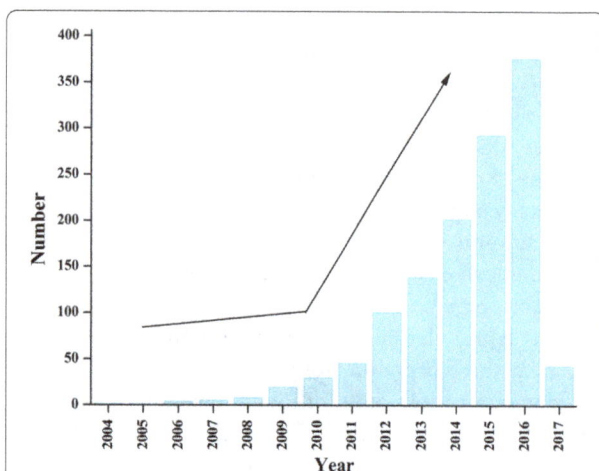

Fig. 1 The number of published papers on DESs starting from 2004 (adapted from ISI Web of Knowledge)

Up to now, many kinds of DESs have been developed with various compounds (Fig. 2). They are generally prepared by mixing a hydrogen bond acceptor (HBA), such as quaternary ammonium salt with a metal salt or HBD at moderate temperature with constant stirring. The first examples of DESs were prepared by heating $ZnCl_2$ with a series of quaternary ammonium salts, in which choline chloride is the HBA obtained at the surprisingly lowest melting point of 23–25 °C (Abbott et al. 2001). In 2014, Abbott et al. proposed a general formula of Cat^+X^-zY for DESs, where Cat^+ refers to ammonium, phosphonium or sulfonium cation and X generally represents a halide anion (Smith et al. 2014). This formula represents four classes of DESs according to the nature of the components used.

Class I: Cat^+X^-zMClx, M = Zn, Sn, Fe, Al, Ga, In.
Class II: $Cat^+X^-zMClx\ yH_2O$, M = Cr, Co, Cu, Ni, Fe.
Class III: Cat^+X^-zRZ, Z = $CONH_2$, COOH, OH.
Class IV: $MClx + RZ = MCl_{x-1}^+RZ + MCl_{x+1}^-$, M = Al, Zn and Z = $CONH_2$, OH.

The main difference between Class II DESs and I is the inherent water, which makes it possible for many hydrated metal halide salts to be used as HBDs to synthesize DESs, since they have relatively lower cost.

Another set of DESs, "natural deep eutectic solvent (NADES)", are developed by utilizing many primary metabolites, such as amino acids, sugars, choline and organic acid (Choi et al. 2011). NMR-based metabolomics analysis indicated that NADES indeed existed as the third liquid other than water and lipid in an organism. For example, a mixture of sucrose, fructose and glucose (1:1:1, molar ratio) forms a clear and uniform liquid at room temperature. They can help plants survive anhydrobiosis and play a role in the synthesis of intracellular macromolecules. These NADES may preserve the linkage between understanding of cellular metabolism and physiology.

Properties
DESs have some special physical traits which are very useful for biocatalytic process, i.e., a wide liquid state range, nonflammability and low volatility (Abbott et al. 2011; Paiva et al. 2014; Smith et al. 2014; Zhang et al. 2012). Low temperature of freezing point (Tf) makes it possible for DESs to remain as a liquid in a biocatalytic process, because most biocatalytic reactions proceed in a liquid environment, in either monophasic or biphasic system. As reported, the T_f of a DES depends on many factors. Increased hydroxyl group of HBD increased the value of T_f. For example, the HBD in ChCl/Gly has three hydroxyl groups and melts at −40 °C, while the HBD in

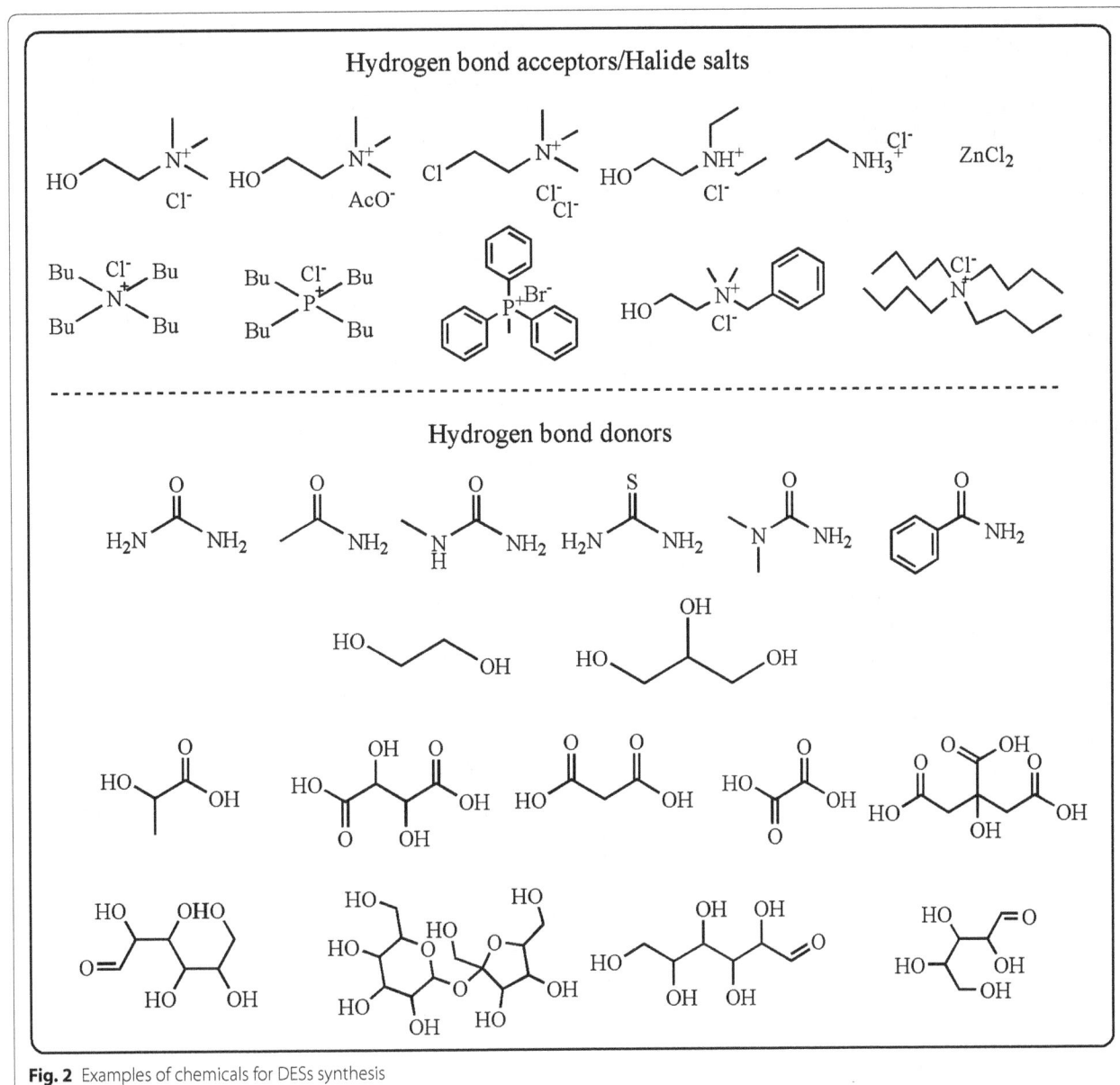

Fig. 2 Examples of chemicals for DESs synthesis

ChCl/EG has two hydroxyl groups and melts at −66 °C (Zhang et al. 2012). DESs with different mole ratios of HBA to HBD also show different T_f but no obvious tendency was observed. The glycerol-based DESs with methyl triphenyl phosphonium bromide as HBA (1:2– 1:4, molar ratio) have a T_f range from −5.5 to 15.8 °C (Shahbaz et al. 2011). Besides, DESs are commonly not flammable and have very low volatility, because their individual components are generally solids or inflammable liquid. These two properties decrease the risk of explosion and burning for an organic reaction if DESs are used as solvents instead of volatile organic solvents, thus making it safe to operate. What is more, density

and viscosity are the other two important characteristics which affect the mass transfer when selecting a solvent for use. Generally, the viscosity of DESs lies in the constitution and is higher than that of the molecular solvent ethanol (Smith et al. 2014). The specific relationship between these factors has been documented at length in another review (Zhang et al. 2012). Most properties change as a function of temperature (Zhao et al. 2015). For instance, as the temperature increases, the density and viscosity decrease.

Two other advantages of DESs over ILs are easier preparation and lower cost. Typical preparation of DESs involves a mixing of the components in a certain molar ratio (e.g., 1:2 of HBA/HBD) with constant stirring at moderate temperature. This simple synthesis is 100% atom economic, requiring no further purification. Besides, the commonly used components of DESs, such as ChCl and glycerol, are all from natural materials. Especially, NADESs are absolutely formed by primary metabolites, such as glucose, malic acid and succinic acids (Dai et al. 2013a). These raw materials exist naturally and have relatively low cost of production. By contrast, the common preparation of ILs has two steps: first alkylation of an amine/phosphine/sulfide to afford an intermediate salt, followed by anion exchange to give the ILs (Hallett and Welton 2011). The complexity of synthesis will spontaneously increase the production cost. Thus, the synthesis of DESs is easier and costs less than that of ILs.

Toxicity

A good question about whether DESs are toxic was proposed in 2013 (Hayyan et al. 2013b). The answer for that is some of them are toxic. Most of them are basically soluble in water. Especially in the cases of Class I and Class II DESs, the metal ion will inevitably leave a trace in the environment, thus exerting pressure on recycling these solvents.

To date, only a few groups have investigated the toxicological properties of DESs in detail (Table 1). Hayyan and co-workers first investigated the toxicity and cytotoxicity of ChCl- and phosphonium-based DESs, of which glycerine, ethylene glycol, triethylene glycol and urea were used as the HBDs (Hayyan et al. 2013a, b). They pointed out that the toxicity and cytotoxicity of different DESs varied with the structure of the components used. ChCl-based DESs were less toxic than phosphonium-based DESs. Afterward, the toxicity of ammonium-based DESs was evaluated in in vitro cell lines and in vivo animal models (Hayyan et al. 2015). In this study, the DESs tested showed higher toxicity than their individual components,which was consistent with other DESs reported previously. Later, some ternary DESs consisting of ChCl, water and natural substance were also examined for their cytotoxicity (Hayyan et al. 2016). Selection of DESs parents is very important for the toxicity of resulting DESs. The IC_{50} of ChCl/Gly/H_2O toward Hela S3 cell was almost 2.4-fold higher than that of ChCl/fructose/H_2O (427 mM vs 177 mM), though in the former DES Gly had higher concentration (1:2:1 vs 5:2:5, respectively). Glycerol is a wildly used sweetener and humectant in the food industry and used in strain preservation; so the toxicity of glycerol-based DESs is expectable to some degree.

However, the toxic DESs can be used for beneficial purposes, e.g., as effective anti-bacterial agents. The DES choline/geranate can inhibit the growth of some bacteria,

fungi and viruses, making it a possible broad-spectrum antimicrobial agent (Zakrewsky et al. 2016).

In our work, a method of bacterial growth inhibition was used to assess the toxicology of four types of DESs based on ChCl, which was coupled with amine-, alcohol-, sugar- and organic acid-based HBDs (Zhao et al. 2015). The organic acid-based DESs significantly inhibited bacterial growth, suggesting the limitation of utilizing organic acid when designing DESs. This finding was in accordance with Hayyan's study, which found that ChCl/MA was greatly toxic toward the tested mammalian cells (Hayyan et al. 2016). The other three types of DESs displayed less toxicity to the tested Gram-positive (*Staphylococcus aureus* and *Listeria monocytogenes*) and Gram-negative (*Escherichia coli* and *Salmonella enteritidis*) bacteria. Interestingly, a further MIC test demonstrated that the two Gram-negative bacteria were less sensitive than Gram-positive bacteria toward the organic acid-based DESs. A mechanism of this difference may be related to the oligosaccharide constituent.

All the examples give us information, that is, reasonable design and post-processing of DESs are very important for us to utilize this solvent. Novel DESs can be made according to the properties of parent components for special use. Some DESs showed toxicity to some extent, but there is limitation for these studies, due to the fact that the tested organisms are different with what we use in a biocatalytic process. For a typical reaction, one can determine the effect of DESs they use on the biocatalyst, under the guidance of previous researches.

Biodegradability

Biodegradability is another important property except toxicity when talking about the "greenness" of DESs. A large proportion of DESs are considered as "readily biodegradable", possibly because most of the components forming DESs are natural products (Radosevic et al. 2015; Zakrewsky et al. 2016). Taking the NADES, for instance, all the individual components come from natural materials and can be metabolized by different kinds of organisms in nature. Glycerol can enter the pathway of glycolysis or gluconeogenesis for final metabolism and the sugar portion can also be directly utilized (Mbous et al. 2017).

Table 2 shows the recent studies which reported the biodegradability of DESs. In Radosevic's work, the closed bottle test (OECD) was used according to the OECD guideline 301 D (1992D), and all the tested ChCl-based DESs (ChCl:Gly, ChCl:Glc, and ChCl:OA) provided a biodegradation level of over 60% after 14 days (Radosevic et al. 2015). The highest biodegradability level was observed up to 96% with ChCl:Gly. These findings for DES biodegradability were in accordance with our

Table 1 The toxicity of some DESs

DESs	Organism	Toxicity comments	References
MTPB/Gly (1:3) MTPB/EG (1:3) MTPB/TEG (1:3)	*Escherichia coli, Staphylococcus aureus, Pseudomonas aeruginosa, Bacillus subtilis, Artemia salina*	All the DESs showed toxic effect on some bacteria The cytotoxicity of DESs was much higher than their individual components	Hayyan et al. (2013a)
ChCl/Gly (1:3) ChCl/EG (1:3) ChCl/U (1:3) ChCl/TEG (1:3)	*Escherichia coli, Staphlococcus aureus, Pseudomonas aeruginosa, Bacillus subtilis, Artemia salina*	All the DESs showed no toxic effect on the bacteria The cytotoxicity of DESs was much higher than their individual components	Hayyan et al. (2013b)
ChCl/Gly (1:3) ChCl/EG (1:3) ChCl/TEG (1:3) ChCl/U (1:3)	PC3, A375, HepG2, HT29, MCF-7, OKF6, H413, ICR mice	The cytotoxicity of DESs varied from various cell lines. The toxic effects of DESs were higher than their individual components	Hayyan et al. (2015)
ChCl or ChOAc/U (1:1) ChCl or ChOAc/Gly (1:1) ChCl or ChOAc/A (1:1) ChCl or ChOAc/EG (1:1)	*Escherichia coli*	0.75 M DES could afford an inhibition index of 72.8–93.8% for the bacterium and was more toxic then the components	Wen et al. (2015)
ChCl/EG (1:2) ChCl/Gly (1:2) ChCl/U (1:2) DAC/EG (1:2) EAC/Gly (1:2) DAC/MA (1:1) DAC/ZnN (1:1) DAC/ZnCl$_2$ (1:2)	*Aspergillus niger, Cyprinus carpio* fish	Metal salt-containing DESs were most toxic than others. DAC-based DESs were less toxic than ChCl-based DESs	Juneidi et al. (2015)
20 kinds of NADESs	*Staphylococcus aureus, Listeria monocytogenes, Escherichia coli, Salmonella enteritidis*	All the DESs except for acid-containing DESs showed no toxic effect on the bacteria	Zhao et al. (2015)
ChCl/ZnCl$_2$ (1:2) ChCl/U (1:2) ChCl/Gly (1:3) ChCl/EG (1:3) ChCl/DEG (1:2) ChCl/TEG (1:3) ChCl/Fru (2:1) ChCl/Glc (2:1) ChCl/PTSA (1:3) ChCl/MA (1:1)	*Phanerochaete chrysosporium, Aspergillus niger, Lentinus tigrinus, Candida cylindracea*	ZnCl$_2$, PTSA and MA DESs had the most toxic effect	Juneidi et al. (2016)
ChCl/Fru/water (5:2:5) ChCl/Glc/water (5:2:5) ChCl/Suc/water (4:1:4) ChCl/Gly/water (1:2:1) ChCl/malonic acid (1:1)	HelaS3, CaOV3, B16F10, MCF-7	NADESs except malonic acid as HBD are less toxic than DESs	Hayyan et al. (2016)
Choline and geranate (1:2)	*M. tuberculosis, S. aureus, P. aeruginosa* et al.	The DES is so toxic that it can act as a broad-spectrum antiseptic agent	Zakrewsky et al. (2016)
ChCl/Fru (2:1) ChCl/Glc (2:1) DAC/TEG (1:3)	HelaS3, PC3, AGS, A375, MCF-7, WRL-68	The DESs (ChCl/Fru and ChCl/Glc, 98 mM $\leq EC_{50} \leq$ 516 mM) were less toxic than DAC/TEG(34 mM $\leq EC_{50} \leq$ 120 mM)	Mbous et al. (2017)

study, in which all the ChCl-based DESs had degradability of over 69.3% after 28 days (Zhao et al. 2015). However, Wen et al. found a different conclusion (Wen et al. 2015). They found that only two (ChCl/U and ChCl/A) of the eight DESs tested could be regarded as readily biodegradable. The two DESs showed a degradability close to 80%; nevertheless, the others provided a degradability below 50% in contrast. In the same year, Juneidi et al. also presented a comprehensive study on the biodegradability for cholinium-based DESs (Juneidi et al. 2015). It is

worth mentioning that this work involved in the toxicity and biodegradability of DESs-containing a metal salt and a hydrated metal salt (i.e., ZnCl$_2$ and ZnN). There was a significant difference between the degradability of DAC:ZnN and DAC:ZnCl (about 80% vs 62%). Even the biodegradability of the former was better than that of the conventional IL [BMPyr][NTf$_2$] (77%). The inherent structure of HBAs and HBDs is the basic factor which determines the biodegradability of different DESs, i.e., the number of hydroxy groups (Radosevic et al. 2015).

Table 2 The biodegradability of some DESs

DESs	Assay method	Resource of microorganism	DES concentration (mg/L)	Comments	Reference
ChCl/Glc (2:1) ChCl/OA (1:1) ChCl/Gly (1:2)	Closed bottle test	Effluent from an urban wastewater treatment plant	100	Over 60% of biodegradation level after 14 days	Radosevic et al. (2015)
20 kinds of NADESs	Closed bottle test	Fresh lake water	3	All DESs had a biodegradation level over 69.3% after 28 days. The acid-based DESs were degraded slower than others	Zhao et al. (2015)
ChCl or ChOAc/U (1:1) ChCl or ChOAc/Gly (1:1) ChCl or ChOAc/A (1:1) ChCl or ChOAc/EG (1:1)	Closed bottle test	Activated sludge from wastewater treatment plant	4	Only ChCl/U and ChCl/A were readily biodegradable	Wen et al. (2015)
ChCl/EG (1:2) ChCl/Gly (1:2) ChCl/U (1:2) DAC/EG (1:2) EAC/Gly (1:2) DAC/MA (1:1) DAC/ZnN (1:1) DAC/ZnCl$_2$ (1:2)	Closed bottle test	Wastewater from secondary effluent treatment plant	5	All DESs were referred to as readily biodegradable. ChCl-based DESs had higher biodegradability than DAC-based DESs	Juneidi et al. (2015)

For a special DES, the difference in result between various reports should be attributed to the difference of the measure condition, the molar ratio of DESs and the wastewater microorganisms.

In summary, DESs exhibit relatively low toxicity toward organisms in a laboratory scale and could be classified as biodegradable green solvents. These benign characteristics may attract more attention of researchers to shift from the classical imidazolium and pyridinium ILs to DESs. More organic chemistry reactions could be explored in such solvents.

The application of DESs in biocatalysis

Recent decades have witnessed the dramatic publications of ILs in chemical synthesis with isolated enzymes or whole cells (Wang et al. 2012). By comparison, only dozens of studies about the biocatalysis in DESs have been published. The following parts will introduce the recent applications of DESs in biocatalytic reactions, in which DESs played different roles such as solvent, co-solvent or substrate.

DESs as solvent for enzyme catalysis

As a new generation of solvent, DESs have distinctive characteristics, e.g., abundant hydrogen bond and low T_f. The components of DESs may display inhibitory effect on enzyme activity. For example, urea could induce the unfolding of enzymes and further make them inactive (Attri et al. 2011). Even so, some hydrolases still exhibited better activity in DES compared with ILs and catalyzed a lot of hydrolysis and transesterification reactions as well as Henry reaction and aldol reaction in this special solvent.

Lipase

Lipases, which are commonly investigated in an organic environment, are the popular choices for enzymatic reaction carried out in DESs (Table 3). In 2008, Kazlauskas et al. first tried the lipase-catalyzed biotransformation reaction in DESs (Fig. 3) (Gorke et al. 2008). Four lipases (iCALB, CALB, CALA and PCL) were investigated for their transesterification activity in eight DES-containing systems. In glycerol-containing DESs, all the enzymes showed conversions of ethyl valerate to varying degrees. Especially in ChCl/Gly, PCL exhibited the lowest conversion (22%), still much higher than that in toluene (5.0%). Interestingly, the alcohol component, either ethylene glycol or glycerol, could compete with 1-butanol. When the concentration of the substrate 1-butanol was much lower than the EG component (400 mM vs 10 M), the CALB-catalyzed reaction still displayed a nearly equal amount of product esters. The reason for this may be the high hydrogen bond network, which hinders the reaction of the alcohol components with the substrate. Additionally, the iCALB exhibited great aminolysis activity in ChCl:Gly-containing system (52 μmol h^{-1} mg^{-1}), which was five times higher than in BMIM[BF$_4$] (9 μmol h^{-1} mg^{-1}). This example opened the possibilities for exploring various biocatalytic reactions in DES-containing solvents.

Table 3 Examples of lipase-catalyzed reactions in DESs

Enzyme	DES	Substrate	Product	Comments	Reference
iCALB, CALB, CALA and PCL	ChCl/Gly (1:2) ChCl/U (1:2) EAC/Gly (1:1.5)	Ethyl valerate with 1-butano	Butyl valerate	ChCl/Gly showed good compatibility with all the lipases	Gorke et al. (2008)
iCALB	ChCl/U (1:2) ChOAc/Gly	Miglyol oil 812	Triglyceride	High yield showed the potential of DES as solvent in the biodiesel synthesis	Zhao et al. (2011a)
iCALB	ChCl/Gly (1:2) ChCl/U (1:2)	Vinyl ester and alcohols	Esters	Some HBDs could compete with the substrate	Durand et al. (2012)
Novozyme 435	ChCl/Gly (1:2)	Soybean oil	Biodiesel	This work expanded the substrate spectrum of biodiesel synthesis	Zhao et al. (2013)
Novozyme 435	ChCl/Gly (1:2) ChCl/U (1:2)	Phenolic esters	Phenolic esters	Water content in DES–water mixtures makes great difference on reaction efficiency	Durand et al. (2013)
iCALB	ChCl/U	Phenolic esters	Phenolic esters	First investigated the effect of water activity and U content on product yields	Durand et al. (2014)
Lipozyme CalB L Novozym 435	ChCl/U (1:2) ChCl/GlyZ (1:1)	Oleic acid and decanol	Decyl oleate	Esters product could be easily separated from the aqueous reaction mixtures	Kleiner and Schörken (2015)
Novozyme 435	ChCl/U (1:2) ChCl/Glc	Glucose and vinyl hexanoate	Glucose-6-O-hexanoate	Glucose component in DESs can act as substrate	Pöhnlein et al. (2015)
Lipozyme TLIM, Novozym 435	ChAc/U (2:1)	Glucose with fatty acid vinyl esters; methyl glucoside with fatty acids	Sugar fatty acid esters	Utilization of combination of ILs and DESs	Zhao et al. (2016)
CALB, Alcalase-CLEA, PPL	ChCl/Gly (1:1.5)	Aromatic aldehydes and ketones	Aldol products	First tested the lipase-catalyzed aldol reaction in DES	Gonzalez-Martinez et al. (2016)
Lipase from *Candida rugosa*	ChCl/U/Gly (1:1:1)	p-Nitrophenyl palmitate	p-Nitrophenol	Glycerol-containing DESs enhance the activity and stability more than urea-based DESs. The effects of DESs on activity and stability of lipase were partially correlated with the solvatochromic parameters. For example, the stability of lipase was correlated with hydrogen bond acidity of DESs mixtures	Kim et al. (2016)
Thermomyces lanuginosus lipase *Pseudozyma antarctica* lipase B	ChCl/U (1:2) ChCl/Gly (1:2)	Rapeseed oil and cooking oil	Biodiesel	Improved the additional value of cooking oil	Kleiner et al. (2016)
Lipozyme TLIM, Novozym 435	ChAc/U (2:1)	Glucose with fatty acid vinyl esters; methyl glucoside with fatty acids	Glucose-based fatty acid esters	Utilization of combination of ILs and DESs	Zhao et al. (2016)

Table 3 continued

Enzyme	DES	Substrate	Product	Comments	Reference
Lipase AS	ChCl/Gly (1:2)	Aldehydes	Nitroalcohols	Addition of water could improve enzyme activity and inhibit DES-catalyzed reaction	Tian et al. (2016)
Burkholderia cepacia lipase	ChCl/EG (1:2)	*p*-Nitrophenyl palmitate	*p*-Nitrophenol	Significantly improved enzyme activity	Juneidi et al. (2017)
Lipase from ANL	ChCl/Gly (1:3) *Burkholderia cepacia* lipase	Dihydromyricetin	Dihydromyricetin-16-acetate	Enhancing substrate solubility	Cao et al. (2017)
iCALB	ChCl/different sugars (1:1)	Fatty acid esters	Glycolipids	Sugar can serve as HBD and substrate	Siebenhaller et al. (2017)
Lipase G	ChCl/xylitol (1:1)	Glyceryl trioleate	Epoxidized vegetable oils	DES stabilized the enzyme	Zhou et al. (2017a)
iCALB	ChCl/Gly (1:2)	Benzoic acid and glycerol	α-Monobenzoate glycerol	Water as co-solvent enzyme remained active in high concentration of DES (92%,v/v)	Guajardo et al. (2017)
PPL	ChCl/U (1:2)	Amines with aryl halides	N-aryl amines	DES acted as catalyst as well as solvent	Pant and Shankarling (2017)

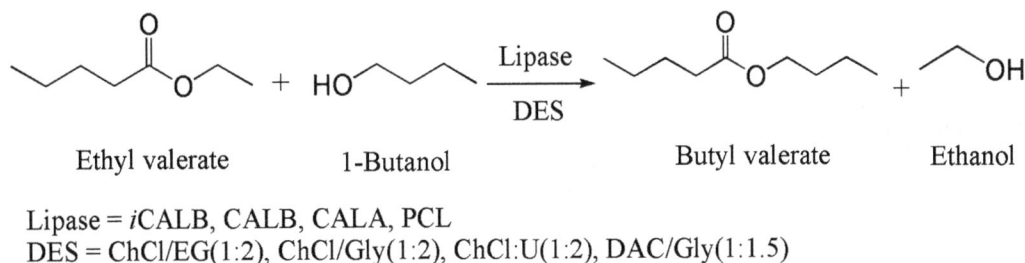

Fig. 3 Lipase-catalyzed transesterification of ethyl valerate to butyl valerate in DESs (adapted from Gorke et al. 2008)

In a following research by Durand (2012), the iCALB-catalyzed transesterification of vinyl ester was used to study the advantages and limitations of seven kinds of DESs (Fig. 4). The same competing phenomenon was observed between the DES component and the alcoholysis reaction. When the reaction occurred in ChCl:U or ChCl:Gly-containing system, alcohol substrate with different chain lengths showed no influence on the conversion and selectivity. A preliminary grinding of the immobilized enzymes could improve specific surface area in DESs, enhancing the reaction efficiency. Recently, they has intended to improve the lipase-catalyzed reaction efficiency in ChCl-based DESs (Durand et al. 2014). As mentioned earlier, albeit the liquid state of DESs at room temperature, their viscosity should be considered for the mass transfer once being used in a biocatalytic reaction. In this work, the effect of water content and urea concentration was systematically assessed. The thermodynamic activity of water (a_w) was used as the study object. When the ChCl/water was lower than 1.75 in the ternary mixture, water trended to be bound to the salt and could not affect the lipase activity. In cases of higher a_w, excess water molecules could affect its role as the solvent, leading to the occurrence of side reactions. When the water content in the mixture was 1 and 1.5 mol, the urea content showed no significant effect on the lipase stability. However, the denaturation effect of urea increased with the increase of urea content at higher water content.

In some cases, DES can play a bifunctional role, that is, substrate and solvent, in lipase-catalyzed reactions. Several DESs with sugars (e.g., arabinose, xylose, glucose and mannose) as HBD was used as substrate other than solvent to form glycolipid, which was detected by MS-based methods (Fig. 5) (Siebenhaller et al. 2017). This approach of using DESs exhibits an apparent advantage that high concentration of the sugar substrate can push the reaction forward in the product direction.

Lipase-catalyzed promiscuous reactions, such as carbon–carbon bond formation, Henry and Aldol reactions, have attracted some groups' attention (Gonzalez-Martinez et al. 2016; Guan et al. 2015; Tian et al. 2016; Zhou et al. 2017b). Gonzalez-Martinez and co-workers investigated the lipase-catalyzed aldol reactions in DES-containing system in 2016 (Gonzalez-Martinez et al. 2016). A model reaction between 4-nitrobenzaldehyde and acetone was set up to examine the feasibility of DES as a reaction medium for the promiscuous aldol process (Fig. 6a). Three hydrolases, CALB, Alcalase-CLEA and PPL, were explored. Compared to the other two enzymes, PPL displayed great activity at different reaction temperatures (30–60 °C) in a system containing ChCl:Gly. The ratio of ChCl to Gly (1:1.5 or 1:2, molar ratio) posed little effect on the conversion and product composition. For this, the ratio was not probably the main aspect that affected the reaction efficiency under the optimized conditions. Different mole ratios of the HBA to HBD may

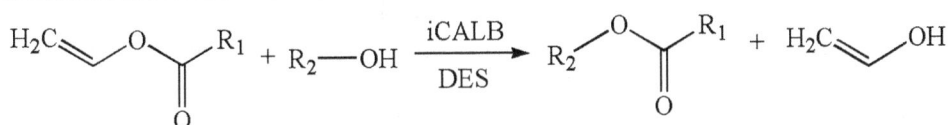

Fig. 4 Lipase-catalyzed transesterification of vinyl ester in DESs (adapted from Durand et al. 2012)

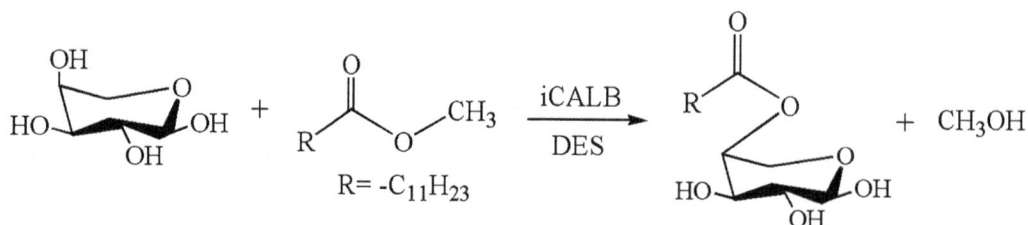

Fig. 5 Lipase-catalyzed production of glycolipids in DESs (adapted from Siebenhaller et al. 2017)

a

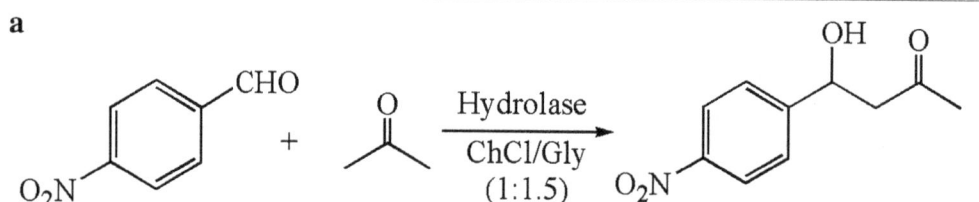

Hydrolase = CALB, Alcalase-CLEA, PPL

b

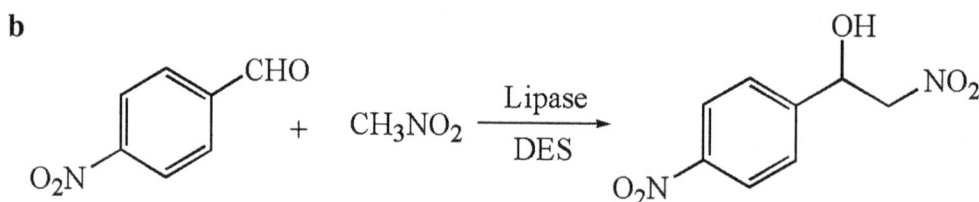

Lipase = lipase from *Aspergillus niger*

DES = ChCl/Gly, ChCl/EG, ChCl/U with different molar ratio

Fig. 6 Lipase-catalyzed promiscuous reactions in DESs (adapted from (**a**) Gonzalez-Martinez et al. 2016; (**b**) Tian et al. 2016)

change the hydrogen bonding formation, further leading to the change of the stereochemical structure of the enzyme, which affects the reaction in turn. Secondly, they tested the suitability of the ChCl:Gly-containing system by exploring different substrate groups, i.e., substituted benzaldehydes with acetone, cyclopentanone and cyclohexanone. In most cases, satisfactory results were obtained. At the same year, Tian et al. explored the Henry reaction catalyzed by lipase AS in ChCl/Gly(1:2)-containing media (Fig. 6b) (Tian et al. 2016). Water as a co-solvent was important in the reaction, not only improving enzyme activity, but also inhibiting spontaneous reaction. More recently, a chemoenzymatic epoxidation of alkenes catalyzed by CALB was reported. DES, ChCl/sorbitol, efficiently stabilized the biocatalyst, allowing the enzyme

to remain active in an oxidative solvent system (Zhou et al. 2017b). The idea of trying lipase-catalyzed promiscuous reaction is of great significance to discover new functions for enzymes in DESs and synthesize many valuable chemicals.

Protease

Proteases also find promising applications in DESs-containing systems. To explore the feasibility of DES in biocatalysis, Zhao and co-workers evaluated the protease-catalyzed transesterification activities in glycerol-based DESs based on choline salt (chloride or acetate form) (Fig. 7a) (Zhao et al. 2011b). This work indicated again the key role of water in the reaction system. While water content increased from 2% (v/v) to 4%, the activity of immobilized subtilisin exhibited a remarkable 1.8-fold increase (from 0.50 to 0.90 μmol min^{-1} g^{-1}) in choline acetate/glycerol (1:1.5). This phenomenon was more obvious for immobilized α-chymotrypsin. Under reaction conditions with addition of water to the DESs solution, the biocatalytic efficiencies of these two proteases were better than those in t-butanol. Apparently, certain DES could activate the protease activity to some extent. This report also demonstrated that immobilization materials could function with water molecules to lower a_w to stabilize the forward reaction.

Except transesterification reaction, proteases-catalyzed peptides synthesis in DESs also proved to be successful. Different ChCl-based DESs using glycerol, urea, xylitol and isosorbide as HBDs were used as solvents for the production of peptides catalyzed by α-chymotrypsin (Maugeri and Leitner 2013). High productivities of 20 g L^{-1} h^{-1} were afforded at the optimal conditions with ChCl/glycerol and addition of 10–30% (v/v) water. The α-chymotrypsin showed good recycling ability in this DES solution system. Recently, the immobilized papain on a magnetic material was successfully used for the synthesis of N-(benzyloxycarbonyl)-alanyl-glutamine (Z-Ala-Gln) in ChCl/urea (1:2) as well, the yield of which reached about 71.5% (Fig. 7b) (Cao et al. 2015). In this example, more glutamine (Gln) dissolved in DES buffer remarkably improved the yield and a high Gln/Z-Ala-OMe ratio of 3–4 could effectively inhibit the side reaction to form Z-Ala-OH.

Epoxide hydrolase

Epoxide hydrolases (EH), another important hydrolase, catalyze the enantioselective hydrolysis of epoxides to the corresponding diols, which are important chiral synthetic intermediates. Up to now, there are only two examples of using DES in EH-catalyzed reactions. The first example was reported in 2008 (Gorke et al. 2008). By the addition of 25% (v/v) ChCl/Gly to the buffer, the conversion of styrene oxide catalyzed by EHAD1 increased from 4.6 to 92%. No change in the enantioselectivity was observed with the increase of conversion. The other one was carried out by Widersten et al. (Fig. 8) (Lindberg et al. 2010). Ethane diol, glycerol and urea were used as the hydrogen bond donors. The DES ChCl/Gly still exhibited a superior solvent property for this reaction than the other two DESs. Surprisingly, the regioselectivity of this hydrolysis reaction varied with the alteration of solvents. More

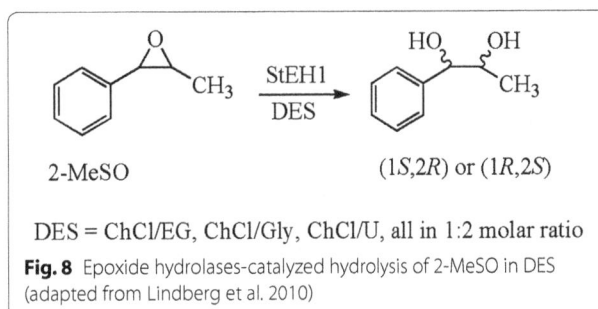

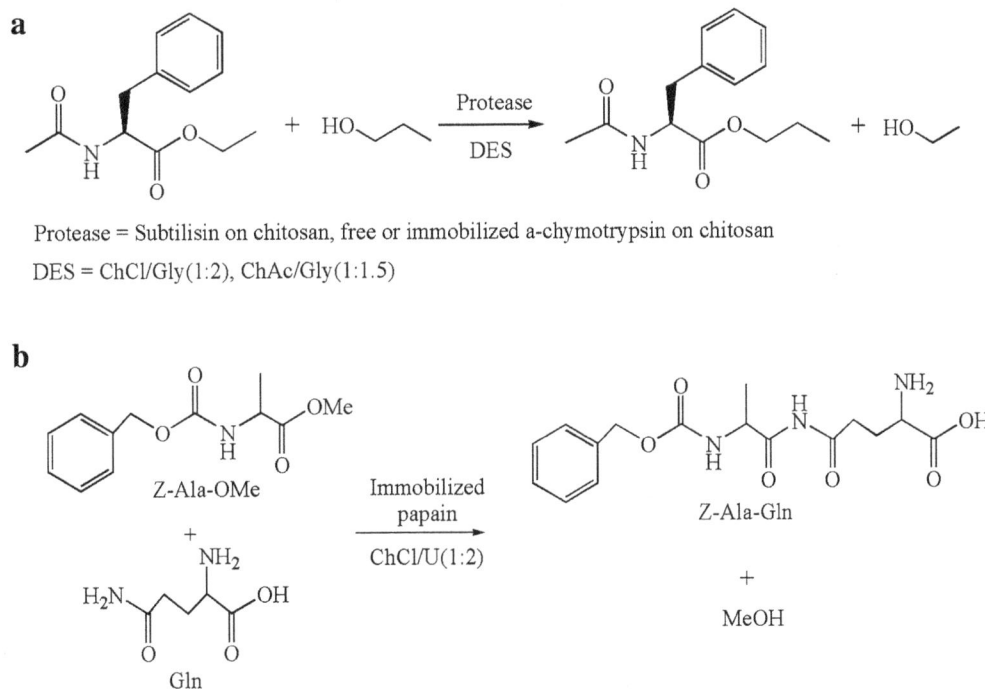

DES = ChCl/EG, ChCl/Gly, ChCl/U, all in 1:2 molar ratio

Fig. 8 Epoxide hydrolases-catalyzed hydrolysis of 2-MeSO in DES (adapted from Lindberg et al. 2010)

Protease = Subtilisin on chitosan, free or immobilized a-chymotrypsin on chitosan

DES = ChCl/Gly(1:2), ChAc/Gly(1:1.5)

Fig. 7 Protease-catalyzed reactions in DESs (adapted from (**a**) Zhao et al. 2011b) (**b**) Cao et al. 2015;

amounts of (1*R*, 2*R*)-2-MeSO (2-methylstyrene oxide) diols were produced when the diol-and triol-containing DESs were added into the buffer as co-solvents, while the effect of glycerol on the StEH1 catalysis was the least. Many researchers considered that the decrease of enzymatic efficiency in higher ILs concentrations could be attributed to the denaturation of enzyme. However, in this work, the authors proposed that higher DES concentrations could cause the destabilization of enzyme–substrate or reaction intermediate complexes according to steady-state kinetic parameters of StEH1 at different DES concentrations. In addition, the substance concentration with the addition of DES had a 1.5-fold improvement compared to the DES-free buffer. This enhancement, to some degree, was related to the hydrogen bond in the solvent environment. The excellent dissolution-promoting ability of DES has already been used in nature product extraction (Dai et al. 2013b).

Other enzymes

In other examples of enzymatic reactions, many enzymes, such as haloalkane dehalogenases (Stepankova et al. 2014), benzaldehyde lyase (Maugeri and Domínguez de María 2014) and phospholipase D (Yang and Duan 2016), also showed improved reaction efficiency in the DES-containing system (Table 4). For the diglycosidase from *Acremonium* sp. DSM24697, the deglycosylation of hesperidin happened when ChCl combined with urea, glycerol and ethylene glycol was added into the buffer, whereas the enzyme activity was inhibited by DES to some extent (Weiz et al. 2016). The apparent advantage of this approach is the enhanced substrate solubility by DES. Such a property has been successfully applied in the extraction of the natural product [e.g., flavonoid (Zhao et al. 2015) and grape skin phenolics (Cvjetko Bubalo et al. 2016)]. This case is beneficial for the modification of many natural products, which have good solubility in DES.

DES as solvent for whole-cell biocatalysis

Whole-cell biocatalysis using bacteria or fungi as catalyst, e.g., *Acetobacter* sp. and recombinant *Saccharomyces cerevisiae*, has been successfully applied in some organic reactions (Gangu et al. 2009; Wachtmeister and Rother 2016; Xu et al. 2016). In comparison with isolated enzymes, whole-cell catalysts, either in the form of natural organisms or genetically recombinant strains, have a series of advantages, such as lower cost than purified enzymes (Tufvesson et al. 2011), perfect cofactor regeneration system (Hummel and Gröger 2014), applicability in untraditional media (Dennewald and Weuster-Botz 2012) and feasibility of multi-step cascade reactions in one cell (Chen et al. 2015; Peters et al. 2017).

Generally, a monophasic system forms when DES is added to aqueous buffer due to its good solubility in water. As a result, the DES-containing medium will serve as an efficient buffer system for a whole cell-catalyzed process. DES can improve, to some degree, the substrate concentration in the reaction system and further increase the reaction efficiency.

The first example of whole-cell biocatalysis in DES was the reduction of ethyl acetoacetate catalyzed by baker's yeast (Fig. 9) (Table 5) (Maugeri and de Maria 2014). The yeast worked well when ChCl/Gly (1:2) was added into an aqueous buffer, and even enantioselectivity had a complete inverse when altering the proportion of DES used. The tested ChCl/Gly may go into the cell and inhibit the enantioselective enzyme activity or affect the configuration of partial enzymes. A similar enantioselectivity inversion by the addition of DES was further expanded to a series of arylpropanone substrates (Vitale et al. 2016). Both the examples used wild-type baker's yeast as a biocatalyst. It is worth understanding how DESs affect the intracellular enzymes. Recombinant *E. coli* cells expressing oxidoreductases were also investigated for their catalytic reduction of ketones in DES-containing system (Muller et al. 2015). The cells performed surprisingly even in an 80% (v/v) DES-containing system for a broad range of aromatic substrates (Fig. 10).

In short, genetically modified recombinant hosts overexpressing different kinds of enzymes with different functions, especially ketoreductase (Ni and Xu 2012), open the way to the industrial production of many valuable compounds. The cost-effective, nonhazardous DESs will present greatly stimulative potential in these biocatalytic processes.

DES as extractive agents in biotransformation

The separation of the product or unreacted substrate from a reaction should be considered in an industrial catalytic process, especially the biocatalytic reaction, regardless of the utilization of isolated or immobilized biocatalyst. DESs also show great ability to separate the desired compounds from a reaction mixture. Many products are traditionally extracted using organic liquids, such as ethyl acetate, *n*-butyl alcohol, *n*-hexane and isopropyl ether.

As mentioned earlier, DESs are often biomass derived and formed by mixing of ChCl and a hydrogen bond donor. Krystof et al. (2013) successfully separated 5-hydroxymethylfurfural (HMF) esters with HMF after the lipase-catalyzed (trans)esterification reaction, which exhibited a hydrogen donor character by using ChCl-based DES. High ester purity (>99%) and efficiency (up to >90% HMF ester recovery) were observed. This finding paved a new path for the HMF chemistry. Inspired by that, ChCl-based DES was also utilized by Qin et al.

Table 4 Other examples of enzymatic reactions in DESs

Catalyst	DES	Substrate	Comments	References
Haloalkane dehalogenases	ChCl/EG (1:2)	1-Iodohexane	Improved enzyme thermostability and substrate solubility	Stepankova et al. (2014)
Benzaldehyde lyase	ChCl/Gly (1:2)	Butyraldehyde; valeraldehyde; benzaldehyde; 2-furaldehyde	Improved e.e.	Maugeri and Domínguez de María (2014)
Phospholipase D	ChCl/EG (1:2)	Phosphatidylcholine with L-serine	>90% yield of phosphatidylserine	Yang and Duan (2016)
Diglycosidase	ChCl/Gly (1:2); ChCl/EG (1:2)	Hesperidin	Enhanced substrate solubility	Weiz et al. (2016)
Horseradish peroxidase; cytochrome c	ChCl or EAC with U, Gly and EG (1:1.5,1:2)	Guaiacol	Enhancing the functional stability of protein	Papadopoulou et al. (2016)
Chondroitinases ABCl	ChCl/Gly (1:2)	Chondroitin	Improving thermal stability remarkably	Daneshjou et al. (2017)
β-D-glucosidase	ChCl/EG (2:1)	Daidzin	Application of DES in the synthesis of bioactive compound	Cheng and Zhang (2017)
Bovine liver catalase	ChCl/EG (1:2)	Hydrogen peroxide	DES could change the Km and Kcat of enzyme in DES-containing solution. Structure test found that the 3D structure was influenced by the addition of DES	Harifi-Mood et al. (2017)

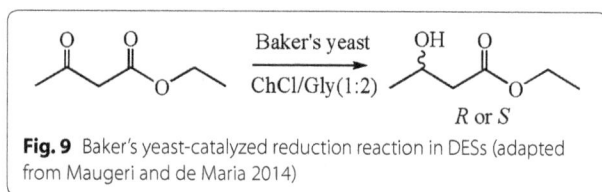

Fig. 9 Baker's yeast-catalyzed reduction reaction in DESs (adapted from Maugeri and de Maria 2014)

to separate HMF and 2,5-diformylfuran (DFF) from the reaction mixtures, and the purity of the latter reached up to 97% (Qin et al. 2015). The two examples discussed above to some extent were based on the characteristics of the hydrogen bond donor of the separated chemical compounds. Maybe, the formation of a new hydrogen bond existed between the DES and target compounds.

Hydrophobic DES containing quaternary ammonium salt/decanoic acid mixture could extract volatile carboxylic acids from aqueous solution (van Osch et al. 2015). The DESs extracted acetic acid, propionic acid and butyric acid more efficiently than a traditional solvent like trioctylamine. For example, $DecA:N_{8881}$-Cl extracted twice as much acetic acid as trioctylamine (38.0% vs 18.6%). $DecA:N_{7777}$-Cl extracted 91.5% of the butyric acid. This kind of DES costs less than some ammonium and phosphonium ionic liquids for extracting butyric acid (Blahušiak et al. 2013; Marták and Schlosser 2016). However, another challenge about the separation of acid from DES is put forward subsequently. The traditional distillation method could facilitate the process owing to the extremely low volatility and high thermal stability of DESs (Wu et al. 2015).

DESs function in biomass pretreatment

DES pretreatment of biomass is a good alternative of IL-based processes. Currently, most attention has been focused on the use of ILs, especially imidazole- and choline-based ILs (Sheldon 2016), for preprocessing cellulose-based biomass to improve the following enzymatic hydrolysis process indirectly. DESs are able to dissolve biopolymers like polysaccharides by breaking the supermolecular structure formed by intermolecular hydrogen bonds (Ren et al. 2016). For example, ChCl/imidazole can be used to pretreat corncob at a relatively low temperature of 80 °C and the final glucose yield reached 92.3% after the enzymatic hydrolysis (Procentese et al. 2015). Due to the diversity of HBDs, researchers have more choice of using different HBDs in conjunction with ChCl for pretreatment; thus, different results are obtained normally, even for the same biomass like corncob (Zhang et al. 2016). Considering the relative cheapness of ILs, it is useful to develop more cellulase-compatible DESs, making it possible to omit the separation of DESs from the system.

Influence of DESs on biocatalysis

As shown in the last chapter, many positive examples show the great promising future of DESs in biocatalysis. It is of interest to elucidate under what circumstances and how the biocatalysts retain their biological function and stability in DESs.

Effect of the characteristics of DESs on the biocatalytic process

The tunability of DES contributes to much more possibility of designing vast kinds of DES, thus bringing various effect on a biocatalytic reaction system. The selection of HBAs and HBDs make a big difference in the whole process. For instance, ChCl/Gly is usually used as a good solvent/co-solvent for enzymatic reaction, while the performance of ChCl/MA is not satisfactory (Durand et al.

Table 5 Examples of whole-cell biocatalytic reactions in DES

Microbial	DES	Substrate	Product	Comments	References
Baker's yeast	ChCl/Gly (1:2)	Ethyl acetoacetate	Ethyl 3-hydroxybutyrate	Increasing the DES content switched the enantioselectivity	Maugeri and de Maria (2014)
Acetobacter sp. CCTCC M209061	ChCl/U (1:2)	3-Chloropropiophenone	(S)-3-Chloro-1-phenylpropanol	Combination of ILs and DESs in the biphasic system effectively improved the substrate concentration	Xu et al. (2015)
Recombinant E. coli over-expressing ADH	ChCl/Gly (1:2)	Ketones	Alcohols	The e.e. of many aromatic substrates increased by the addition of DES	Muller et al. (2015)
Baker's yeast	ChCl/Gly (1:2)	Aryl-containing ketones	Aryl-containing alcohols	(S)-oxidoreductases of baker's yeast was possibly inhibited by DES	Vitale et al. (2016)
Recombinant E. coli CCZU-T15	ChCl/Gly (1:2)	4-chloro-3-oxobutanoate	(S)-4-chloro-3-hydroxybutanoate	Addition of Tween-80 improved significantly substrate concentration from 2 to 3 M	Dai et al. (2017)
Lysinibacillus fusiformis CGMCC 1347	ChOAc/U (1:1) ChOAc/EG (1:1) ChCl/Lac (4:1) ChCl/Raf (11:2)	Isoeugenol	Vanillin	NADESs were first used in whole-cell biocatalysis	Yang et al. (2017)

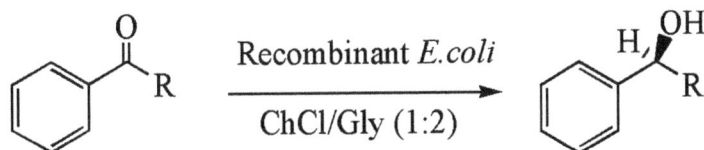

Fig. 10 Recombinant *E. coli*-catalyzed reduction reactions in DES (adapted from Muller et al. 2015)

2012). Different HBDs also form DESs with different viscosities (Zhang et al. 2012), which affects the mass transfer of all the reactants (substrate, product, catalyst, etc.) and further changes the reaction rate. At this point, the hydrogen bond is another concern posed by various HBDs. There is a difference of just one hydroxyl group between Gly and EG; however, the resulting DESs showed remarkable difference on the lipase-catalyzed reaction (Durand et al. 2012).

Hydrogen bonding of DESs is the main power which makes them distinctive with their individual components. This special force can activate the enzyme by increasing the enzyme affinity with the substrate (Juneidi et al. 2017). It is also necessary to consider the ratio of HBA/HBD, which might affect the solvatochromic parameters and the hydrogen bond formation between the reaction mixtures (Kim et al. 2016). A molecular dynamics simulation of lipase in ChCl/U firstly proved that the hydrogen bond between DES components could hinder the attack of U to the enzyme function domains, thus resulting in a stabilized enzyme (Monhemi et al. 2014).

In addition, DES concentration and water content are two interacting factors with significant influence on the reaction efficiency. A high amount of water not only decreased the viscosity of the DES solution, but also enabled the contact between the enzyme molecule and substrate (Guajardo et al. 2017). But too much water might have a destructive effect on the hydrogen bond network of DES and weaken the benefit of DES.

Influence of DESs on biocatalyst structure and activity

Firstly, DESs can change the secondary structure of enzymes. Take the horseradish peroxidase (HRP) for example (Wu et al. 2014): higher HBD molar ratio (e.g., from 2:1 to 1:2) in ChCl-based DES increased the α-helix content in the HRP. More α-helix and less β-sheet in the structure were beneficial for their activity and stability. The improvement in activity may result from the looser tertiary structure of HRP. A recent report confirmed that the ChCl component had a disruptive influence on the α-helix, which would cause a decrease of the enzyme activity. However, the glycerol component could markedly increase the α-helix content and decrease the

β-sheet content, leading to a more stable enzyme. Thus, a trade-off effect between HBA and HBD, either activation or deactivation, can be observed for a certain DES.

Secondly, DESs can affect the 3-D structure of enzymes in an aqueous environment. The functional configuration of enzymes is essentially related to the folding and unfolding degree. Several factors including hydrophobicity, solvent polarity and hydrogen bond characteristics of ILs can affect separately or together the protein stability and activity (Weingartner et al. 2012). Similarly, the concentration of DESs in an aqueous solution had significant effect on the unfolding and refolding process of the enzyme, which could be determined by spectroscopy technologies such as intrinsic fluorescence and CD spectroscopy (Esquembre et al. 2013). The two methods can effectively estimate the folded degree of proteins compared to the native-like secondary and tertiary structures. A redshift of the fluorescence emission spectra of tryptophan residues in lysozyme was observed when tryptophan side chain was exposed to a polar surrounding of water molecules. In neat ChCl/U and ChCl/Gly solution, the emission maximum of lysozyme shifted from 335 nm (observed in buffer) to 332 and 329 nm, respectively (Esquembre et al. 2013). However, higher concentration of DESs can cause the irreversible unfolding of lysozyme at high temperatures. Meanwhile, the protein accumulation in low content of DESs buffer at room temperature was interestingly reversible.

In whole-cell biocatalysis, DESs increased the permeability of cells, lowering resistance to mass transfer resistance, thereby improving the reaction efficiency (Xu et al. 2016). ILs similarly can improve the permeability of cells (Xiao et al. 2012). In other cases, DESs can enhance the affordable substrate concentration for a cell catalyst, thereby improving the catalytic rate and yield of the product (Xu et al. 2016). The worst situation is that DESs possibly react with the cell membrane and induce cell apoptosis, finally leading to cell death.

Conclusion and perspective

Deep eutectic solvents are easier to prepare with different kinds of biocompatible and naturally occurring constituents. Varying the component and ratio yields DESs with different T_f, polarity, density and viscosity. Some DESs show toxic profile toward laboratory organisms, which depends on the used components, test conditions and organisms. However, most DESs can be considered as "readily biodegradable" solvents. DESs can work as solvents, co-solvents or extracting solvents in a specific biocatalytic reaction. Remarkably, as solvents, DESs (e.g., Gly-based DESs) can activate and stabilize the enzyme, thus achieving a high reaction efficiency. Multiple hydrolases (lipase, protease and epoxide hydrolase) and other enzymes exhibit great catalytic performance in these solvents, which displays their vast potential to replace ILs and organic solvents in biocatalytic reactions. An in-depth understanding of how DESs activate and stabilize enzymes will promote the application of DESs in biocatalysis to step off laboratory scale.

However, DESs in biocatalysis needs researchers to fulfill their full physical–chemical characterization and the important toxicity data. The plausible relationship between the structure and function of enzymes and DESs will be an interesting field to explore. Additionally, the downstream separation of target product from DESs is also a big challenge. More utilization of environmental-friendly and economical solvents in biocatalysis will embody the greenness and sustainability of green chemistry.

Abbreviations
ChCl: choline chloride; MTPB: methyltriphenylphosphonium bromide; DAC: *N,N*-diethyl ethanol ammonium chloride; Gly: glycerol; EG: ethylene glycol; DEG: diethylene glycol; TEG: triethylene glycol; U: urea; A: acetamide; OA: oxalic acid; MA: malonic acid; PTSA: para-toluene sulfonic acid; Glc: glucose; Lac: lactose; Raf: raffinose; Fru: fructose; Suc: sucrose; ZnN: zinc nitrate hexahydrate; Choline and geranate: choline bicarbonate/geranic acid; PC3: human prostate cancer cell line; A375: human malignant melanoma cell line; HepG2: human liver hepatocellular cell line; HT29: human colon adenocarcinoma cell line; MCF-7: human breast cancer cell line; OKF6: human oral keratinocyte cell line; H413: carcinoma-derived human oral keratinocyte cells; HelaS3: human cervical cancer cell line; CaOV3: human ovarian cancer cell line; B16F10: mouse skin cancer cel line; AGS: human gastric cancer cell line; WRL-68: human hepatocyte cell line; CALA *Candida antarctica* lipase A iCALB: *Candida antarctica* lipase B immobilized on acrylic resin; CALB: *Candida antarctica* lipase B; CRL: *Candida rugosa* lipase; PCL: *Pseudomonas cepacia* lipase; PPL: *Porcine pancreas* lipase; Alcalase-CLEA: protease from *Bacillus licheniformis*; Lipase AS: lipase from *Aspergillus niger*; StEH1: potato epoxide hydrolase; EHAD1: epoxide hydrolase from *Agrobacterium radiobacter* AD1; BMIM[Tf$_2$N]: 1-butyl-3-methylimidazolium bis(trifluoromethanesulfonyl) imide; [BMPyr][NTf2]: 1-butyl-1-methylpyrrolidinium Bis(trifluoromethanesulfonyl)imide; [Bmim] HSO4: 1-butyl-3-methylimidazolium hydrosulfate.

Authors' contributions
PX and WYL drafted the manuscript. GWZ, MHZ and NL modified the manuscript. All authors read and approved the final manuscript.

Author details
[1] Laboratory of Applied Biocatalysis, School of Food Sciences and Engineering, South China University of Technology, Guangzhou 510640, China. [2] State Key Laboratory of Pulp and Paper Engineering, South China University of Technology, Guangzhou 510640, China. [3] State Key Laboratory of Bioreactor Engineering, East China University of Science and Technology, Shanghai 200237, China.

Acknowledgements
We wish to thank the National Natural Science Foundation of China (21676104; 21336002; 21376096), the Open Funding Project of the State Key Laboratory of Bioreactor Engineering, and the Program of State Key Laboratory of Pulp and Paper Engineering (2017ZD05) for partially funding this work. We also thank Professor Romas Kazlauskas for editing this manuscript.

Competing interests
The authors declare that they have no competing interests.

References
(1992D) OECD 301D Guidelines for testing of chemicals closed bottle test. Organisation of economic cooperation and development, Paris

Abbott AP, Capper G, Davies DL, Munro HL, Rasheed RK, Tambyrajah V (2001) Preparation of novel, moisture-stable, Lewis-acidic ionic liquids containing quaternary ammonium salts with functional side chains. Chem Commun 19:2010–2011

Abbott AP, Harris RC, Ryder KS, D'Agostino C, Gladden LF, Mantle MD (2011) Glycerol eutectics as sustainable solvent systems. Green Chem 13(1):82–90

Anastas P, Eghbali N (2010) Green chemistry: principles and practice. Chem Soc Rev 39(1):301–312

Anastas PT, Warner JC (1998) Principles of green chemistry. green chemistry: theory and practice, Oxford University Press, New York

Atilhan M, Aparicio S (2016) Deep eutectic solvents on the surface of face centered cubic metals. J Phys Chem C 120(19):10400–10409

Attri P, Venkatesu P, Kumar A, Byrne N (2011) A protic ionic liquid attenuates the deleterious actions of urea on alpha-chymotrypsin. Phys Chem Chem Phys 13(38):17023–17026

Blahušiak M, Schlosser Š, Marták J (2013) Extraction of butyric acid with a solvent containing ammonium ionic liquid. Sep Purif Technol 119:102–111

Bommarius AS, Paye MF (2013) Stabilizing biocatalysts. Chem Soc Rev 42(15):6534–6565

Cao SL, Xu H, Li XH, Lou WY, Zong MH (2015) Papain@magnetic nanocrystalline cellulose nanobiocatalyst: a highly efficient biocatalyst for dipeptide biosynthesis in deep eutectic solvents. ACS Sustain Chem Eng 3(7):1589–1599

Cao SL, Deng X, Xu P, Huang ZX, Zhou J, Li X, Zong M, Lou W (2017) Highly efficient enzymatic acylation of dihydromyricetin by the immobilized lipase with deep eutectic solvents as co-solvent. J Agric Food Chem 65(10):2084–2088

Carriazo D, Serrano MC, Gutierrez MC, Ferrer ML, del Monte F (2012) Deep-eutectic solvents playing multiple roles in the synthesis of polymers and related materials. Chem Soc Rev 41(14):4996–5014

Chen FF, Liu YY, Zheng GW, Xu JH (2015) Asymmetric amination of secondary alcohols by using a redox-neutral two-enzyme cascade. ChemCatChem 7(23):3838–3841

Cheng QB, Zhang LW (2017) Highly efficient enzymatic preparation of daidzein in deep eutectic solvents. Molecules 22(1):186

Choi YH, van Spronsen J, Dai YT, Verberne M, Hollmann F, Arends I, Witkamp GJ, Verpoorte R (2011) Are natural deep eutectic solvents the missing link in understanding cellular metabolism and physiology? Plant Physiol 156(4):1701–1705

Cvjetko Bubalo M, Ćurko N, Tomašević M, Kovačević Ganić K, Radojčić Redovniković I (2016) Green extraction of grape skin phenolics by using deep eutectic solvents. Food Chem 200:159–166

Dai Y, van Spronsen J, Witkamp G-J, Verpoorte R, Choi YH (2013a) Natural deep eutectic solvents as new potential media for green technology. Anal Chim Acta 766:61–68

Dai YT, Witkamp GJ, Verpoorte R, Choi YH (2013b) Natural deep eutectic solvents as a new extraction media for phenolic metabolites in *Carthamus tinctorius* L. Anal Chem 85(13):6272–6278

Dai Y, Huan B, Zhang HS, He YC (2017) Effective biotransformation of ethyl 4-chloro-3-oxobutanoate into ethyl (S)-4-chloro-3-hydroxybutanoate by recombinant *E-coli* CCZU-T15 whole cells in ChCl Gly-water media. Appl Biochem Biotechnol 181(4):1347–1359

Daneshjou S, Khodaverdian S, Dabirmanesh B, Rahimi F, Daneshjoo S, Ghazi F, Khajeh K (2017) Improvement of chondroitinases ABCI stability in natural deep eutectic solvents. J Mol Liq 227:21–25

Dennewald D, Weuster-Botz D (2012) Ionic liquids and whole-cell–catalyzed processes ionic liquids in biotransformations and organocatalysis. Wiley, New York, pp 261–314

Durand E, Lecomte J, Baréa B, Piombo G, Dubreucq E, Villeneuve P (2012) Evaluation of deep eutectic solvents as new media for *Candida antarctica* B lipase catalyzed reactions. Process Biochem 47(12):2081–2089

Durand E, Lecomte J, Barea B, Dubreucq E, Lortie R, Villeneuve P (2013) Evaluation of deep eutectic solvent–water binary mixtures for lipase-catalyzed lipophilization of phenolic acids. Green Chem 15(8):2275–2282

Durand E, Lecomte J, Barea B, Villeneuve P (2014) Towards a better understanding of how to improve lipase-catalyzed reactions using deep eutectic solvents based on choline chloride. Eur J Lipid Sci Technol 116(1):16–23

Esquembre R, Sanz JM, Wall JG, del Monte F, Mateo CR, Ferrer ML (2013) Thermal unfolding and refolding of lysozyme in deep eutectic solvents and their aqueous dilutions. Phys Chem Chem Phys 15(27):11248–11256

Gangu SA, Weatherley LR, Scurto AM (2009) Whole-cell biocatalysis with ionic liquids. Curr Org Chem 13(13):1242–1258

Garcia G, Aparicio S, Ullah R, Atilhan M (2015) Deep eutectic solvents: physicochemical properties and gas separation applications. Energy Fuels 29(4):2616–2644

Gonzalez-Martinez D, Gotor V, Gotor-Fernandez V (2016) Application of deep eutectic solvents in promiscuous lipase-catalysed aldol reactions. Eur J Org Chem 8:1513–1519

Gorke JT, Srienc F, Kazlauskas RJ (2008) Hydrolase-catalyzed biotransformations in deep eutectic solvents. Chem Commun 10:1235–1237

Guajardo N, Domínguez de María P, Ahumada K, Schrebler RA, Ramírez-Tagle R, Crespo FA, Carlesi C (2017) Water as cosolvent: non viscous deep eutectic solvents for efficient lipase-catalyzed esterifications. ChemCatChem 9(8):1393–1396

Guan Z, Li LY, He YH (2015) Hydrolase-catalyzed asymmetric carbon-carbon bond formation in organic synthesis. RSC Adv 5(22):16801–16814

Hallett JP, Welton T (2011) Room-temperature ionic liquids: solvents for synthesis and catalysis. 2. Chem Rev 111(5):3508–3576

Hammond OS, Bowron DT, Edler KJ (2016) Liquid structure of the choline chloride-urea deep eutectic solvent (reline) from neutron diffraction and atomistic modelling. Green Chem 18(9):2736–2744

Harifi-Mood AR, Ghobadi R, Divsalar A (2017) The effect of deep eutectic solvents on catalytic function and structure of bovine liver catalase. Int J Biol Macromol 95:115–120

Hayyan M, Hashim MA, Al-Saadi MA, Hayyan A, AlNashef IM, Mirghani MES (2013a) Assessment of cytotoxicity and toxicity for phosphonium-based deep eutectic solvents. Chemosphere 93(2):455–459

Hayyan M, Hashim MA, Hayyan A, Al-Saadi MA, AlNashef IM, Mirghani MES, Saheed OK (2013b) Are deep eutectic solvents benign or toxic? Chemosphere 90(7):2193–2195

Hayyan M, Looi CY, Hayyan A, Wong WF, Hashim MA (2015) In vitro and in vivo toxicity profiling of ammonium-based deep eutectic solvents. PLoS ONE 10(2):e0117934

Hayyan M, Mbous YP, Looi CY, Wong WF, Hayyan A, Salleh Z, Mohd-Ali O (2016) Natural deep eutectic solvents: cytotoxic profile. Springerplus 5:913

Hummel W, Gröger H (2014) Strategies for regeneration of nicotinamide coenzymes emphasizing self-sufficient closed-loop recycling systems. J Biotechnol 191:22–31

Juneidi I, Hayyan M, Hashim MA (2015) Evaluation of toxicity and biodegradability for cholinium-based deep eutectic solvents. RSC Adv 5(102):83636–83647

Juneidi I, Hayyan M, Ali OM (2016) Toxicity profile of choline chloride-based deep eutectic solvents for fungi and *Cyprinus carpio* fish. Environ Sci Pollut R 23(8):7648–7659

Juneidi I, Hayyan M, Hashim MA, Hayyan A (2017) Pure and aqueous deep eutectic solvents for a lipase-catalysed hydrolysis reaction. Biochem Eng J 117((Part A)):129–138

Kim SH, Park S, Yu H, Kim JH, Kim HJ, Yang YH, Kim YH, Kim KJ, Kan E, Lee SH (2016) Effect of deep eutectic solvent mixtures on lipase activity and stability. J Mol Catal B-Enzym 128:65–72

Kleiner B, Schörken U (2015) Native lipase dissolved in hydrophilic green solvents: a versatile 2-phase reaction system for high yield ester synthesis. Eur J Lipid Sci Technol 117(2):167–177

Kleiner B, Fleischer P, Schörken U (2016) Biocatalytic synthesis of biodiesel utilizing deep eutectic solvents: a two-step-one-pot approach with free lipases suitable for acidic and used oil processing. Process Biochem 51(11):1808–1816

Krystof M, Perez-Sanchez M, de Maria PD (2013) Lipase-catalyzed (Trans)esterification of 5-hydroxy-methylfurfural and separation from HMF esters using deep-eutectic solvents. Chemsuschem 6(4):630–634

Lindberg D, Revenga MD, Widersten M (2010) Deep eutectic solvents (DESs) are viable cosolvents for enzyme-catalyzed epoxide hydrolysis. J Biotechnol 147(3–4):169–171

Marták J, Schlosser Š (2016) New mechanism and model of butyric acid extraction by phosphonium ionic liquid. J Chem Eng Data 61(9):2979–2996

Maugeri Z, de Maria PD (2014) Whole-cell biocatalysis in deep-eutectic-solvents/aqueous mixtures. Chemcatchem 6(6):1535–1537

Maugeri Z, Domínguez de María P (2014) Benzaldehyde lyase (BAL)-catalyzed enantioselective CC bond formation in deep-eutectic-solvents–buffer mixtures. J Mol Catal B Enzym 107:120–123

Maugeri Z, Leitner W, Domínguez de María P (2013) Chymotrypsin-catalyzed peptide synthesis in deep eutectic solvents. Eur J Org Chem 20:4223–4228

Mbous YP, Hayyan M, Wong WF, Looi CY, Hashim MA (2017) Unraveling the cytotoxicity and metabolic pathways of binary natural deep eutectic solvent systems. Sci Rep 7:41257

Monhemi H, Housaindokht MR, Moosavi-Movahedi AA, Bozorgmehr MR (2014) How a protein can remain stable in a solvent with high content of urea: insights from molecular dynamics simulation of *Candida antarctica* lipase B in urea: choline chloride deep eutectic solvent. Phys Chem Chem Phys 16(28):14882–14893

Muller CR, Lavandera I, Gotor-Fernandez V, de Maria P (2015) Performance of recombinant-whole-cell-catalyzed reductions in deep-eutectic-solvent–aqueous-media mixtures. Chemcatchem 7(17):2654–2659

Ni Y, Xu JH (2012) Biocatalytic ketone reduction: a green and efficient access to enantiopure alcohols. Biotechnol Adv 30(6):1279–1288

Paiva A, Craveiro R, Aroso I, Martins M, Reis RL, Duarte ARC (2014) Natural deep eutectic solvents –solvents for the 21st century. ACS Sustain Chem Eng 2(5):1063–1071

Pant PL, Shankarling GS (2017) Deep eutectic solvent/lipase: two environmentally benign and recyclable media for efficient synthesis of *N*-aryl amines. Catal Lett 147(6):1371–1378

Papadopoulou AA, Efstathiadou E, Patila M, Polydera AC, Stamatis H (2016) Deep eutectic solvents as media for peroxidation reactions catalyzed by heme-dependent biocatalysts. Ind Eng Chem Res 55(18):5145–5151

Peters C, Rudroff F, Mihovilovic MD, Bornscheuer U (2017) Fusion proteins of an enoate reductase and a Baeyer-Villiger monooxygenase facilitate the synthesis of chiral lactones. Biol Chem 398(1):31–37

Petkovic M, Hartmann DO, Adamova G, Seddon KR, Rebelo LPN, Pereira CS (2012) Unravelling the mechanism of toxicity of alkyltributylphosphonium chlorides in *Aspergillus nidulans* conidia. New J Chem 36(1):56–63

Pöhnlein M, Ulrich J, Kirschhöfer F, Nusser M, Muhle-Goll C, Kannengiesser B, Brenner-Weiß G, Luy B, Liese A, Syldatk C, Hausmann R (2015) Lipase-catalyzed synthesis of glucose-6-O-hexanoate in deep eutectic solvents. Eur J Lipid Sci Technol 117(2):161–166

Potdar MK, Kelso GF, Schwarz L, Zhang CF, Hearn MTW (2015) Recent developments in chemical synthesis with biocatalysts in ionic liquids. Molecules 20(9):16788–16816

Procentese A, Johnson E, Orr V, Campanile AG, Wood JA, Marzocchella A, Rehmann L (2015) Deep eutectic solvent pretreatment and subsequent saccharification of corncob. Bioresour Technol 192:31–36

Qin YZ, Li YM, Zong MH, Wu H, Li N (2015) Enzyme-catalyzed selective oxidation of 5-hydroxymethylfurfural (HMF) and separation of HMF and 2,5-diformylfuran using deep eutectic solvents. Green Chem 17(7):3718–3722

Radosevic K, Bubalo MC, Srcek VG, Grgas D, Dragicevic TL, Redovnikovic IR (2015) Evaluation of toxicity and biodegradability of choline chloride based deep eutectic solvents. Ecotoxicol Environ Saf 112:46–53

Ren HW, Chen CM, Wang QH, Zhao DS, Guo SH (2016) The properties of choline chloride-based deep eutectic solvents and their performance in the dissolution of cellulose. BioResources 11(2):5435–5451

Roosen C, Muller P, Greiner L (2008) Ionic liquids in biotechnology: applications and perspectives for biotransformations. Appl Microbiol Biotechnol 81(4):607–614

Savile CK, Janey JM, Mundorff EC, Moore JC, Tam S, Jarvis WR, Colbeck JC, Kreber A, Fleitz FJ, Brands J, Devine PN, Huisman GW, Hughes GJ (2010) Biocatalytic asymmetric synthesis of chiral amines from ketones applied to sitagliptin manufacture. Science 329(5989):305

Shahbaz K, Mjalli FS, Hashim MA, AlNashef IM (2011) Using deep eutectic solvents based on methyl triphenyl phosphonium bromide for the removal of glycerol from palm-oil-based biodiesel. Energy Fuels 25(6):2671–2678

Sheldon RA (2016) Biocatalysis and biomass conversion in alternative reaction media. Chem Eur J 22(37):12983–12998

Siebenhaller S, Muhle-Goll C, Luy B, Kirschhöfer F, Brenner-Weiss G, Hiller E, Günther M, Rupp S, Zibek S, Syldatk C (2017) Sustainable enzymatic synthesis of glycolipids in a deep eutectic solvent system. J Mol Catal B Enzym. doi:10.1016/j.molcatb.2017.01.015

Smith EL, Abbott AP, Ryder KS (2014) Deep eutectic solvents (DESs) and their applications. Chem Rev 114(21):11060–11082

Stepankova V, Vanacek P, Damborsky J, Chaloupkova R (2014) Comparison of catalysis by haloalkane dehalogenases in aqueous solutions of deep eutectic and organic solvents. Green Chem 16(5):2754–2761

Thuy Pham TP, Cho C-W, Yun Y-S (2010) Environmental fate and toxicity of ionic liquids: a review. Water Res 44(2):352–372

Tian XM, Zhang SQ, Zheng LY (2016) Enzyme-catalyzed henry reaction in choline chloride-based deep eutectic solvents. J Microbiol Biotechnol 26(1):80–88

Tufvesson P, Lima-Ramos J, Nordblad M, Woodley JM (2011) Guidelines and cost analysis for catalyst production in biocatalytic processes. Org Process Res Dev 15(1):266–274

van Osch D, Zubeir LF, van den Bruinhorst A, Rocha MAA, Kroon MC (2015) Hydrophobic deep eutectic solvents as water-immiscible extractants. Green Chem 17(9):4518–4521

van Rantwijk F, Sheldon RA (2007) Biocatalysis in ionic liquids. Chem Rev 107(6):2757–2785

Vitale P, Abbinante VM, Perna FM, Salomone A, Cardellicchio C, Capriati V (2016) Unveiling the hidden performance of whole cells in the asymmetric bioreduction of aryl-containing ketones in aqueous deep eutectic solvents. Adv Synth Catal 358:1–10

Wachtmeister J, Rother D (2016) Recent advances in whole cell biocatalysis techniques bridging from investigative to industrial scale. Curr Opin Biotechnol 42:169–177

Wang Q, Yao X, Tang S, Lu X, Zhang X, Zhang S (2012) Urea as an efficient and reusable catalyst for the glycolysis of poly(ethylene terephthalate) wastes and the role of hydrogen bond in this process. Green Chem 14(9):2559–2566

Weingartner H, Cabrele C, Herrmann C (2012) How ionic liquids can help to stabilize native proteins. Phys Chem Chem Phys 14(2):415–426

Weiz G, Braun L, Lopez R, de María PD, Breccia JD (2016) Enzymatic deglycosylation of flavonoids in deep eutectic solvents–aqueous mixtures: paving the way for sustainable flavonoid chemistry. J Mol Catal B Enzym 130:70–73

Wen Q, Chen JX, Tang YL, Wang J, Yang Z (2015) Assessing the toxicity and biodegradability of deep eutectic solvents. Chemosphere 132:63–69

Wu BP, Wen Q, Xu H, Yang Z (2014) Insights into the impact of deep eutectic solvents on horseradish peroxidase: activity, stability and structure. J Mol Catal B Enzym 101:101–107

Wu X, Li G, Yang H, Zhou H (2015) Study on extraction and separation of butyric acid from clostridium tyrobutyricum fermentation broth in PEG/Na$_2$SO$_4$ aqueous two-phase system. Fluid Phase Equilib 403:36–42

Xiao ZJ, Du PX, Lou WY, Wu H, Zong MH (2012) Using water-miscible ionic liquids to improve the biocatalytic anti-Prelog asymmetric reduction of prochiral ketones with whole cells of Acetobacter sp. CCTCC M209061. Chem Eng Sci 84:695–705

Xu P, Xu Y, Li XF, Zhao BY, Zong MH, Lou WY (2015) Enhancing asymmetric reduction of 3-chloropropiophenone with immobilized Acetobacter sp. CCTCC M209061 cells by using deep eutectic solvents as cosolvents. ACS Sustain Chem Eng 3(4):718–724

Xu P, Zheng GW, Du PX, Zong MH, Lou WY (2016) Whole-cell biocatalytic processes with ionic liquids. ACS Sustain Chem Eng 4(2):371–386

Yang SL, Duan ZQ (2016) Insight into enzymatic synthesis of phosphatidylserine in deep eutectic solvents. Catal Commun 82:16–19

Yang TX, Zhao LQ, Wang J, Song GL, Liu HM, Cheng H, Yang Z (2017) Improving whole-cell biocatalysis by addition of deep eutectic solvents and natural deep eutectic solvents. ACS Sustain Chem Eng 5(7):5713–5722

Zakrewsky M, Banerjee A, Apte S, Kern TL, Jones MR, Del Sesto RE, Koppisch AT, Fox DT, Mitragotri S (2016) Choline and geranate deep eutectic solvent as a broad-spectrum antiseptic agent for preventive and therapeutic applications. Adv Healthc Mater 5(11):1282–1289

Zhang Q, De Oliveira Vigier K, Royer S, Jerome F (2012) Deep eutectic solvents: syntheses, properties and applications. Chem Soc Rev 41(21):7108–7146

Zhang CW, Xia SQ, Ma PS (2016) Facile pretreatment of lignocellulosic biomass using deep eutectic solvents. Bioresour Technol 219:1–5

Zhao DB, Liao YC, Zhang ZD (2007) Toxicity of ionic liquids. Clean Soil Air Water 35(1):42–48

Zhao H, Baker GA, Holmes S (2011a) New eutectic ionic liquids for lipase activation and enzymatic preparation of biodiesel. Org Biomol Chem 9(6):1908–1916

Zhao H, Baker GA, Holmes S (2011b) Protease activation in glycerol-based deep eutectic solvents. J Mol Catal B Enzym 72(3–4):163–167

Zhao H, Zhang C, Crittle TD (2013) Choline-based deep eutectic solvents for enzymatic preparation of biodiesel from soybean oil. J Mol Catal B Enzym 85–86:243–247

Zhao BY, Xu P, Yang FX, Wu H, Zong MH, Lou WY (2015) Biocompatible deep eutectic solvents based on choline chloride: characterization and Application to the extraction of rutin from Sophora japonica. ACS Sustain Chem Eng 3(11):2746–2755

Zhao KH, Cai YZ, Lin XS, Xiong J, Halling P, Yang Z (2016) Enzymatic synthesis of glucose-based fatty acid esters in bisolvent systems containing ionic liquids or deep eutectic solvents. Molecules 21(10):1294

Zhou P, Wang X, Zeng C, Wang W, Yang B, Hollmann F, Wang Y (2017a) Deep eutectic solvents enable more robust chemoenzymatic epoxidation reactions. ChemCatChem 9(6):934–936

Zhou PF, Wang XP, Yang B, Hollmann F, Wang YH (2017b) Chemoenzymatic epoxidation of alkenes with Candida antarctica lipase B and hydrogen peroxide in deep eutectic solvents. RSC Adv 7(21):12518–12523

Application of methanol and sweet potato vine hydrolysate as enhancers of citric acid production by *Aspergillus niger*

Daobing Yu[1], Yanke Shi[1], Qun Wang[2], Xin Zhang[1*] and Yuhua Zhao[2*]

Abstract

Background: Agricultural waste is as an alternative low-cost carbon source or beneficial additives which catch most people's eyes. In addition, methanol and sweet potato vine hydrolysate (SVH) have been reported as the efficient enhancers of fermentation according to some reports. The objective of the present study was to confirm SVH as an efficient additive in CA production and explore the synergistic effects of methanol and SVH in fermentation reactions.

Results: The optimal fermentation conditions resulted in a maximum citric acid concentration of 3.729 g/L. The final citric acid concentration under the optimized conditions was increased by 3.6-fold over the original conditions, 0.49-fold over the optimized conditions without methanol, and 1.8-fold over the optimized conditions in the absence of SVH. Kinetic analysis showed that Q_p, $Y_{p/s}$, and $Y_{x/s}$ in the optimized systems were significantly improved compared with those obtained in the absence of methanol or SVH. Further, scanning electron microscopy (SEM) revealed that methanol stress promoted the formation of conidiophores, while SVH could neutralize the effect and prolong Aspergillus niger vegetative growth. Cell viability analysis also showed that SVH might eliminate the harmful effects of methanol and enhance cell membrane integrity.

Conclusions: SVH was a superior additive for organic acid fermentation, and the combination of methanol and SVH displayed a significant synergistic effect. The research provides a preliminary theoretical basis for SVH practical application in the fermentation industry.

Keywords: Citric acid, Methanol, Sweet potato vines hydrolysate, Synergistic effect, *Aspergillus niger*

Background

Citric acid (CA) is a versatile organic acid with a broad range of uses in the food, pharmaceutical, beverage, and cosmetic industries because of its mild sour flavor, water solubility, strong complexation, and high level of biological safety. Currently, the global production of CA has reached 1.7 million tons per year, produced almost entirely through fermentation, with an annual growth rate of 5% (Kana et al. 2012). The major method for manufacturing CA relies on the submerged fermentation of starch-based or sucrose-based feedstock by the fungus *Aspergillus niger*. Karaffa and Kubicek (2003) summarized the important role of glycolysis, the process of excretion and transport of CA, and the critical fermentation variables for CA accumulation. Finally, they proposed that the type and concentration of the carbon source is probably the most crucial parameter for successful citric acid production. A yield of 95 kg CA per 100 kg was even achieved of supplied sugar using best fermentation strains. However, high energy coupled with raw material costs has pushed CA production into an unprofitable market (Dhillon et al. 2011).

In the past few years, the use of agricultural waste as an alternative low-cost carbon source or beneficial additives has received considerable attention. For example, carob pod (Roukas 1998), kiwi fruit peel (Hang and Woodams

*Correspondence: zhangxins@126.com; yhzhao225@zju.edu.cn
[1] College of Forestry and Biotechnology, Zhejiang Agriculture and Forestry University, Lin'an 311300, Zhejiang, People's Republic of China
[2] College of Life Sciences, Zhejiang University, Hangzhou 310058, Zhejiang, People's Republic of China

1998), pineapple waste (Kumar et al. 2003), jackfruit waste (Angumeenal and Venkappayya 2005), sugar cane bagasse (Ali and Haq 2005), orange peel (Rivas et al. 2008), and apple pomace (Dhillon et al. 2011; Ali et al. 2015) have all been utilized for CA production by *Aspergillus* sp. Sweet potato (*Ipomoea batatas*) is the world's seventh most important food crop after rice, wheat, potatoes, corn, barley, and cassava (FAO 1997), with a production exceeding 105 million metric tons per year worldwide (Ishida et al. 2000). However, only the tuberous roots of sweet potato are used as food in most areas, whereas the vines are discarded as agricultural waste (Ishida et al. 2000), causing environmental pollution and wasting resources. Sweet potato vines contain nutritional and functionally valuable components and represent a considerable source of carbon, nitrogen, and energy (Ishida et al. 2000). However, there are few relevant reports on the effective application of sweet potato vines in bio-industry, apart from previous research confirming that the combination of molasses and SVH was good alternative carbon source in lipid fermentation (Shen et al. 2013, 2015).

A preliminary study revealed that CA production was markedly increased when 1% methanol and 10% SVH were added to basal medium under standard fermentation conditions. The objective of the present study was to confirm SVH as an efficient additive in CA production and explore the synergistic effects of methanol and SVH in fermentation reactions. To achieve the best fermentation results, the response surface methodology (RSM) was used to optimize key variables, such as temperature, pH, inoculum quantity, aeration condition, and fermentation period. In addition, the mechanism of methanol- and SVH-enhanced CA production was assessed by cultivation dynamics, SEM, and flow cytometry (FCM) of *A. niger* cells under different incubation conditions. The research will provide a preliminary theoretical basis for SVH practical application in the fermentation industry.

Methods
Microorganism and medium
The fungus *Aspergillus niger* sp. ZJUY previously isolated in our laboratory was used in the present study. The PDA medium for harvesting spores contained 200.0 g/L peeled potato, 20.0 g/L glucose, and 20.0 g/L agar (pH not adjusted). The CA-fermentation basal medium contained 50.0 g/L potato starch supplemented with 20 μL/L α-amylase (20,000 U/mL), 50.0 g/L glucose, 3.0 g/L NH_4Cl, 0.2 g/L $MgSO_4 \cdot 7H_2O$ (pH 6.3). SVH was prepared as described in Zhan et al. (2013). A single-factor analytic approach was adopted to determine the optimal concentration of methanol and SVH. RSM was utilized to optimize the CA-fermentation conditions as described below.

Cultivation
ZJUY was inoculated on PDA agar and cultured at 28 °C for 96 h. Spores were eluted with 20 mL 0.1% (v/v) Tween-80, of which approximately 15 mL was filtered through lens paper, transferred to a sterilized 50-mL flat-top centrifuge tube, and then separated by centrifugation (2700×g, 5 min). The supernatant was removed and the spores resuspended in 20 mL of 0.1% (v/v) Tween-80, and then centrifuged again. The resulting pellet was resuspended in 3 mL sterile water and diluted to 1×10^9 CFU/mL in seed broth. Fermentation experiments were performed in a 250-mL Erlenmeyer flask containing 100 mL medium and cultured at 37 °C at 200 rpm for 4 days.

Determination of citric acid concentration
Citric acid concentration was determined with an Agilent 1200 Series high-performance liquid chromatography (HPLC) instrument equipped with a UV/Vis detector and Eclipse Plus C18 column (250 × 4.6 mm × 5 μm; Agilent Technologies, Santa Clara, CA, USA). The processed broth was diluted 10-fold in the phosphate buffer (25 mM, pH 2.4), filtered through a 0.22-μm membrane, and injected into 2.0-mL autosampler vials. CA was separated with a mobile phase composed of methanol and phosphate buffer (25 mM, pH 2.4) at a 1:9 ratio (v/v) with a flow rate of 1.0 mL/min at 30 °C. The injection volume was 20 μL and we performed three replicates of each trial. CA quantitation was performed at the wavelength of maximum absorbance for each analyte ($\lambda = 210$ nm; A_{210}) obtained from UV spectrophotometry spectra determination (Rodrigues et al. 2007). CA was identified by comparing its retention time with that of a standard substance, and its concentration was quantified using an external standard calibration, calculated by the following equations:

$$C = k \times F \tag{1}$$

$$F = 9 \times 10^{-4} \times S - 0.0010 \tag{2}$$

where, k is the dilution ration ($k = 10$), F is the linear regression function ($R^2 = 1.0000$), S is the peak area of CA at A_{210}, and C is the CA concentration (g/L).

Determination of the optimal levels of fermentation factors
Factors previously identified to affect CA production were analyzed in single-factor experiments, including temperature (°C; 20, 25, 30, 35, 40, 45), pH (3.5, 4.0, 4.5, 5.0, 5.5, 6.0, 6.5, 7.0), inoculum quantity (%; 0.5, 1, 3, 5, 8, 10, 12, 15), rotational speed (rpm; 80, 100, 120, 140, 160, 180, 200, 200, 220), fermentation period (d; 3, 4, 5, 6, 7, 8, 9, 10), methanol content (%; 1, 2, 3, 4, 5, 6, 7, 8), and SVH concentration (%; 1, 3, 5, 8, 10, 12, 15, 20). All trials were performed in triplicate as described above.

Table 1 Plackett–Burman design for 7 variables and 12 trials

Trial	Variable							CA (g/L)
	pH	Temperature (°C)	Inoculum quantity (%)	Rotational speed (rpm)	Fermentation period (d)	Methanol content (%)	SVH (%)	
1	6 (−1)	44 (+1)	15 (+1)	125 (+1)	5 (−1)	6 (−1)	10 (−1)	1.917
2	7.5 (+1)	44	12 (−1)	125	6.25 (+1)	7.5 (+1)	10	1.856
3	7.5	44	12	100 (−1)	5	7.5	10	1.755
4	7.5	35 (−1)	12	100	6.25	6	12.5 (+1)	2.072
5	7.5	35	15	125	5	7.5	12.5	2.050
6	6	35	12	125	5	7.5	12.5	2.000
7	6	35	12	100	5	6	10	1.784
8	6	44	12	125	6.25	6	12.5	2.166
9	7.5	35	15	125	6.25	6	10	2.105
10	6	44	15	100	6.25	7.5	12.5	2.057
11	6	35	15	100	6.25	7.5	10	2.396
12	7.5	44	15	100	5	6	12.5	2.004

The (−1) indicates the low level

(+1) indicates the high level

Factorial design and optimization study of CA production

Plackett–Burman design

For screening purposes, seven independent variables were screened in 12 combinations organized according to the Plackett–Burman design (Table 1). All experiments were performed in triplicate and the average CA concentration was treated as the response. The optimal level of each single-factor experiment became the low level, while the high level was 1.25 times the low level. The main effect of each variable was calculated as the difference between the average of measurements made at the high setting (+) and the average of measurements observed at the low setting (−) for that factor.

Central composite design

Fermentation factors affecting CA production were optimized with CCD in Design-Expert software, version 10.0, (Stat-Ease Inc., Minneapolis, MN, USA) using 50 experimental runs and 5 variables. Each factor was examined at five different levels: relatively low (−), low (−), basal (0), high (+), and relatively high (++) (Table 2). The CCD results were fitted to a second-order polynomial model as follows:

$$Y = \beta_0 + \beta_1 A + \beta_2 B + \beta_3 C + \beta_4 D \\ + \beta_5 E + \beta_{12} AB + \beta_{13} AC + \beta_{14} AD \\ + \beta_{15} AE + \beta_{23} BC + \beta_{24} BD + \beta_{25} BE \\ + \beta_{34} CD + \beta_{35} CE + \beta_{45} DE + \beta_{11} A^2 \\ + \beta_{22} B^2 + \beta_{33} C^2 + \beta_{44} D^2 + \beta_{55} E^2 \tag{3}$$

where, Y is the dependent variable (CA concentration), A is the initial pH, B is temperature (°C), C is inoculum

quantity (%), D is fermentation period (d), E is SVH concentration (%), β_0 is the regression coefficient at the center point, and β is the estimated coefficient for each term of the response surface model.

Citrate synthase activity analysis

Aspergillus niger cell-free extracts were prepared as described by Kobayashiet al. (2013), and the protein concentrations determined with a Bradford Protein Assay Kit. CS activity was measured as previously described by Srere (1966) with some modifications. The reaction mixture contained 50 mM Tris–HCl (pH 8.0), 500 mM $MgCl_2$, 5.0 mM 5,5′-dithiobis-(2-nitrobenzoic acid) (DTNB), 20 mM acetyl-CoA, 50 mM oxaloacetate, and crude enzyme solution in a total volume of 1.0 mL. One unit (U) was defined as the amount of enzyme catalyzing the liberation of 1 μmol of CoA-SH per min.

Kinetic characterization of fermentation

Kinetic parameters of the CA-fermentation process were determined according to Pirt (1975). Kinetic parameters: Q_p = grams of CA produced/L/h, $Y_{p/s}$ = grams of CA produced/gram of substrate consumed, $Y_{x/s}$ = grams of cells/gram of substrate utilized, Q_s = grams of substrate consumed/L/h (Ali 2007).

Observations of the mycelial morphology

Scanning electron microscope was used to observe changes in mycelial morphology under different cultural conditions. For this, the culture samples were centrifuged, washed, and the insoluble fractions subsequently resuspended in sterile water and fixed in 2.5%

Table 2 Independent variables and levels of variation in CCD

Trial	Variable					CA (g/L)
	pH	Temperature (°C)	Inoculum quantity (%)	Fermentation period (d)	SVH (%)	
1	4 (−)	35 (+)	12 (−)	7 (+)	12 (+)	2.011
2	6 (+)	35	18 (+)	7	10 (−)	2.485
3	6	25 (−)	18	7	12	2.032
4	5 (0)	30 (0)	15 (0)	6 (0)	13.4 (++)	2.426
5	6	25	12	7	10	2.292
6	5	30	15	6	11 (0)	2.588
7	4	35	18	5 (−)	10	2.216
8	4	35	12	5	12	2.579
9	6	35	12	7	10	2.617
10	4	25	18	7	12	2.284
11	4	35	18	7	10	2.263
12	4	35	18	5	12	2.211
13	6	25	18	5	12	2.852
14	5	30	8 (−)	6	11	2.775
15	5	30	15	6	11	2.803
16	5	42 (++)	15	6	11	2.581
17	6	25	12	7	12	2.466
18	4	25	18	5	12	2.027
19	5	30	15	6	11	2.716
20	5	30	15	6	11	2.685
21	6	25	18	5	10	2.245
22	4	35	18	7	12	2.062
23	5	30	15	6	11	2.788
24	4	25	12	7	12	2.264
25	4	25	12	5	10	2.303
26	5	30	22 (++)	6	11	2.855
27	6	35	18	5	10	2.916
28	6	35	18	5	12	3.682
29	6	25	12	5	10	2.413
30	4	25	12	7	10	2.226
31	6	35	18	7	12	2.591
32	4	25	18	5	10	1.985
33	4	25	12	5	12	2.148
34	6	25	12	5	12	2.583
35	6	35	12	7	12	2.383
36	5	30	15	8.4 (++)	11	2.286
37	2.6 (−)	30	15	6	11	1.981
38	5	18 (−)	15	6	11	2.132
39	6	35	12	5	10	2.851
40	7.4 (++)	30	15	6	11	2.171
41	5	30	15	6	11	2.741
42	5	30	15	6	8.6 (−)	2.733
43	5	30	15	3.6 (−)	11	2.078
44	5	30	15	6	11	2.407
45	5	30	15	6	11	2.593
46	4	25	18	7	10	2.541
47	4	35	12	7	10	2.628
48	4	35	12	5	10	2.619

Table 2 continued

Trial	Variable					CA (g/L)
	pH	Temperature (°C)	Inoculum quantity (%)	Fermentation period (d)	SVH (%)	
49	6	35	12	5	12	2.882
50	6	25	18	7	10	1.953

Each factor was examined at 5 levels including (−), (−), (0), (+) and (++) which indicates the highest level

glutaraldehyde at 4 °C overnight. The fixed sample was washed separately with 1% osmium and PBS three times and then dehydrated with an ethanol gradient. The dehydrated samples were dried with critical point drier, and sputter-coated with gold before SEM analysis.

Detection of cell viability

Filtered broth from 24-h cultures was centrifuged to collected spores, which were then diluted to approximately 10^6 spores/mL. The spores were subsequently fixed with 0.2% oxymethylene, washed twice with PBS buffer, and then stained with 30 μg/mL propidium iodide (PI; Beyotime, China) at room temperature for 30 min. The stained spheroblasts were pelleted by centrifuging for 2 min at $16,000 \times g$ at 4 °C, washed, and then resuspended in 500 μL PBS buffer. Flow cytometry (FCM) coupled with PI was used to monitor *A. niger* membrane integrity and sorted at a rate of approximately 500 events per second, with 30,000 total events detected in each run. Flow cytometry was carried out using a BD FACSCalibur instrument (Becton–Dickinson, USA) fitted with a 15 mW argon ion laser for excitation (488 nm), while monitoring with three different emission channels (530/30, 585/42, and 670 nm Lp). BD CellQuest Pro software (Becton–Dickinson, USA) was used for instrument control, data acquisition, and data analysis.

Results

Single-factor optimization of fermentation

Single-factor optimization was used to confirm the effects of temperature, pH, inoculum quantity, rotational speed, fermentation period, methanol content, and SVH concentration on CA production (Fig. 1). Notably, temperature showed a significant effect on CA concentration according to normal distribution with minor variation. The maximum CA concentration was observed at 35 °C (Fig. 1a), which was significantly decreased when the fermentation temperature was higher than 45 °C or lower than 20 °C. This phenomenon was correlated with biomass production (data not shown). Similar findings were reported by Karthikeyan and Sivakumar (2010).

Rault et al. (2009) specified the effects of fermentation pH on the physiological-state dynamics of *Lactobacillus bulgaricus* CFL1. The cells maintained a more vigorous physiological state when pH was controlled at 5,

with higher viability and steady acidification activity, but these characteristics fluctuated and declined at pH 6 during the fermentation process. Moreover, pH also played an important role in CA accumulation (Fig. 1b), which increased with initial pH from 3.5 to 6.0, and a highest value (2.618 g/L) was achieved at an initial pH of 6.0. However, the CA concentration decreased dramatically once the initial pH exceeded 7.0, consistent with results from Roukas (1998) on the effect of initial pH on CA production from carob pods by surface fermentation.

Analysis of various *A. niger* inoculation levels on CA fermentation revealed that the maximum CA concentration was obtained with an inoculation quantity of 12% (Fig. 1c), equivalent to a final spore concentration of 10^8 CFU/mL, similar to previous findings (Vandenberghe et al. 2000). An optimal inoculum level is critical for CA accumulation because low inoculums may provide inadequate biomass and limit CA formation, whereas excessive levels could generate too much biomass to retain sufficient nutrients necessary for CA production (Sabu et al. 2006).

In the bioreactor, agitation replaced the rotational shaking, but these techniques have similar effects on CA production. We selected the small-scale flask-shaking fermentation method for analysis. Rotational speed is known to affect both the mobility of fermentation broths and the rate of oxygen transfer (Amanullah et al. 1998). The maximum specific growth rate of the culture in the exponential phase was closely related to the amount of dissolved oxygen. As shown in Fig. 1d, the peak of CA concentration appeared at 100 rpm and decreased when the rotation speed exceeded the optimal level, perhaps due to the excessive fungal growth.

Different fermentation periods were compared to determine the optimal incubation time. Notably, the highest CA concentration was obtained at 5 days of culture in our study. Further incubation resulted in a sharp decreased in CA production (Fig. 1e), in agreement with results from Karthikeyan and Sivakumar (2010). However, the optimal period varied from material to material. For instance, the best fermentation time for *A. niger* was 5 days when cassava bagasse was used as a substrate (Vandenberghe et al. 2000), 4 days with jack fruit carpel fiber (Angumeenal and Venkappayya 2005) and orange peel autohydrolysate (Rivas et al. 2008).

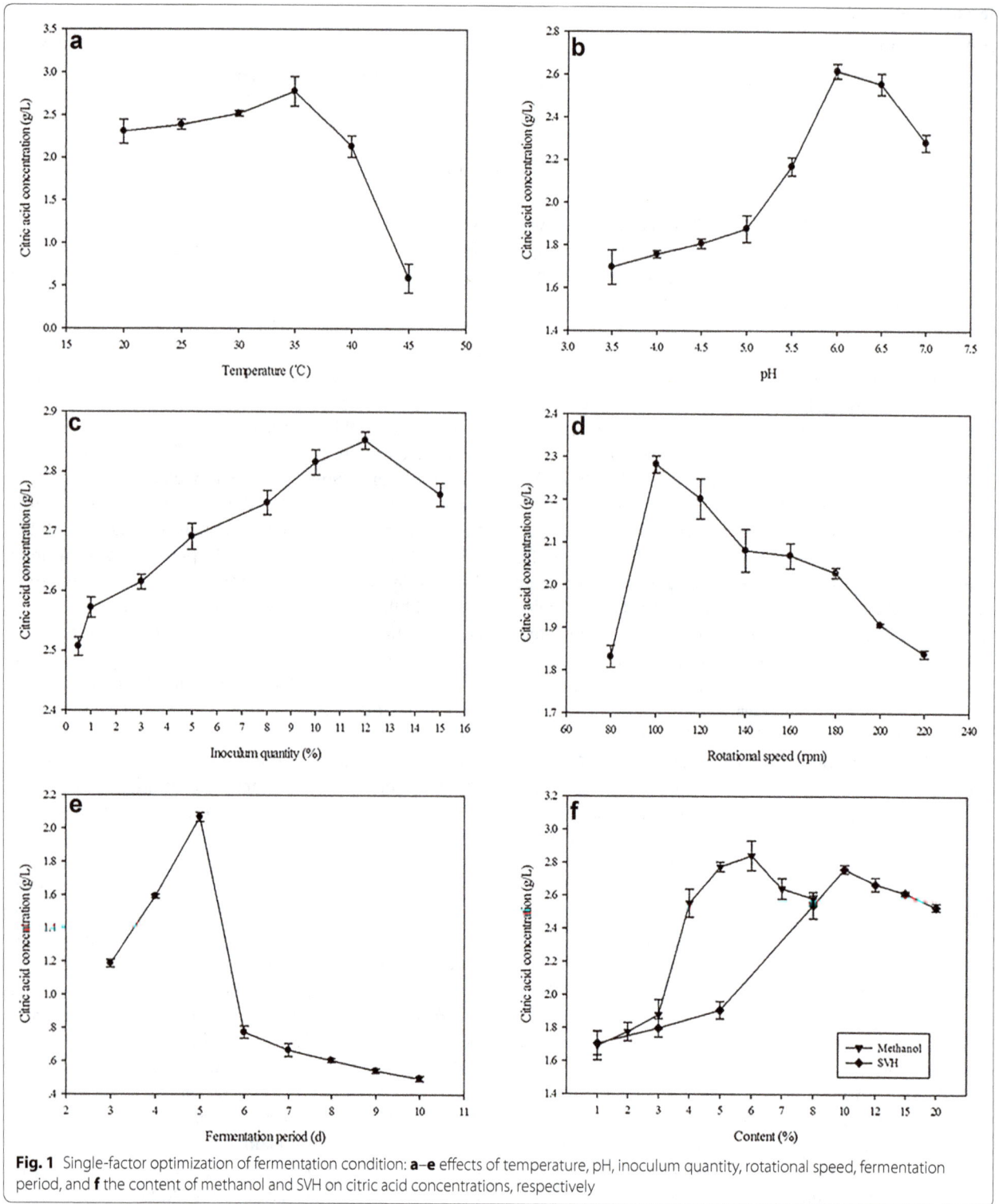

Fig. 1 Single-factor optimization of fermentation condition: **a–e** effects of temperature, pH, inoculum quantity, rotational speed, fermentation period, and **f** the content of methanol and SVH on citric acid concentrations, respectively

Methanol is known to boost CA production by *A. niger* (Navaratnam et al. 1998), likely by stimulating its excretion by increasing cell membrane permeability, which can reduce the mass transfer resistance of the membrane and strengthen the catalysis of the cell, without damaging intracellular organic structures or causing cell lysis. (Rivas et al. 2008). Moreover, 2-oxoglutarate dehydrogenase activity was low, whereas that of pyruvate carboxylase was high in the presence of methanol. Maddox et al. (1986) also observed strong correlations between CA production and the activities of these two enzymes. In the present study, the maximum threshold of CA concentration was obtained with 6% methanol (Fig. 1f).

Figure 1f depicts the effect of SVH addition on CA production, which showed a steady increase as SVH increased from 1 to 10%, and the highest concentration (2.762 g/L) was achieved with 10% SVH. However, the gradually increasing trend was reversed with SVH concentrations greater than 10%.

Evaluation of significant factors affecting CA concentration

The main effect of each variable on CA concentration was estimated as the difference between the average CA concentrations at the high (+1) and low (−1) levels. As shown in Table 3, all variables had significant effects on CA production during the fermentation period with the exception of rotational speed (from 100 to 125 rpm) and methanol content (from 6 to 7.5%), whereas fermentation period affected CA accumulation the most, followed by inoculation quantity, temperature, and SVH ($p < 0.05$).

Optimization of fermentation conditions by CCD

Each independent variable was investigated at five levels according to the CCD. Table 2 represents the design matrix of the coded variables together, according to its result on CA concentration. Best-fit models were determined by quadratic regression. The modified quadratic model was a highly significant ($p < 0.001$) representation of the actual relationships between the responses and significant variables. ANOVA was used to evaluate the significance of the coefficients of the modified quadratic

model (Table 4), and a second-order polynomial function was fitted to the experimental results using Design-Expert 10.0 software as follows:

$$
\begin{aligned}
Y = {} & 2.66 + 0.11A + 0.11B - 0.033C - 0.051D \\
& - 0.021E + 0.064AB + 0.024AC - 0.097AD \\
& + 0.069AE - 2.031 \times 10^{-3}BC - 0.054BD \\
& - 0.050BE + 8.281 \times 10^{-3}CD + 0.034CE \\
& - 0.051DE - 0.11A^2 - 0.058B^2 + 0.023C^2 \\
& - 0.089D^2 - 0.019E^2
\end{aligned}
\tag{4}
$$

Solving the model according to the data obtained from Table 2 revealed an optimal response under the following conditions: 18% inoculum quantity, 12% SVH, 7.5% methanol, initial pH of 6.0, 125 r/min rotational speed, and culturing at 35 °C for 5 days. These factors produced a final CA concentration of 3.673 g/L, which was very similar to the predicted condition of Trial 28 in the CCD experiment.

Effects of the interaction of tested variables on the CA concentration were clearly represented in the three-dimensional response surface plots (Fig. 2). The results were analyzed by ANOVA which illustrated that the interactions of initial pH-fermentation temperature, initial pH-fermentation period, and initial pH-SVH significantly affected CA production ($p < 0.05$).

Biochemical behaviors of A. niger in different fermentation systems

To confirm the model's validity and explore the synergistic effects of adding both methanol and SVH to the CA-fermentation system, an experiment was performed under the predicted optimized conditions in the presence or absence of methanol and SVH. The original fermentation conditions were used as the control: initial pH 6.3, 37 °C fermentation temperature, 1% inoculum quantity, and 200 rpm rotational speed, with a 4-day fermentation period.

Notably, the addition of SVH and methanol enhanced CA production by A. niger using glucose and potato starch as the substrates (Fig. 3a). The optimized conditions produced a CA concentration of 3.729 g/L, which is higher than the CA concentrations of 1.2 mg/g and 0.45 g/L obtained using sugar cane bagasse (Ali and Haq 2005) and apple pomace (Ali et al. 2015), respectively, and close to the theoretically predicted concentration of 3.673 g/L. These results demonstrated the accuracy and applicability of the CCD model as a useful optimization method for biotechnological applications. The final CA concentration produced by this approach was 4.6-fold higher than the control, 0.49-fold higher than that produced in the absence of methanol, and

Table 3 Analysis of variance (ANOVA) for Plackett–Burman design

Source	Sum of squares	df	F value	p value Prob > F	
Model	0.34	10	600.73	0.0317	significant
A-initial pH	0.019	1	337.99	0.0346	
B-temperature	0.035	1	628.85	0.0254	
C-inoculation quantity	0.067	1	1187.6	0.0185	
E-fermentation period	0.11	1	1929.24	0.0145	
F-methanol content	3.63E−04	1	6.44	0.2389	
G-SVH	0.024	1	424.99	0.0309	

Table 4 ANOVA and coefficient values for CCD

Source	Sum of squares	df	F value	p value Prob > F	
Model	3.26	20	5.11	<0.0001	Significant
A-initial pH	0.5	1	15.49	0.0005	
B-fermentation temperature	0.52	1	16.32	0.0004	
C-inoculum quantity	0.047	1	1.47	0.2351	
D-fermentation period	0.11	1	3.57	0.0690	
E-SVH	0.02	1	0.62	0.4391	
AB	0.13	1	4.16	0.0506	
AC	0.018	1	0.57	0.4579	
AD	0.3	1	9.39	0.0047	
AE	0.15	1	4.72	0.0382	
BC	1.32E−04	1	4.13E−03	0.9492	
BD	0.093	1	2.9	0.0995	
BE	0.079	1	2.47	0.1270	
CD	2.20E−03	1	0.069	0.7952	
CE	0.036	1	1.13	0.2975	
DE	0.083	1	2.6	0.1177	
A^2	0.64	1	20.1	0.0001	
B^2	0.19	1	5.84	0.0222	
C^2	0.03	1	0.93	0.3434	
D^2	0.44	1	13.71	0.0009	
E^2	0.019	1	0.6	0.4461	
Residual	0.93	29			
Lack of fit	0.81	22	2.13	0.1544	Not significant
Pure error	0.12	7			
Cor total	4.19	49			

Coefficient of determination (R^2) = 0.7788, CV = 7.31%. The "Pred R-Squared" of 0.2324 was not as close to the "Adj R-Squared" of 0.6262 as one might normally expect. The "Adeq Precision" value of 10.000 indicated an adequate signal

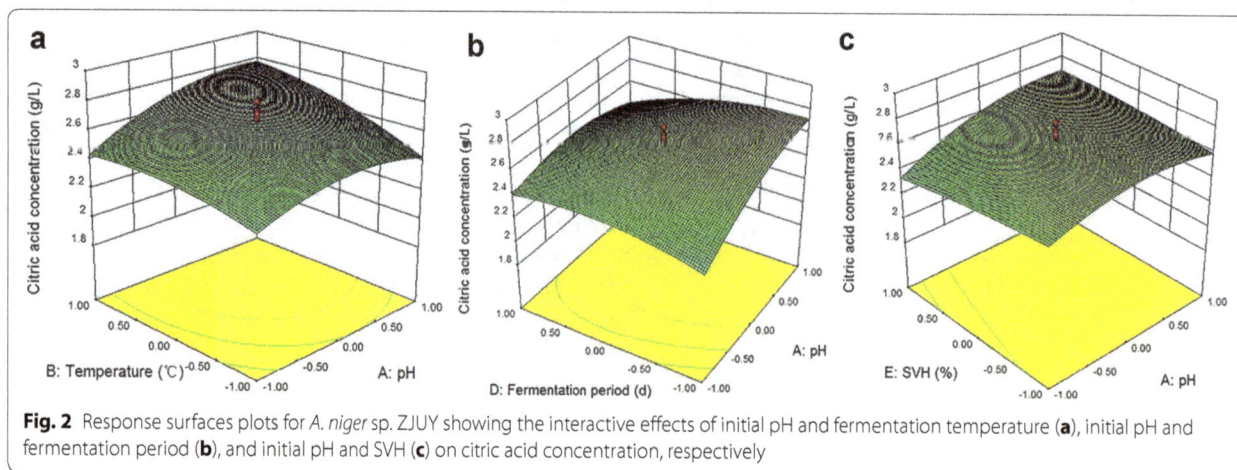

Fig. 2 Response surfaces plots for A. niger sp. ZJUY showing the interactive effects of initial pH and fermentation temperature (**a**), initial pH and fermentation period (**b**), and initial pH and SVH (**c**) on citric acid concentration, respectively

1.8-fold higher than that produced without SVH, indicating SVH is a cost-effective additive in CA production, and its combination with the permeabilization agent methanol shows an apparent synergistic effect in CA accumulation.

In addition, the pH of the four different cultivation systems behaved similarly with prolonged incubation times (Fig. 3b). All pH values dramatically declined at the initial phase, and then flattened out. It was also noted that pH in systems with SVH was obviously higher than others

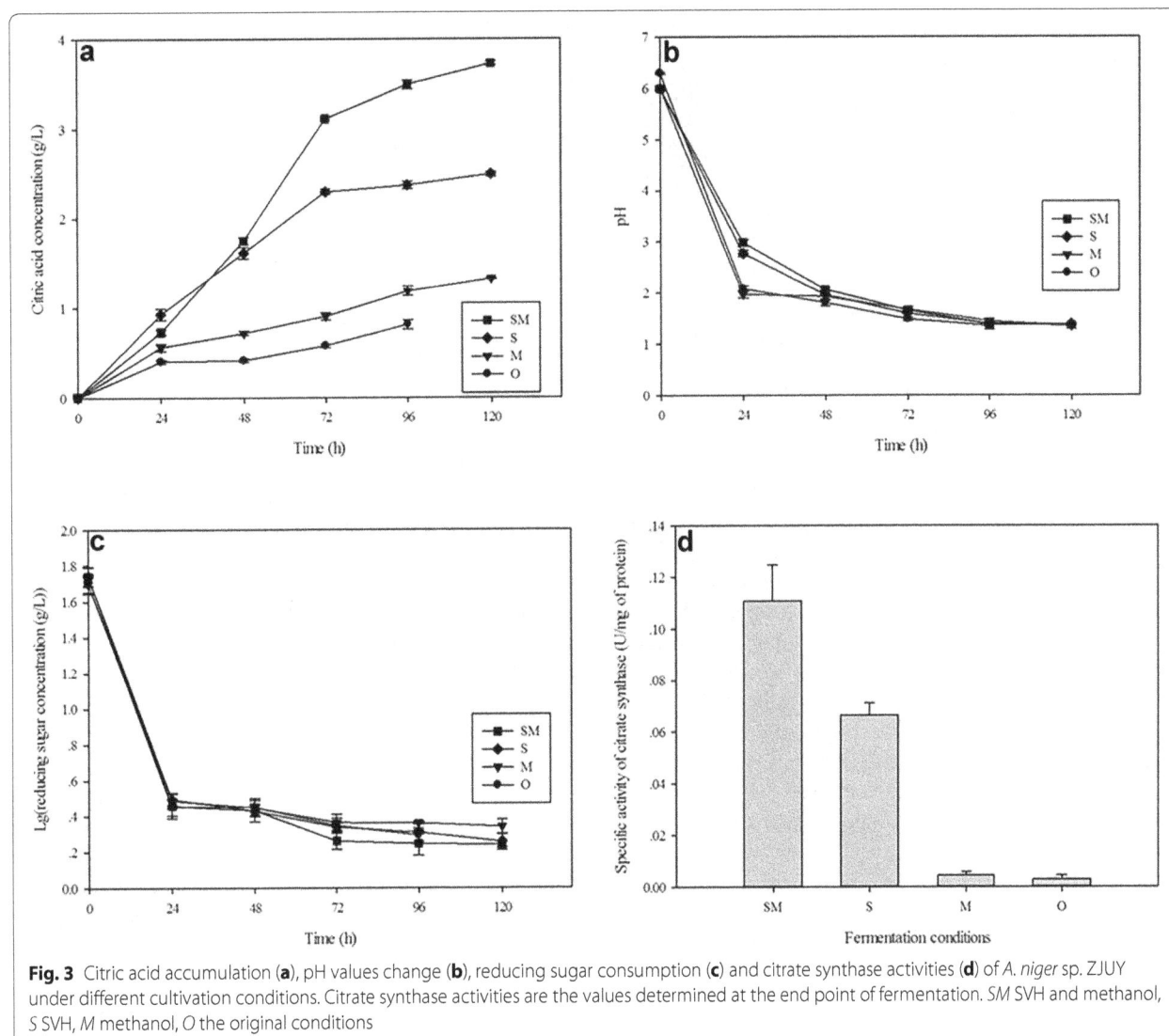

Fig. 3 Citric acid accumulation (**a**), pH values change (**b**), reducing sugar consumption (**c**) and citrate synthase activities (**d**) of *A. niger* sp. ZJUY under different cultivation conditions. Citrate synthase activities are the values determined at the end point of fermentation. *SM* SVH and methanol, *S* SVH, *M* methanol, *O* the original conditions

during the first 48 h, after which the pH curve gradually reduced with that of the others. The influence of pH on CA production has been confirmed by some researchers, and Aravantinos-Zafiris et al. (1994) observed a dramatic increase in CA production when the pH was increased from 3 to 4 and an optimum range fell from pH 4–6. However, CA production can considerably increase the acidity of the fermentation broth, consequently limiting the microorganism's capacity to ferment sugars. To offset this, Rivas et al. (2008) added $CaCO_3$ in the fermentation system to neutralize acidification and obtained positive results. In the present study, pH continuously decreased with increased CA concentration. Finally, the ultimate pH values were stabilized to approximately 1.3. During the fermentation period, SVH neutralized the acidification caused by CA release in the early stages, and maintained system a suitable pH to conduct CA fermentation.

Thus, the alleviation of acidification may be one means by which SVH enhances CA accumulation.

Figure 3c showed that SVH could promote decomposition and utilization of the reducing sugar. The reducing sugar residues in the systems with SVH were generally below that observed in the absence of SVH, and methanol accelerated reducing sugar consumption. Citrate synthase (CS) is directly related to CA synthesis, and the fluctuation of citroyl synthetase activities perfectly matched with CA production under different fermentation systems. The strains cultivated in the optimized conditions displayed a highest CS activity (0.11 U/mg), which was somewhat higher than that in the absence of methanol, whereas CS showed the lowest activities in the original conditions (Fig. 3d). This phenomenon indicated SVH could improve CS activity and promote CA production by *A. niger*.

Table 5 Comparison of kinetic parameters for CA fermentation by *A. niger* sp. ZJUY under different fermentation systems

Kinetic parameters	Control	SM			S			M		
		24 h	48 h	72 h	24 h	48 h	72 h	24 h	48 h	72 h
Q_p	0.001244	0.0426	0.0496	0.0384	0.0286	0.0285	0.02	0.006617	0.007201	0.008691
$Y_{p/s}$ (g/g)	0.1516	1.7454	0.4612	0.5023	0.3139	0.2879	0.263	0.1194	0.0907	0.1598
$Y_{x/s}$ (g cells/g)	10.4660	8.0882	7.8851	8.5778	3.2607	2.6796	1.6608	2.647	4.6332	6.3582
Q_s	0.008421	0.0322	0.108	0.0764	0.0928	0.099	0.0761	0.0565	0.0796	0.0542
LSD	0.5359	1.0517	1.1139	1.6851	1.1123	1.3153	1.1788	0.1975	0.6522	0.4612
Significance level	HS	HS	HS	HS	HS	HS	S	HS	HS	HS

Significant differences were determined by one-way ANOVA followed by LSD post hoc test using software SPSS, version 16.0, (SPSS Inc., Chicago, IL, USA). Differences were considered to be significant at $p < 0.05$

HS highly significance, *S* significance*SM* SVH and methanol, *S* SVH, *M* methanol

Analysis of fermentation kinetics

A comparison of kinetic parameters relating to the effects of SVH and methanol added 24, 48, or 72 h after inoculation on CA production by *A. niger* showed significant improvements in Q_p, $Y_{p/s}$, and $Y_{x/s}$ over those obtained in the system without methanol or SVH. However, when the culture was monitored for $Y_{x/s}$, the value of control samples were significantly higher than that of other systems, demonstrating that the control was mostly attributed to vegetative growth during this stage, and accumulated relatively high biomass. In this research, all treatments with methanol, SVH, or both showed higher Q_s than that of control, and higher Q_s value of SM and S systems matched exactly with higher citrate synthase activities and the CA output presented in Table 5.

SEM and flow cytometry of *A. niger* cells under the different incubation conditions

Scanning electron microscope of the *A. niger* surface showed that all mycelium produced conidiophores, except for those incubated under the original conditions, in which only spindly mycelium were observed, and methanol stress could promote the formation of conidiophores in the process of fermentation. It is also worth mentioning that the amount of conidiophores was markedly reduced in the system containing SVH, as compared to those where SVH was absent (Fig. 4), suggesting that SVH could relieve the toxic effect of methanol by maintaining *A. niger* in a prolonged vegetative growth phase, and SVH-methanol played a better synergistic effect on CA production. Moreover, *A. niger* cell membrane integrity showed that the death rates of spores in the absence of SVH were slightly higher than other systems during the early phase, and the death rates of spores decreased from 48.21% (absence of SVH) to 33.97% (absence of methanol) to 20.35% (optimized conditions) at the terminal stage. The phenomenon suggested that SVH could effectively maintain membrane integrity. Further,

when the fluorescence intensity reached from 10 to 100, the death rate decreased in the beginning and then increased in the absence of SVH (blue-broken line), likely since spores being detected were in different life cycles, wherein methanol promotes apoptosis of spore-bearing and new generated spores. Overall, SVH may prevent the mycelium from the toxic effect of methanol and sustain the cell membrane integrity (Fig. 5).

Discussion

Bioconversion of the biomass from agro-industrial wastes to produce beneficial fermentation products has received increasing attention in recent years. SVH has been proved to be an effective nutrient supplier for single cell oil production by *Trichosporon fermentans* in previous studies. With SVH as substrate, lipid production was markedly improved compared with that in the original medium (Zhan et al. 2013). Our preliminary experiments indicated that SVH could also be a superior additive for organic acid fermentation by *A. niger*. In the present study, the CA-fermentation conditions were optimized based on RSM, and the positive effect of SVH on CA accumulation was explored. Results showed that a 12% SVH addition could effectively promote CA production and the interaction of initial pH-fermentation temperature, pH-fermentation period, and pH-SVH significantly affected CA productivity ($p < 0.05$). Zhan et al. (2013) have reported that SVH is abundant in organic components and contains a relatively high amount of minerals; the organic components include reducing sugars, some crude polysaccharides, as well as little nitrogen and soluble protein. The considerable concentration (40.83 g/L) of reducing sugars in SVH can be used as a good supplement to glucose during the accumulation of CA. Khare et al. (1995) claimed that the organic nitrogen sources—such as peptone—did not significantly increase CA yield. However, in contrast, Zhang et al. (1999) investigated the relationship between CA output and the amount of

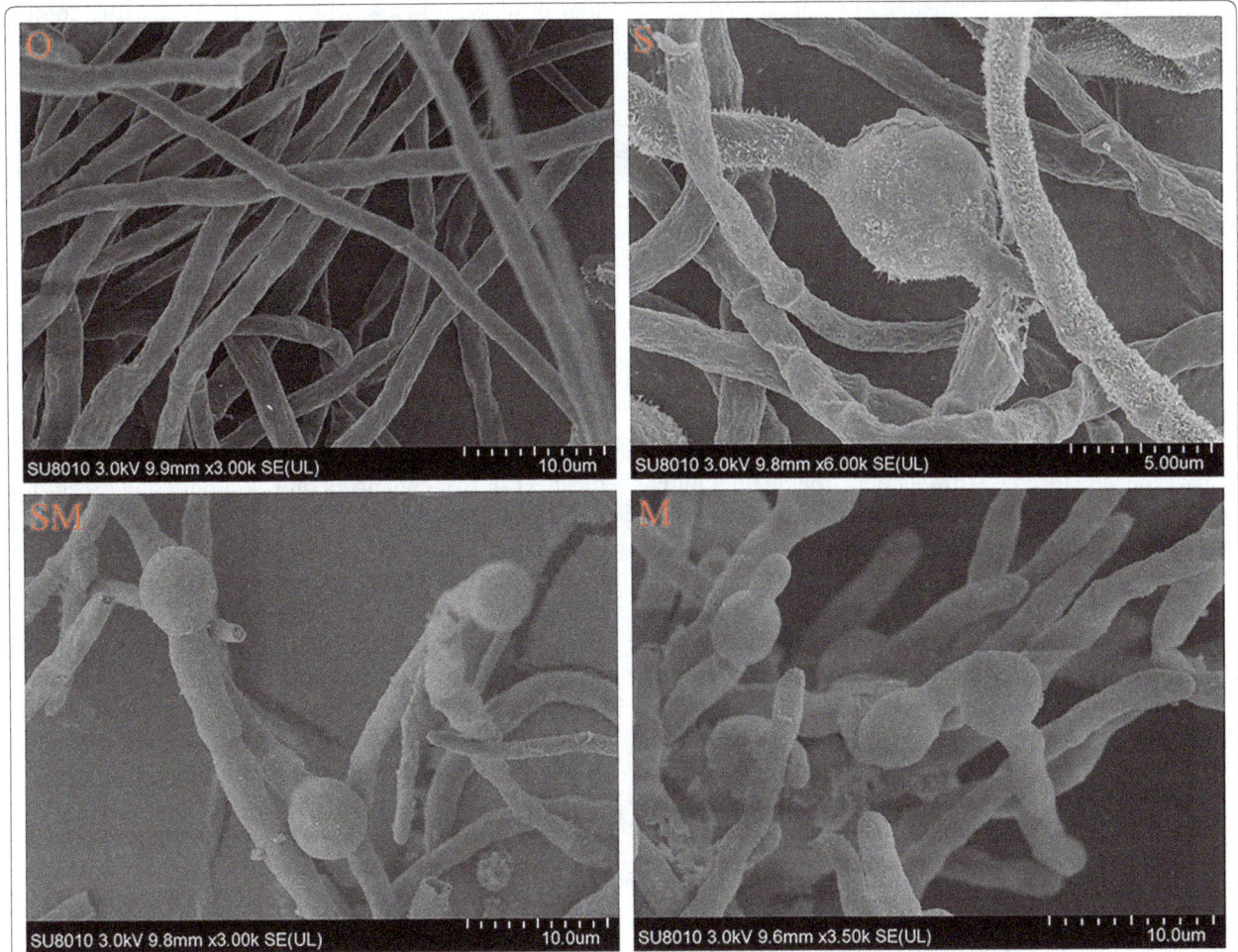

Fig. 4 The scanning electron microscope (SEM) graphs of *A. niger* sp. ZJUY cell surface in absence of both SVH and methanol (O), presence of SVH (S), presence of both SVH and methanol (SM), and presence of methanol (M)

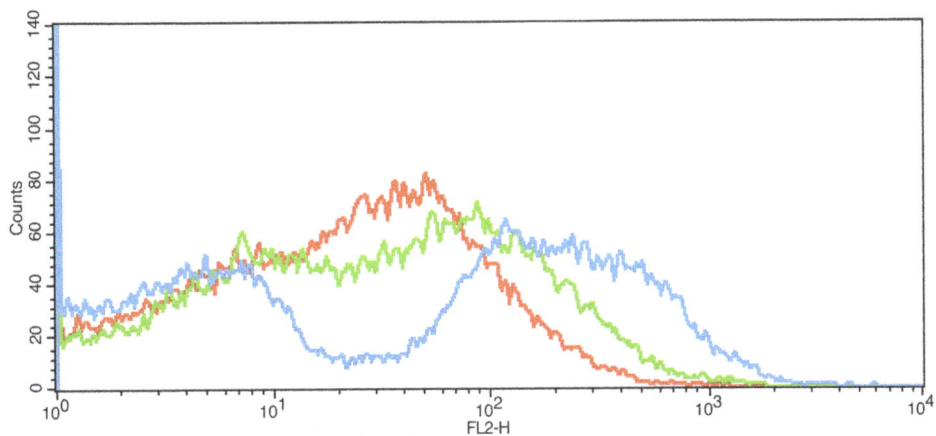

Fig. 5 Cell viability was detected by flow cytometry of *A. niger* sp. ZJUY cells, grown on methanol (*blue-broken line*), SVH (*green-broken line*), and methanol supplemented with SVH (*red-broken line*)

organic nitrogen in the fermentation medium, and found that the CA output increased by 27% with the optimum amounts of some amino acid (e.g., Thr, Met) or when enzymatic protein hydrolysate from corn was added to low quality sweet potato powder during fermentation. Our findings were consistent with the latter, in that the amount of total nitrogen in the SVH was 0.94 g/L, indicating that SVH could be an excellent media supplement to offset the absence of certain essential amino acid required for promoting the growth of fungus and activating the citrate synthase system. Thus, the addition of SVH could elevate the CA production to some extent. Besides, SVH afforded some essential trace elements (Ca^{2+}, Mg^{2+}, Fe^{2+}, Mn^{2+}, Cu^{2+}, etc.) for viable microorganisms. Ali and Haq (2005) found that copper sulfate remarkably enhanced CA production. The distinct nutritional properties of SVH might explain its positive effect on CA accumulation, indicating its great potential for use in the fermentation industry.

Methanol could enhance the CA secretion by altering membrane permeability. Mirbagheri et al. (2011) proposed that the permeabilization process only alters the intracellular lipid bilayer, while leaving the outer membrane intact. However, methanol still posed as a toxic threat to cell growth to some extent. Our SEM analyses showed that the largest number of conidiospores distributed above the mycelium in systems containing only methanol. Moreover, flow cytometry of spores under different fermentation conditions showed a decreased viability as compared to others during the early and late phase, which indicated that methanol had toxic effects on living cells and the presence of methanol accelerated *A. niger* reproductive growth.

This study verified the speculative synergistic effect of methanol and SVH as enhancers in CA production. The optimal methanol-SVH fermentation system resulted in a maximum citric acid concentration of 3.729 g/L, increased by 3.6-fold over the original condition, 0.49-fold over the optimized conditions without methanol, and 1.8-fold over the optimized conditions in the absence of SVH. The mechanisms of how methanol and SVH cooperatively increased CA production were initially explored from the perspective of fermentation dynamics, morphology, and cell viability. Fermentation kinetic studies showed that all the values for the SVH-methanol system were significantly improved with respect to Q_p, $Y_{p/s}$, and $Y_{x/s}$ over those obtained in systems without methanol or SVH. This improvement confirmed that the combination of SVH and methanol increased uptake of the carbon source and its subsequent assimilation. However, it was noted that when the culture was monitored for $Y_{x/s}$, the value of controls were significantly higher than

that of other systems, which demonstrated the control mainly conducted vegetative growth during this stage, produced much biomass but accumulated little CA. Higher Q_s values indicate a higher biomass conversion and productivity. In our study, all treatments added with methanol, SVH, or both showed higher Q_s than controls, and the higher Q_s value of the SM and S systems perfectly matched with higher citrate synthase activities and CA accumulation.

Morphology and cell viability studies showed that SVH could effectively relieve methanol toxic effect on growth of *A. niger* and maintain high CA productivity. In the SVH-added system, the death rate of *A. niger* cells was fairly low and only few conidiospores were observed. These results might illustrate the remarkable synergistic effect of two enhancers on CA production except for nutritional factor.

In our preliminary experiment, we employed SVH as a sole carbon source in mineral medium with starch and glucose as control. However, CA production was relatively lower than the otherwise, and when SVH was added as a nutrient supplement instead of the same amount of carbon source, CA production was considerably improved. The results suggested that SVH acted as a beneficial additive rather than alternative carbon sources in our study, maybe due to its poor carbon source and low C/N. A similar phenomenon has also been found by Shen et al. (2015), who showed that the co-fermentation of molasses and SVH increased the lipid output by 35% when 10% SVH was added, compared to the poor lipid accumulation on pure molasses by *T. fermentans*.

The CA yield of this work was significantly lower when compared with those of many other reports. For example, Roukas achieved a maximum CA concentration of 85.5 g/L from carob pod extract in surface fermentation (Roukas 1998), Hang obtained a high CA concentration of 100 g/kg in solid-state fermentation with kiwi fruit peel as only carbon source (Hang and Woodams 1998), then the successful cases were also repeated by Angumeenal (73 g/L), Dhillon (44.9 g/kg), and Vandenberghe (88.1 g/kg) (Angumeenal and Venkappayya 2005; Dhillon et al. 2011; Vandenberghe et al. 2000). However, the weak CA production in our study was mostly attributed to differences in the fermenting strain. The main purpose of this study was to explore a new way for the resource utilization of sweet potato vine waste. Certainly, application of industrial strains in CA fermentation could be more meaningful, while the fact that SVH can be used as a beneficial additive to enhance CA production of a weak production strain actually indicated its broad applicability in the fermentation industry.

Application of methanol and sweet potato vine hydrolysate as enhancers of citric acid production by Aspergillus...

43

Conclusions

The main purpose of this study was not to produce citric acid in amount but to explore a new way for the resource utilization of agricultural waste-sweet potato vine. What's more is, we found that the combination of methanol and SVH displayed a significant synergistic effect on CA accumulation even with a weaker strain, which suggested its great application potential in fermentation.

Authors' contributions
DY wrote the experiment; DY and YS conducted the experiment; QW conceived the research; XZ and YZ provided experimental materials and guidance. All authors read and approved the final manuscript.

Competing interests
The authors declare that the research was conducted in the absence of any commercial or financial relationships that could be construed as a potential competing interests.

Funding
We gratefully acknowledge financial support from the National Natural Science Foundation of China (Grant Nos. 41271335, 31470191, and 41671314), the Major State Basic Research Development Program of China (973 Program) (Grant No. 2015CB150502).

References
Ali S (2007) Application of kaolin to improve citric acid production by a thermophilic Aspergillus niger. Appl Microbiol Biotechnol 73(4):755–762

Ali S, Haq IU (2005) Role of different additives and metallic micro minerals on the enhanced citric acid production by Aspergillus niger MNNG-115 using different carbohydrate materials. J Basic Microbiol 45(1):3–11

Ali SR, Anwar Z, Irshad M, Mukhtar S, Warraich NT (2015) Bio-synthesis of citric acid from single and co-culture-based fermentation technology using agro-wastes. J Radiat Res Appl Sci 9(1):57–62

Amanullah A, Tuttiett B, Nienow AW (1998) Agitator speed and dissolved oxygen effects in xanthan fermentations. Biotechnol Bioeng 57(2):198–210

Angumeenal A, Venkappayya D (2005) Artrocarpus heterophyllus—a potential substrate for citric acid biosynthesis using Aspergillus niger. LWT Food Sci Technol 38(1):89–93

Aravantinos-Zafiris G, Tzia C, Oreopoulou V, Thomopoulos CD (1994) Fermenta tion of orange processing wastes for citric acid production. J Sci Food Agric 65(1):117–120

Dhillon GS, Brar SK, Verma M, Tyagi RD (2011) Apple pomace ultrafiltration sludge—a novel substrate for fungal bioproduction of citric acid: optimisation studies. Food Chem 128(4):864–871

FAO (1997) FAO production yearbook 1996, vol 50. Collection FAO: Statistiques (FAO); Coleccion FAO: Estadistica (FAO)

Hang Y, Woodams E (1998) Microbial production of citric acid by solid fermentation of kiwifruit peel. J Food Sci 52:226–227

Ishida H, Suzuno H, Sugiyama N, Innami S, Tadokoro T, Maekawa A (2000) Nutritive evaluation on chemical components of leaves, stalks and stems of sweet potatoes (Ipomoea batatas poir). Food Chem 68(3):359–367

Kana EBG, Olokeb JK, Lateefb A, Oyebanjib A (2012) Comparative evaluation of artificial neural network coupled genetic algorithm and response surface methodology for modeling and optimization of citric acid production by Aspergillus niger MCBN297. Chem Eng 27:397–402

Karaffa L, Kubicek CP (2003) Aspergillus niger citric acid accumulation: do we understand this well working black box? Appl Microbiol Biotechnol 61(3):189–196

Karthikeyan A, Sivakumar N (2010) Citric acid production by Koji fermentation using banana peel as a novel substrate. Biores Technol 101(14):5552–5556

Khare SK, Jha K, Gandhi AP (1995) Citric acid production from Okara (soyresidue) by solid-state fermentation. Biores Technol 54(3):323–325

Kobayashi K, Hattori T, Honda Y, Kirimura K (2013) Gene identification and functional analysis of methylcitrate synthase in citric acid-producing Aspergillus niger WU-2223L. Biosci Biotechnol Biochem 77(7):1492–1498

Kumar D, Jain V, Shanker G, Srivastava A (2003) Utilisation of fruits waste for citric acid production by solid state fermentation. Process Biochem 38(12):1725–1729

Maddox I, Hossain M, Brooks J (1986) The effect of methanol on citric acid production from galactose by Aspergillus niger. Appl Microbiol Biotechnol 23(3–4):203–205

Mirbagheri M, Nahvi I, Emtiazi G, Darvishi F (2011) Enhanced production of citric acid in Yarrowia lipolytica by Triton X-100. Appl Biochem Biotechnol 165(3–4):1068–1074

Navaratnam P, Arasaratnam V, Balasubramaniam K (1998) Channelling of glucose by methanol for citric acid production from Aspergillus niger. World J Microbiol Biotechnol 14(4):559–563

Pirt SJ (1975) Principles of microbe and cell cultivation. Blackwell Scientific, New York

Rault A, Bouix M, Béal C (2009) Fermentation pH influences the physiologicalstate dynamics of Lactobacillus bulgaricus CFL1 during pH-controlled culture. Appl Environ Microbiol 75(13):4374–4381

Rivas B, Torrado A, Torre P, Converti A, Domínguez JM (2008) Submerged citric acid fermentation on orange peel autohydrolysate. J Agric Food Chem 56(7):2380–2387

Rodrigues CI, Marta L, Maia R, Miranda M, Ribeirinho M, Máguas C (2007) Application of solid-phase extraction to brewed coffee caffeine and organic acid determination by UV/HPLC. J Food Compos Anal 20(5):440–448

Roukas T (1998) Carob pod: a new substrate for citric acid production by Aspergillus niger. Appl Biochem Biotechnol 74(1):43–53

Sabu A, Augur C, Swati C, Pandey A (2006) Tannase production by Lactobacillus sp. ASR-S1 under solid-state fermentation. Process Biochem 41(3):575–580

Shen Q, Lin H, Zhan J, Wang Q, Zhao Y (2013) Sweetpotato vines hydrolysate induces glycerol to be an effective substrate for lipid production of Trichosporon fermentans. Biores Technol 136:725–729

Shen Q, Lin H, Wang Q, Fan X, Yang Y, Zhao Y (2015) Sweetpotato vines hydrolysate promotes single cell oils production of trichosporon fermentans in high-density molasses fermentation. Biores Technol 176:249–256

Srere PA (1966) Citrate-condensing enzyme-oxalacetate binary complex. J Biol Chem 241(9):2157–2165

Vandenberghe LP, Soccol CR, Pandey A, Lebeault J-M (2000) Solid-state fermentation for the synthesis of citric acid by Aspergillus niger. Biores Technol 74(2):175–178

Zhan J, Lin H, Shen Q, Zhou Q, Zhao Y (2013) Potential utilization of waste sweetpotato vines hydrolysate as a new source for single cell oils production by Trichosporon fermentans. Biores Technol 135:622–629

Zhang W, Wang J, Li H et al (1999) Effect of the organic nitrogen sources and citric acid fermentation. J Hebei Univ 20(2):157–162

4

Microbial transformation of artemisinin by *Aspergillus terreus*

Hongchang Yu[1], Baowu Zhu[1] and Yulian Zhan[1,2]*

Abstract

Background: Artemisinin (**1**) and its derivatives are now being widely used as antimalarial drugs, and they also exhibited good antitumor activities. So there has been much interest in the structural modification of artemisinin and its derivatives because of their effective bioactivities. The microbial transformation is a promising route to obtain artemisinin derivatives. The present study focuses on the microbial transformation of artemisinin by *Aspergillus terreus*.

Results: During 6 days at 28 °C and 180 rpm, *Aspergillus terreus* transformed artemisinin to two products. They were identified as 1-deoxyartemisinin (**2**) and 4α-hydroxy-1-deoxyartemisinin (**3**) on the basis of their spectroscopic data.

Conclusions: The microbial transformation of artemisinin by *Aspergillus terreus* was investigated, and two products (1-deoxyartemisinin and 4α-hydroxy-1-deoxyartemisinin) were obtained. This study is the first to report on the microbial transformation of artemisinin by *Aspergillus terreus*.

Keywords: Microbial transformation, Artemisinin, *Aspergillus terreus*

Background

Artemisinin (Fig. 1) **1** (qinghaosu) is a sesquiterpene lactone and its structure was determined by X-ray analysis (Liu et al. 1979). Artemisinin and its derivatives such as dihydroartemisinin, artemether, artesunate, and arteether are now being widely used as antimalaria drugs. In some reports, artemisinin derivatives also exhibited good antitumor activities (Wu et al. 2004; Efferth et al. 2001, 2004; Singh and Lai 2001). There has been much interest in the structural modification of artemisinin and its derivatives because of their effective bioactivities. In this study, we report the microbial transformation of **1** by *Aspergillus terreus*, and two products were obtained.

Methods

General

^{1}H NMR (nuclear magnetic resonance) and ^{13}C NMR spectra were recorded in CDCl$_3$ (chloroform-d) on a Bruker AV 500 MHz spectrometer. Chemical shifts were reported in ppm (δ), and *J* values were reported in Hz.

Microorganism

Aspergillus terreus strain ZYL050009 was isolated from soil samples collected from the yew planting base at Guilin, China. The isolate was identified by amplification of the nuclear ribosomal internal transcribed spacer (ITS) region, using the primers ITS1 (5′-TCC GTA GGT GAA CCT GCG G-3′) and ITS 4 (5′-TCC TCC GCT TAT TGA TAT GC-3′) (White et al. 1989). The amplicons were sequenced, and alignments were performed using BLASTN algorithm. Reference sequences with the highest identity were selected and imported into the open source software MEGA 7.0 (Kumar et al. 2016). For the phylogenetic analysis, tree constructions were done with the software MEGA 7.0 using the neighbor-joining method (Hall 2013). Bootstrap analysis was done using 1000-times resampled data.

Medium

All culture and microbial transformation experiments were performed in the following medium: potato infusion is made by boiling 200 g of sliced potatoes in 1 L deionized water for 30 min and then filtering the broth through cheesecloth. Deionized water is added such that the total volume of the suspension is 1 L. 20 g dextrose is

*Correspondence: zhanyulian@hotmail.com
[1] School of Life and Environmental Sciences, Guilin University of Electronic Technology, Guilin 541004, People's Republic of China
Full list of author information is available at the end of the article

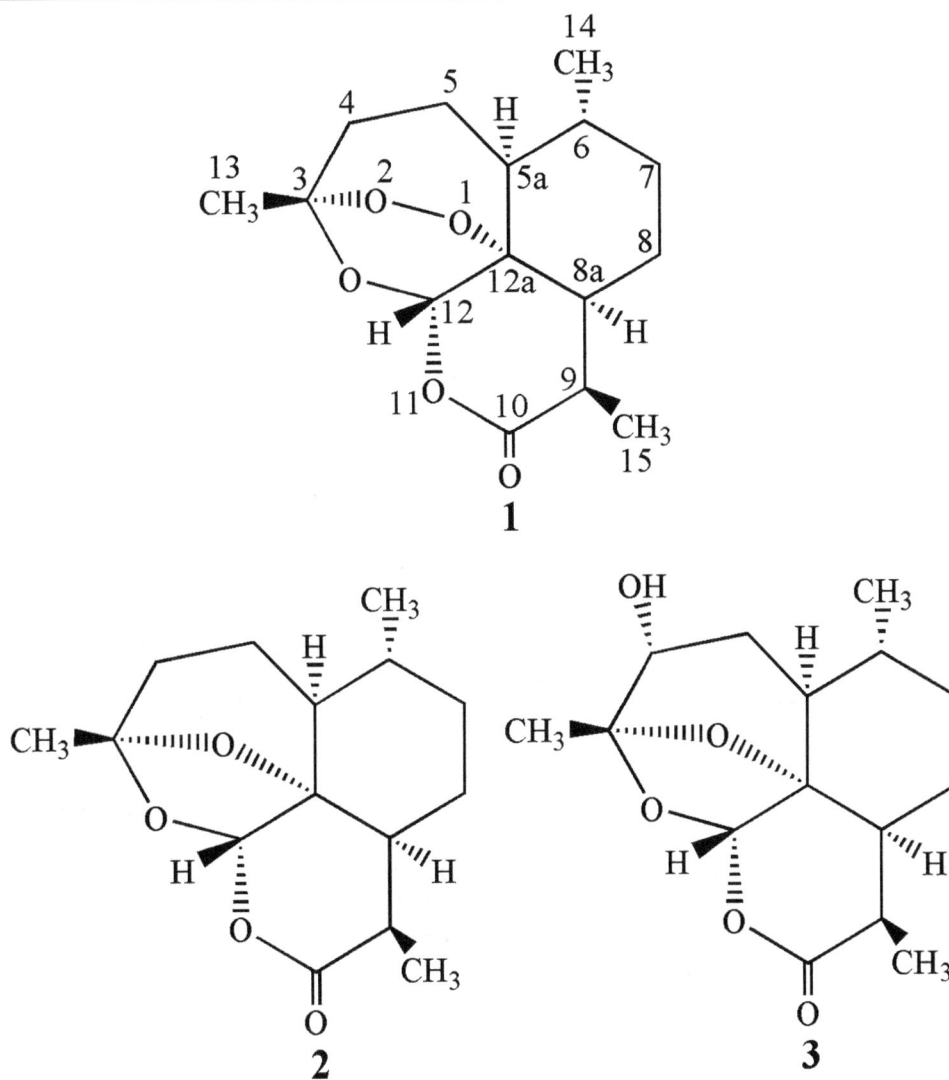

Fig. 1 Structures of artemisinin and two products from microbial transformation of artemisinin by *Aspergillus terreus*

then added and the medium is sterilized by autoclaving at 121 °C for 30 min.

Microbial transformation of artemisinin (1) by *Aspergillus terreus*

Well-developed fungal mycelia were transferred into 250-mL Erlenmeyer flasks containing 60 mL of medium from the surface of agar slants. Cultures were grown for 48 h on a rotary shaker at 28 °C with shaking at 180 rpm, and used to inoculate 51 250-mL shake flasks that contained 60 mL of medium. The cultures were then incubated for 48 h using the same conditions as before. Artemisinin (Mediplantex, Vietnam) was dissolved in acetone (25 mg/mL), filter-sterilized, and 0.4 mL of this solution was added to each flask. A total of 510 mg of artemisinin was transformed. The cultures were incubated for additional 6 days at 28 °C while shaking at 180 rpm. The mycelia were separated by filtration and discarded. The filtrate was extracted three times with an equal volume of ethyl acetate (EtOAc). The extract was evaporated to dryness under vacuum to afford a residue.

Chromatographic conditions

A total of 1.01 g of residue was obtained from the broth. The residue was purified by silica gel column chromatography, using a petroleum ether (60–90 °C)–acetone mobile phase in a gradient mode, eluting with 10–30% acetone.

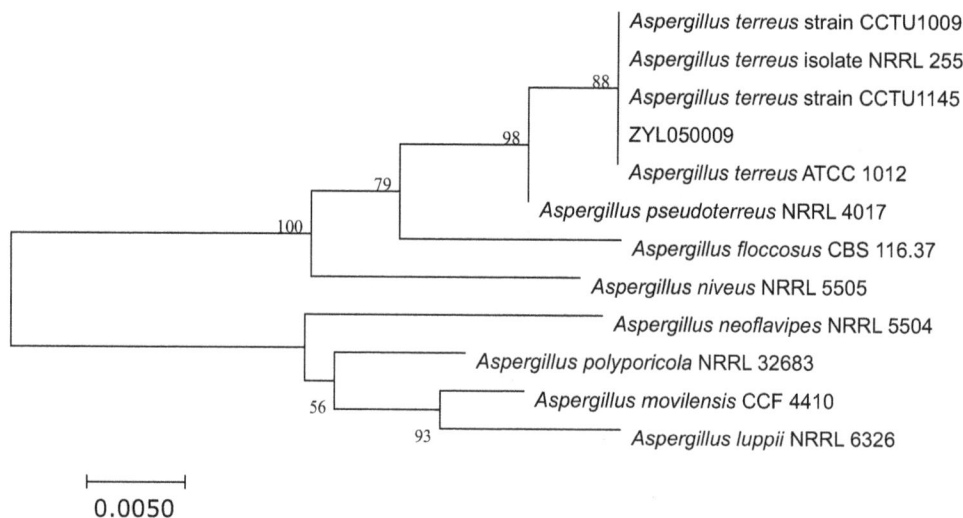

Fig. 2 Phylogenetic tree based on the ITS sequences was generated using the neighbor-joining method and the MEGA7.0 software. All sequences data were retrieved from the GenBank database. Bootstrap values, expressed as percentages of 1000 replications, are given at branching points

Results and discussion

We performed DNA sequencing for species identification. We isolated genomic DNA sample from the strain and sequenced the amplified ITS regions. A BLAST search of ITS rDNA sequences available in the GenBank database showed that 583-bp ITS from the strain shared 99% match with *A. terreus* ATCC 1012 (NR_131276.1), *A. terreus* strain CCTU1145 (GenBank: KY053120.1), *A. terreus* isolate NRRL 255 (GenBank: EF669586.1), and *A. terreus* strain CCTU1009 (GenBank: KY053112.1) (Fig. 2). The isolate was determined as *A. terreus*.

The microbial transformation of artemisinin by *A. terreus* resulted in 20 mg of **2** (yield 3.9%), 39 mg of **3** (yield 7.6%).

The structures of products were identified on the basis of their spectroscopic data. Data of ^{1}H and ^{13}C NMR spectra of product **2** were in agreement with the reported literatures' data (Lee et al. 1989; Gaur et al. 2014). So product **2** was identified as 1-deoxyartemisinin (Fig. 1). The comparison of the ^{1}H and ^{13}C NMR data of product **3** with those of 4α-hydroxy-1-deoxyartemisinin (Parshikov et al. 2004; Zhan et al. 2015) was in complete agreement. Therefore, product **3** was confirmed to be 4α-hydroxy-1-deoxyartemisinin (Fig. 1).

1-deoxyartemisinin (**2**) Colorless needles (from acetone); ^{1}H-NMR (CDCl$_3$, 500 MHz) δ 0.94 (3H, d, $J = 5.6$ Hz, Me-14), 0.99 (1H, m, H-8), 1.08 (1H, m, H-7), 1.19 (3H, d, $J = 7.2$ Hz, Me-15), 1.25 (3H, m, H-5, H-6, H-5a), 1.52 (3H, s, Me-13), 1.63 (1H, m, H-4), 1.78 (2H, m, H-4, H-7), 1.90 (2H, m, H-5, H-8), 2.00 (1H, m, H-8a), 3.18 (1H, m, H-9), 5.69 (1H, s, H-12); ^{13}C NMR (CDCl$_3$, 125 MHz) δ 171.8 (s, C-10), 109.2 (s, C-3), 99.6 (d, C-12),

82.4 (s, C-12a), 44.6 (d, C-5a), 42.4 (d, C-8a), 35.4 (d, C-6), 34.0 (t, C-4), 33.5 (t, C-7), 32.8 (d, C-9), 23.9 (q, C-13), 23.5 (t, C-8), 22.0 (t, C-5), 18.6 (q, C-14), 12.6 (q, C-15).

4α-hydroxy-1-deoxyartemisinin (**3**) Colorless needles (from acetone); ^{1}H-NMR (CDCl$_3$, 500 MHz) δ 5.64 (1H, s, H-12), 3.62 (1H, brs, H-4β), 3.20 (1H, m, H-9), 2.06 (1H, m, H-8a), 1.99 (1H, m, H-5α), 1.93 (1H, m, H-8α), 1.82 (1H, m, H-7α), 1.58 (3H, s, Me-13), 1.54 (1H, m, H-5a), 1.50 (1H, m, H-5β), 1.29 (1H, m, H-6), 1.21 (3H, d, $J = 7.2$ Hz, Me-15), 1.13 (1H, m, H-7β), 1.00 (1H, m, H-8β), 0.93 (3H, d, $J = 6.4$ Hz, Me-14); ^{13}C NMR (CDCl$_3$, 125 MHz) δ 171.3 (s, C-10), 108.9 (s, C-3), 98.9 (d, C-12), 83.0 (s, C-12a), 69.1 (d, C-4), 42.1 (d, C-8a), 40.6 (d, C-5a), 35.1 (d, C-6), 33.4 (t, C-7), 32.7 (d, C-9), 30.3 (t, C-5), 23.5 (t, C-8), 20.5 (q, C-13), 18.4 (q, C-14), 12.6 (q, C-15).

Although artemisinin is effective against chloroquine-resistant parasites, its toxicities (Kamchonwongpaisan et al. 1997) and low solubility (Lin et al. 1989) in water limit its therapeutical use. The modification of artemisinin has been studied by chemical and biological methods in some reports (Gaur et al. 2014; Parshikov et al. 2004, 2006; Zhan et al. 2002; Goswami et al. 2010; Acton 1999). However, synthesis and semisynthesis of artemisinin derivatives are impossible or impracticable because of the complexity of the artemisinin molecule and the chemical lability of the peroxy ring system. Therefore, microbial transformation is a promising route to obtain artemisinin derivatives. There are many reports on microbial transformation of artemisinin by various microorganisms, such as *Aspergillus niger*, *Rhizopus stolonifer*, *Cunninghamella elegans*, *Eurotium amstelodami, Mucor polymorphosporus, Penicillium simplicissimum, Streptomyces griseus* (Zhan et al. 2015; Gaur et al. 2014; Goswami et al. 2010; Liu et al. 2006; Parshikov et al. 2004, 2006; Zhan et al. 2002). Here, we first report the microbial transformation of artemisinin by *A. terreus*.

Conclusions

In this work, we investigated the microbial transformation of artemisinin by *A. terreus*, and obtained two products, 1-deoxyartemisinin and 4α-hydroxy-1-deoxyartemisinin. This is the first report of microbial transformation of artemisinin by *A. terreus*.

Authors' contributions
YZ designed and coordinated the study, performed the data analysis and drafted the manuscript. HY and BZ performed the experiments. All authors read and approved the final manuscript.

Author details
[1] School of Life and Environmental Sciences, Guilin University of Electronic Technology, Guilin 541004, People's Republic of China. [2] State Key Laboratory of Bioreactor Engineering, East China University of Science and Technology, Shanghai 200237, People's Republic of China.

Acknowledgments
Not applicable.

Competing interests
The authors declare that they have no competing interests.

Funding
This work was supported by Open Funding Project of the State Key Laboratory of Bioreactor Engineering (No. 2015OPEN16), and supported by Guangxi Undergraduate Program for Innovation and Entrepreneurship (No. 201510595170).

References
Acton N (1999) Semisynthesis of 3-β-hydroxyartemisinin. J Nat Prod 62:790–793
Efferth T, Dunstan H, Sauerbrey A, Miyachi H, Chitambar CR (2001) The antimalarial artesunate is also active against cancer. Int J Oncol 18:767–773
Efferth T, Benakis A, Romero MR, Tomicic M, Rauh R, Steinbach D, Häfer R, Stamminger T, Oesch F, Kaina B, Marschall M (2004) Enhancement of cytotoxicity of artemisinins toward cancer cells by ferrous iron. Free Radic Biol Med 37:998–1009

Gaur R, Darokar MP, Ajayakumar PV, Shukla RS, Bhakuni RS (2014) In vitro antimalarial studies of novel artemisinin biotransformed products and its derivatives. Phytochemistry 107:135–140
Goswami A, Saikia PP, Barua NC, Bordoloi M, Yadav A, Bora TC, Gogoi BK, Saxena AK, Suri N, Sharma M (2010) Bio-transformation of artemisinin using soil microbe: direct C-acetoxylation of artemisinin at C-9 by *Penicillium simplicissimum*. Bioorg Med Chem Lett 20:359–361
Hall BG (2013) Building phylogenetic trees from molecular data with MEGA. Mol Biol Evol 30:1229–1235
Kamchonwongpaisan S, McKeever P, Hossler P, Ziffer H, Meshnick SR (1997) Artemisinin neurotoxicity: neuropathology in rats and mechanistic studies in vitro. Am J Trop Med Hyg 56:7–12
Kumar S, Stecher G, Tamura K (2016) MEGA7: molecular evolutionary genetics analysis version 7.0 for bigger datasets. Mol Biol Evol 33:1870–1874
Lee IS, El-Sohly HN, Croon EM, Hufford CD (1989) Microbial metabolism studies of the antimalarial sesquiterpene artemisinin. J Nat Prod 52:337–341
Lin AJ, Lee M, Klayman DL (1989) Antimalarial activity of new water-soluble dihydroartemisinin derivatives. 2. Stereospecificity of the ether side-chain. J Med Chem 32:1249–1252
Liu JM, Ni MY, Fan JF, Tu YY, Wu ZH, Wu YL, Zhou WS (1979) Structure and reaction of arteannuin. Acta Chim Sinica 37:129–143
Liu J-H, Chen Y-G, Yu B-Y, Chen Y-J (2006) A novel ketone derivative of artemisinin biotransformed by *Streptomyces griseus* ATCC 13273. Bioorg Med Chem Lett 16:1909–1912
Parshikov IA, Muraleedharan KM, Avery MA, Williamson JS (2004) Transformation of artemisinin by *Cunninghamella elegans*. Appl Microbiol Biotechnol 64:782–786
Parshikov IA, Miriyala B, Muraleedharan KM, Avery MA, Williamson JS (2006) Microbial transformation of artemisinin to 5-hydroxyartemisinin by *Eurotium amstelodami* and *Aspergillus niger*. J Ind Microbiol Biotechnol 33:349–352
Singh NP, Lai H (2001) Selective toxicity of dihydroartemisinin and holo-transferrin toward human breast cancer cells. Life Sci 70:49–56
White TJ, Bruns T, Lee SJWT, Taylor JW (1989) Amplification and direct sequencing of fungal ribosomal RNA genes for phylogenetics. In: Innis MA, Gelfand DH, Sninsky JJ, White TJ (eds) PCR protocols: a guide to methods and applications. Academic Press Inc, San Diego, pp 315–322
Wu GD, Zhou HJ, Wu XH (2004) Apoptosis of human umbilical vein endothelial cells induced by artesunate. Vasc Pharmacol 41:205–212
Zhan J, Zhang Y, Guo H, Han J, Ning L, Guo D (2002) Microbial metabolism of artemisinin by *Mucor polymorphosporus* and *Aspergillus niger*. J Nat Prod 65:1693–1695
Zhan Y, Liu H, Wu Y, Wei P, Chen Z, Williamson JS (2015) Biotransformation of artemisinin by *Aspergillus niger*. Appl Microbiol Biotechnol 99:3443–3446

5

Enhancing transglutaminase production of *Streptomyces mobaraensis* by iterative mutagenesis breeding with atmospheric and room-temperature plasma (ARTP)

Ying Jiang[1] [iD], Yue-Peng Shang[1], Hao Li[1], Chao Zhang[1], Jiang Pan[1], Yun-Peng Bai[1], Chun-Xiu Li[1]* and Jian-He Xu[1,2]*

Abstract

Objectives: To improve the fermentation production of transglutaminase (TGase) from *Streptomyces mobaraensis* for applications in the food industry, the atmospheric and room-temperature plasma (ARTP) mutagenesis was applied to breed *S. mobaraensis* mutants with increased TGase production.

Results: After eight rounds of iterative ARTP mutagenesis, four genetically stable mutants, *Sm5-V1*, *Sm6-V13*, *Sm2-V10*, and *Sm7-V12*, were identified, which showed increased TGase production by 27, 24, 24, and 19%, respectively. The best mutant *Sm5-V1* exhibited a maximum TGase activity of 5.85 U/mL during flask fermentation. Compared to the wild-type strain, the transcription levels of the zymogen TGase genes in the mutants increased significantly as indicated by quantitative real-time PCR, while the gene nucleotide sequences of the mutants did not change at all. It was shown that the overexpression of TGase zymogen gene in the mutants contributes to the increase in TGase production.

Conclusions: ARTP is a potentially efficient tool for microbial mutation breeding to bring some significant changes required for the industrial applications.

Keywords: Atmospheric and room-temperature plasma, Breeding, *Streptomyces mobaraensis*, Transcription level, Transglutaminase

Background

Transglutaminases (TGase, EC 2.3.2.13), also referred to protein-glutamine γ-glutamyltransferases, are enzymes capable of catalyzing acyl-transfer reactions between the γ-carboxamide group of protein or peptide-bound glutamine and ε-amino group of lysine or other primary amines (Zhu et al. 1995; Martins et al. 2014). The covalent modifications of proteins promoted by TGase facilitate extensive applications in food, medicine, and other industries. TGases are widely distributed in prokaryotes and eukaryotes. TGases from animal tissues are Ca^{2+}-dependent enzymes that lead to the precipitation of proteins from food containing casein, soybean globulin, or myosin (Martins et al. 2014). However, the scarcity of resources as well as complexity of further extraction and purification still limits their applications. Microbial transglutaminases (MTG) that were usually found in *Streptomyces* (Duran et al. 1998; Marx et al. 2008) and *Bacillus* (Kobayashi et al. 1998; Soares et al. 2003) species have a wider application due to their advantages as Ca^{2+}-independent activity, thermostability, and broad specificity for acyl donors (Salis et al. 2015). Up to now, MTG for applications in the food processing were mainly produced by fermentation of *Streptomyces mobaraensis* (Yokoyama et al. 2004). TGase from *S. mobaraensis* was known to be secreted as a zymogen and then activated by proteolytic processing to the enzymatically active mature form (Zotzel et al. 2003a, b). Its pro-region is essential for

*Correspondence: chunxiuli@ecust.edu.cn; jianhexu@ecust.edu.cn
[1] State Key Laboratory of Bioreactor Engineering, East China University of Science and Technology, Shanghai 200237, People's Republic of China
Full list of author information is available at the end of the article

efficient protein folding, secretion, and suppression of the enzymatic activity (Yurimoto et al. 2004). Considering the fact that genetically engineered strains are somehow restricted in the food industry, it is more feasible to improve TGase production by mutation breeding.

Microbial mutation breeding by altering the genomes shows great significance for biotechnology researches and applications (Tan et al. 2014; Kumar 2015). Recently, a novel and efficient mutation tool called atmospheric and room-temperature plasma (ARTP) has been successfully employed to cause some significant changes in enzyme activity, biochemical productivity, and metabolism without lethal damages (Guo et al. 2011; Lu et al. 2011; Xu et al. 2012; Wang et al. 2014). Compared to the conventional mutation breeding methods, ARTP mutagenesis shows some distinct advantages, such as low costs, low and controllable plasma temperatures, various active chemical species with a high density, rapid mutation, flexible and secure operations (Laroussi 2005; Zhang et al. 2014, 2015).

In this study, the iterative ARTP mutagenesis was applied to S. mobaraensis for breeding mutants with increased TGase production. Protocols for iterative rapid mutation of S. mobaraensis with helium-driven ARTP system and effective tube screening method of the mutants were established. The transcription levels and the nucleotide sequences of the gene (pro-smtg) encoding TGase zymogen were compared between the mutants and the wild-type strain to explain the reasons for the enhanced TGase production, which may provide a valuable guidance for further researches on mechanisms regarding ARTP effects on the whole cells and the intracellular bio-macromolecules.

Methods
Materials
L-glutamic acid γ-monohydroxamate was purchased from Sigma-Aldrich Co., Ltd. (Shanghai, China). N-α-Carbobenzoxy-L-glutaminyl-glycine (Nα-CBZ-Gln-Gly) was purchased from Civi Chemical Technology Co., Ltd. (Shanghai, China). Reduced glutathione was purchased from Aladdin Reagent Co., Ltd. (Shanghai, China). All-in-One First-Strand cDNA Synthesis SuperMix (One-Step gDNA Removal) and TransStart Tip Green qPCR SuperMix that were used for RT-PCR (Reverse Transcription PCR) and qRT-PCR (quantitative Real-Time PCR) were purchased from TransBionovo Co., Ltd. (Beijing, China). All other reagents were obtained from commercial sources and were of analytical grade.

Strains and media
Escherichia coli DH5α was used as the host strain for the recombinant DNA manipulations. Streptomyces mobaraensis, the wild-type strain (S. mobaraensis ECU7480, stored in our lab) and the mutants generated by ARTP mutagenesis, were grown on a solid medium comprising 20 g/L soluble starch, 3 g/L tryptone, 1 g/L KNO$_3$, 0.5 g/L K$_2$HPO$_4$·3H$_2$O, 0.5 g/L MgSO$_4$·7H$_2$O, 0.5 g/L NaCl, 0.01 g/L FeSO$_4$·7H$_2$O, and 1.5–2.0% (w/v) agar at pH 7.4–7.6 and 30 °C for 5 days. Both tube and flask fermentations were conducted using a seeding medium (20 g/L glycerol, 20 g/L tryptone, 5 g/L yeast extract, 2 g/L MgSO$_4$·7H$_2$O, 2.62 g/L K$_2$HPO$_4$·3H$_2$O, 2 g/L KH$_2$PO$_4$, pH 7.0) at 30 °C for 24 h to prepare the inoculums and a fermentation medium (20 g/L glycerol, 20 g/L tryptone, 5.5 g/L corn steep powder, 5 g/L yeast extract, 2 g/L MgSO$_4$, 2.62 g/L K$_2$HPO$_4$·3H$_2$O, 10 g/L CaCO$_3$, pH 7.2) at 30 °C with 8–10% inoculum size for the production of TGase.

Procedures for iterative mutagenesis with ARTP and directed screening
The procedures for iterative mutation of S. mobaraensis genome with ARTP mutation breeding system purchased from Si Qing Yuan Biotechnology Co., Ltd. (Beijing, China) and the following screening are shown in Fig. 1. In this study, pure helium was used as the working gas at a flow rate of 10 slpm (standard liters per minute). The RF power input was 40 W. The distance (D) between the plasma torch nozzle exit and the sample plate was 4 mm and the plasma jet temperature was below 30 °C. To provide different dosages of the active species in the plasma jet region, 10 μL of the spore suspension was pipetted onto the stainless minidisk and then exposed to ARTP jet for different treatment times ranging from 10 to 60 s. Spores without treatment were used as the control.

After ARTP mutation, the spore suspension was transferred onto solid medium. Single colonies from the solid medium were selected randomly and inoculated into a test tube containing 3 mL fermentation medium for 96 h fermentation. The cell-free supernatant of fermentation broth was withdrawn for TGase activity detection. The top eight mutants with increased TGase production were chosen as the starting strains for the next round of ARTP mutagenesis. After eight rounds of iterative ARTP mutagenesis, flask cultivation was performed to verify the TGase production of selected mutants from shake tube screening. The selected mutants grown on the solid medium were cultivated in 250-mL flasks containing 25 mL seeding medium at 30 °C for 24 h. Then, 2.5 mL of the seeding culture was transferred into 250-mL shake flasks containing 25 mL fermentation medium and cultured at 30 °C. Aliquots of the cell-free supernatant were taken at different time intervals for examination of the TGase activity. Additionally, the genetic stability of the identified mutants was evaluated by eight rounds

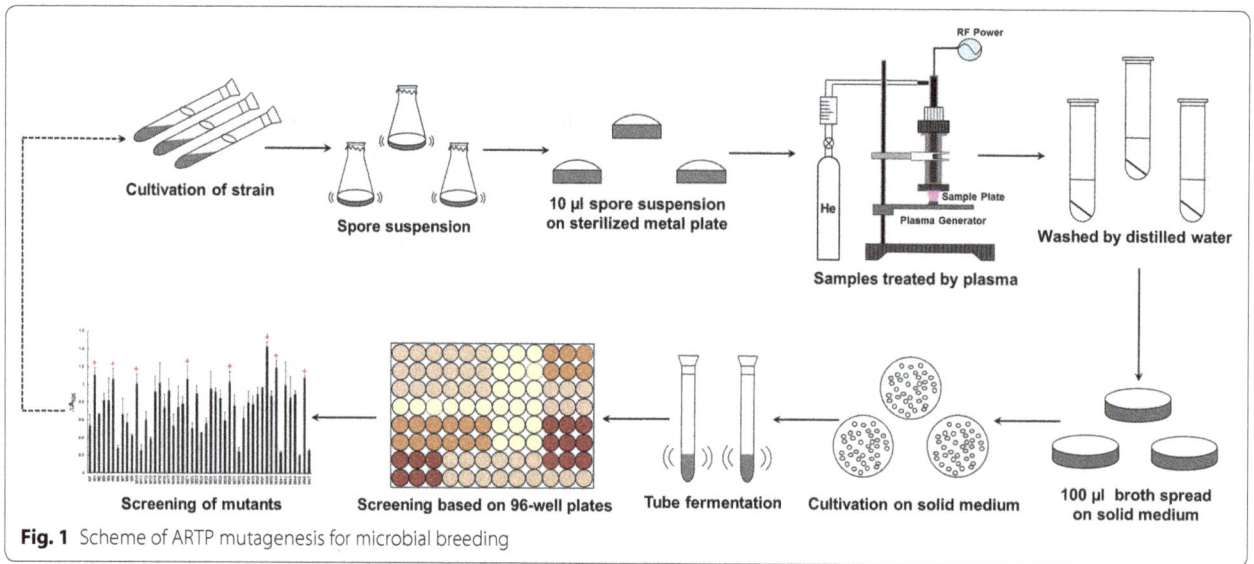

Fig. 1 Scheme of ARTP mutagenesis for microbial breeding

of subcultures. The TGase production of the mutants was examined by shake flask fermentation in every two subcultures.

Evaluation of ARTP mutagenesis of *S. mobaraensis*

The lethal rates of the spores under different treatment times were evaluated according to Eq. 1. The mutation rate and the positive mutation rate were calculated based on Eqs. 2 and 3, respectively.

$$\text{Lethal rate (\%)} = \frac{U - T}{U} \times 100 \tag{1}$$

$$\text{Mutation rate } (R_{\text{M}}) \, (\%) = \frac{M}{T} \times 100 \tag{2}$$

$$\text{Positive mutation rate } (R_{\text{P}}) \, (\%) = \frac{P}{M} \times 100, \tag{3}$$

where U is the total colony count of the sample without treatment, T is the total colony count after treatment with ARTP, M is the total colony count of mutants with different TGase production from the wild-type strain, and P is the total colony count of mutants with increased TGase production than that of the wild-type strain. All the colony numbers were obtained by the colony forming unit (CFU) method on a solid medium.

Activity assay of TGase

The activity of TGase was measured according to the colorimetric hydroxamate procedure (Grossowicz et al. 1950) using Nα-CBZ-Gln-Gly as the substrate. The calibration curve was prepared using L-glutamic acid γ-monohydroxamic acid as a standard. One unit (U)

of TGase activity was defined as the amount of enzyme that catalyzes the formation of 1 μmol L-glutamic acid γ-monohydroxamate per minute at 37 °C.

Quantitative real-time PCR analysis

Total RNAs of *S. mobaraensis* ECU7480 and its mutants were extracted after cultivation for different times. The mycelium pellets were collected by centrifugation and then frozen in liquid nitrogen immediately. The extraction of total RNAs was performed using the SV Total RNA Isolation System (Promega, USA) according to the manufacturer's protocol. The quantity and quality of the isolated RNAs were examined by NanoDrop 2000c UV–Vis spectrophotometer (Thermo Scientific, USA) and agarose gel electrophoresis. Subsequently, the reverse transcription PCR (RT-PCR) was carried out using 1 μg total RNA as the template. The gene transcription level of pro-*smtg* during fermentation process was investigated by quantitative real-time PCR (qRT-PCR) with primers listed in Table 1. The qRT-PCR reaction consisted of an initial denaturation at 94 °C and 40 amplifications cycles of 5 s at 94 °C and 60 s at 64 °C. The target gene transcription level was normalized internally to that of 16S rRNA gene for its transcription during overall growth stages was relatively stable.

Cloning of pro-*smtg* gene from *S. mobaraensis* mutants

Genomic DNAs were obtained from *S. mobaraensis* mutants and the wild-type strain. The primers used for the PCR are listed in Table 1. The PCR procedure was set as follows: (95 °C, 5 min) 1 cycle; (94 °C, 1 min; 65 °C, 30 s; 72 °C, 90 s) 30 cycles; and (72 °C, 10 min) 1 cycle. The amplified PCR products (1200 bp) were ligated into

Table 1 Primers used in this study

Name	Sequences (5′–3′)	Application
16S rRNA_L1	AGCAGCGGAGCATGTGGCTT	16S rRNA gene transcription
16S rRNA_R1	TGCGCTCGTTGCGGGACTTA	
pro-smtg-L1	CATGTCGAGGGACAGGAACA	pro-smtg gene transcription
pro-smtg-R1	TTGCGGAACTTGCTCTCGTA	
pro-smtg-FP	CGGAATTCATGCCGTCCGC AGGC	pro-smtg gene amplification
pro-smtg-RP	CCCAAGCTTTCACGGCCAG CCCTG	

plasmid pMD-19T and sequenced for alignment to that of the wild-type strain.

Results and discussion

Mutation and screening of the mutants

A well-controlled lethal rate is fundamental for effective mutation and screening of the mutants. The lethal rates of *S. mobaraensis* with respect to various treatment times are shown in Fig. 2, which indicated that the lethal rate increased to 66.5, 93.1, and 99.2%, respectively, after treated for 30, 40, and 50 s. When the samples were treated for 50 s or even longer, no spores could survive. According to previous reports (Guo et al. 2011; Hua et al. 2010), a lethal rate of 90% was considered appropriate. Besides, keeping the lethal rate high is necessary for the effective mutation and selection of mutant strains. Therefore, the ARTP treatment time applied in this study was determined as 40 s.

After eight rounds of iterative ARTP mutagenesis, 501 mutants in total were selected for tube-fermentation screening, as shown in Additional file 1: Figure S1. To investigate the accumulative effect on TGase production by iterative ARTP mutagenesis, the mutation and screening results for each round were collected and compared.

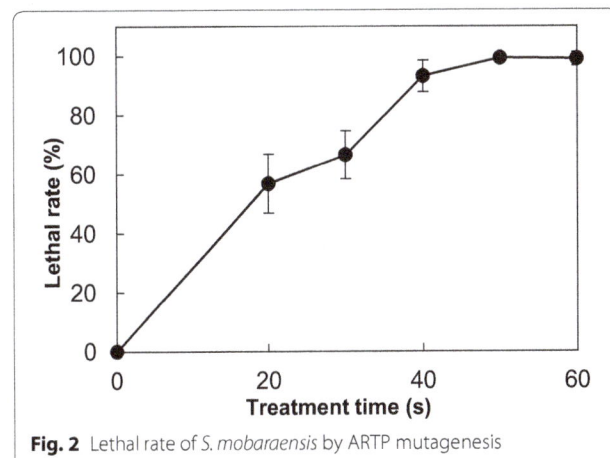

The statistic results shown in Table 2 and Fig. 3 indicated that the proportion of positive mutants was scaled up with the increase of iterative rounds. Moreover, the TGase production of the best mutant for each round presented a fluctuant improving trend. Thus, the increase of iterative rounds might exhibit an accumulative effect on TGase production of the mutants.

TGase production of the mutants

Shake flask fermentation was conducted to verify the potentially positive mutants identified in the tube fermentation. Finally, four mutants, *Sm*5-V1, *Sm*6-V13, *Sm*2-V10, and *Sm*7-V12, were identified in the ARTP breeding process, and the TGase production of these mutants is shown in Table 3. The highest TGase production reached 5.85 U/mL, which represented a 27% increase as compared with the wild-type strain.

The genetic stability is one of the key performance factors for a promising industrial strain because it reflects

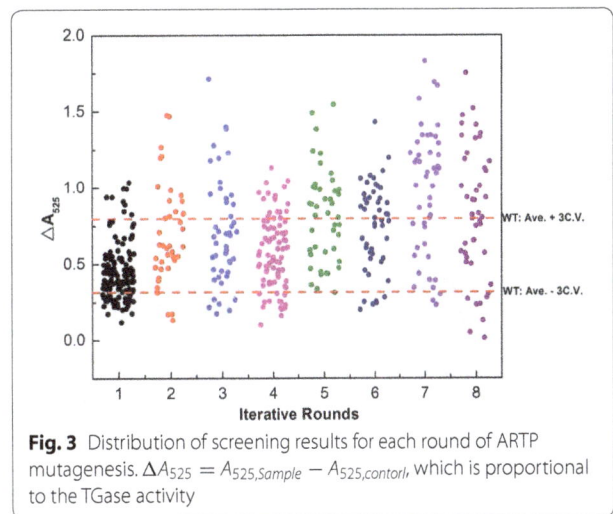

Fig. 3 Distribution of screening results for each round of ARTP mutagenesis. $\Delta A_{525} = A_{525,Sample} - A_{525,contorl}$, which is proportional to the TGase activity

Fig. 2 Lethal rate of *S. mobaraensis* by ARTP mutagenesis

Table 2 Mutation and screening results for each round of ARTP mutagenesis

Iterative round	R_M (%)	R_P (%)	Average ΔA_{525} of mutants
WT	–	–	0.56 ± 0.08
1	32.1	6.4	0.44 ± 0.19
2	48.8	36.6	0.69 ± 0.33
3	45.7	32.6	0.71 ± 0.35
4	34.8	19.1	0.58 ± 0.23
5	56.8	54.6	0.81 ± 0.31
6	62.5	50.0	0.75 ± 0.28
7	74.5	68.1	0.98 ± 0.41
8	69.6	52.2	0.83 ± 0.43

Table 3 TGase production of *S. mobaraensis* wild-type and ARTP mutants

Strain	TGase titer (U/mL)	Relative production (%)
WT	4.60 ± 0.17	100 ± 4
Sm5-V1	5.85 ± 0.21	127 ± 5
Sm6-V13	5.72 ± 0.16	124 ± 4
Sm2-V10	5.68 ± 0.19	124 ± 4
Sm7-V12	5.49 ± 0.13	119 ± 3

the potential mutations at the gene level (Ren et al. 2017). The genetic stability test results shown in Additional file 1:Figure S2 indicated that the TGase production of the identified mutants still remained stable after eight rounds of subcultures. These results proved that ARTP mutagenesis is a promising mutation breeding tool for industrial applications.

Pro-*smtg* gene expression level of the wild-type and mutant strains

The enhanced TGase production caused by ARTP mutagenesis may be attributed to two effects. A direct effect will happen if the structural gene (pro-*smtg*) of TGase zymogen is altered, leading to the change in the specific activity of the protein. Indirect effect happens when the relevant genes regulating the expression of TGase zymogen are altered.

Firstly, the pro-*smtg* gene nucleotide sequences of the four mutants showed no differences from the original sequence of the native enzyme from the wild-type strain, as indicated by gene cloning and sequence alignment shown in Additional file 1: Figure S3. However, qRT-PCR results shown in Fig. 4 indicated that the pro-*smtg* expression levels of the four mutants were remarkably improved as compared to that of the parental strain, which might lead to the increase in TGase production. The analysis from transcription level conducted in this

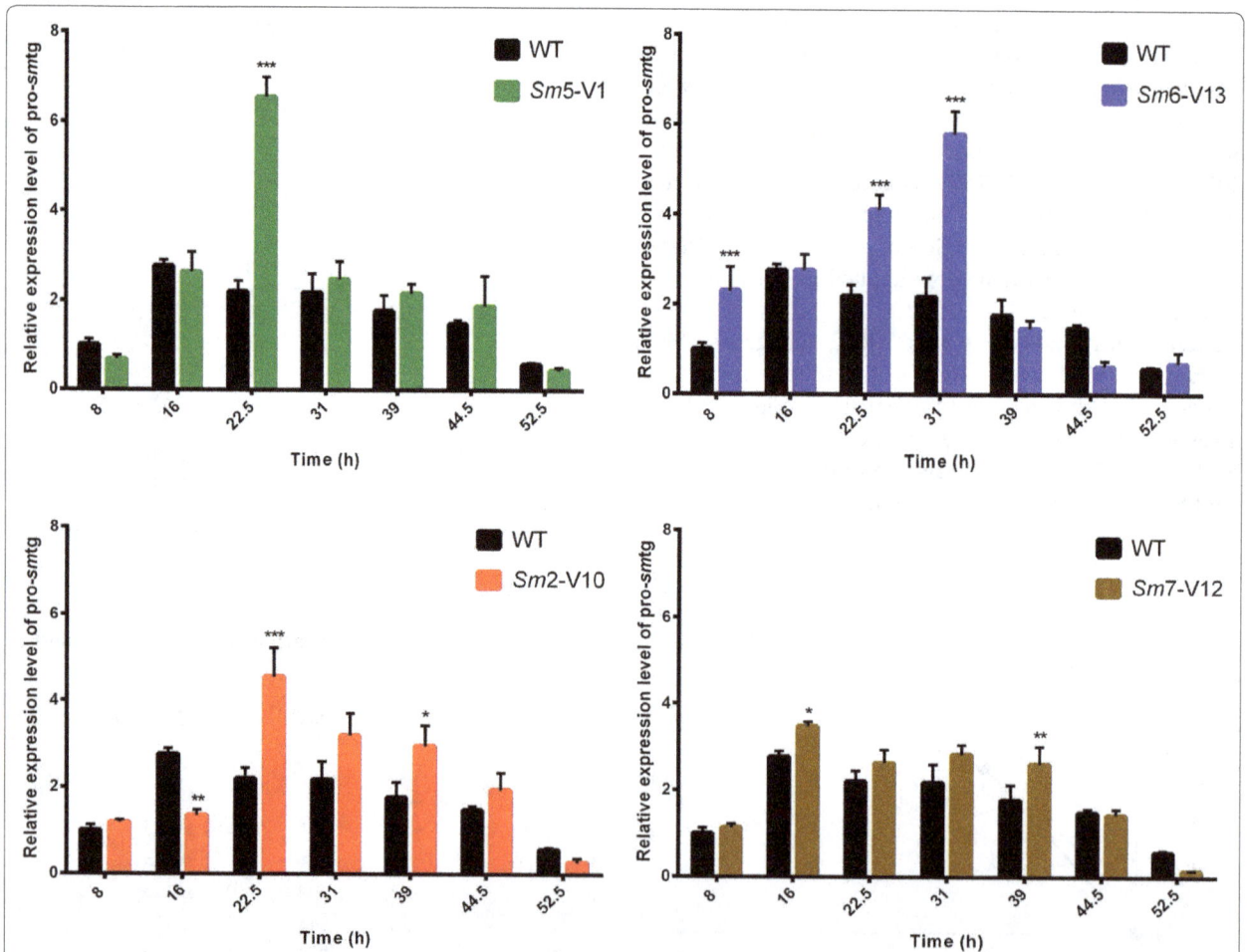

Fig. 4 Transcription analysis of pro-*smtg* gene by qRT-PCR. The pro-*smtg* transcription levels were detected after flask fermentation for different times in wild-type strain (WT) and mutants *Sm5-V1, Sm6-V13, Sm2-V10,* and *Sm7-V12,* which were normalized to that of 16S rRNA gene by the $2^{-\Delta\Delta Ct}$ method. The experiments were performed in three replications

study is of great importance for further investigation of metabolic changes caused by ARTP mutagenesis.

Conclusions

Iterative ARTP mutagenesis was applied to improve the TGase production in *S. mobaraensis* as a novel and efficient mutation tool. As a result, four mutants, *Sm*5-V1, *Sm*6-V13, *Sm*2-V10, and *Sm*7-V12, with increased TGase production by 27, 24, 24, and 19%, respectively, were identified after eight rounds of iterative mutagenesis. The best mutant *Sm*5-V1 showed an enhanced TGase production of 5.85 U/mL. In addition, the results of sequence alignment and qRT-PCR revealed that the expression of TGase zymogen gene increased obviously under the ARTP treatment, while the gene sequence remained unchanged. The analysis of the target gene sequence as well as transcription level conducted in this study would provide a valuable guidance for further researches and applications of ARTP mutagenesis.

Abbreviations
ARTP: atmospheric and room-temperature plasma; TGase: transglutaminase; MTG: microbial transglutaminases; RT-PCR: reverse transcription PCR; qRT-PCR: quantitative real-time PCR.

Authors' contributions
YJ wrote the experiment; YJ and CZ conducted the experiment; YS and HL helped to revise the manuscript; YJ, CL, and JX conceived the research; JP and YB provided experimental materials and guidance. All authors read and approved the final manuscript.

Author details
[1] State Key Laboratory of Bioreactor Engineering, East China University of Science and Technology, Shanghai 200237, People's Republic of China. [2] Shanghai Collaborative Innovation Center for Biomanufacturing Technology, East China University of Science and Technology, Shanghai 200237, People's Republic of China.

Acknowledgements
The authors are thankful to the Prof. Yonghong Wang's research group for providing the ARTP mutation breeding system as well as the operative skills to perform this research work.

Competing interests
The authors declare that the research was conducted in the absence of any commercial or financial relationships that could be construed as a potential competing interest.

Funding
This work is financially sponsored by the National Natural Science Foundation of China (No. 21536004), Shanghai Commission of Science and Technology (No. 15JC1400403), and Shanghai Pujiang Program (No. 15PJ1401200).

References
Duran R, Junqua M, Schmitter JM, Gancet C, Goulas P (1998) Purification, characterisation, and gene cloning of transglutaminase from *Streptoverticillium cinnamoneum* CBS 683.68. Biochimie 80:313–319

Grossowicz N, Wainfan E, Borek E, Waelsch H (1950) The enzymatic formation of hydroxamic acids from glutamine and asparagine. J Biol Chem 187:111–125

Guo T, Tang Y, Xi YL, He AY, Sun BJ, Wu H, Liang DF, Jiang M, Ouyang PK (2011) *Clostridium beijerinckii* mutant obtained by atmospheric pressure glow discharge producing high proportions of butanol and solvent yields.

Hua XF, Wang J, Wu ZJ, Zhang HX, Li HP, Xing XH, Liu Z (2010) A salt tolerant *Enterobacter cloacae* mutant for bioaugmentation of petroleum- and salt-contaminated soil. Biochem Eng J 49:201–206

Kobayashi K, Suzuki SI, Izawa Y, Yokozeki K, Miwa K, Yamanaka S (1998) Transglutaminase in sporulating cells of *Bacillus subtilis*. J Gen Appl Microbiol 44:85–91

Kumar AK (2015) UV mutagenesis treatment for improved production of endoglucanase and β-glucosidase from newly isolated thermotolerant actinomycetes, *Streptomyces griseoaurantiacus*. Bioresour Bioprocess 2:22

Laroussi M (2005) Low temperature plasma-based sterilization: overview and state-of-the-art. Plasma Process Polym 2:391–400

Lu Y, Wang LY, Ma K, Li G, Zhang C, Zhao HX, Lai QH, Li HP, Xing XH (2011) Characteristics of hydrogen production of an *Enterobacter aerogenes* mutant generated by a new atmospheric and room temperature plasma (ARTP). Biochem Eng J 55:17–22

Martins IM, Matos M, Costa R, Silva F, Pascoal A, Estevinho LM, Choupina AB (2014) Transglutaminase: recent achievements and new sources. Appl Microbiol Biotechnol 98:6957–6964

Marx CK, Hertel TC, Pietzsch M (2008) Purification and activation of a recombinant histidine-tagged pro-transglutaminase after soluble expression in *Escherichia coli* and partial characterization of the active enzyme. Enzyme Microb Technol 42:568–575

Ren F, Chen L, Tong QY (2017) Highly improved acarbose production of *Actinomyces* through the combination of ARTP and penicillin susceptible mutant screening. World J Microbiol Biotechnol 33:16

Salis B, Spinetti G, Scaramuzza S, Bossi M, Jotti GS, Tonon G, Crobu D, Schrepfer R (2015) High-level expression of a recombinant active microbial transglutaminase in *Escherichia coli*. BMC Biotechnol 15:84

Soares LHB, Assmann F, Ayub MAZ (2003) Purification and properties of a transglutaminase produced by a *Bacillus circulans* strain isolated from the Amazon environment. Biotechnol Appl Biochem 37:295–299

Tan J, Chu J, Wang YH, Zhuang YP, Zhang SL (2014) High-throughput system for screening of *Monascus purpureus* high-yield strain in pigment production. Bioresour Bioprocess 1:16

Wang Q, Feng LR, Wei L, Li HG, Wang L, Zhou Y, Yu XB (2014) Mutation breeding of lycopene-producing strain *Blakeslea Trispora* by a novel atmospheric and room temperature plasma (ARTP). Appl Biochem Biotechnol 174:452–460

Xu F, Jin HJ, Li HM, Tao L, Wang JP, Lv J, Chen SF (2012) Genome shuffling of *Trichoderma viride* for enhanced cellulase production. Ann Microbiol 62:509–515

Yokoyama K, Nio N, Kikuchi Y (2004) Properties and applications of microbial transglutaminase. Appl Microbiol Biotechnol 64:447–454

Yurimoto H, Yamane M, Kikuchi Y, Matsui H, Kato N, Sakai Y (2004) The propeptide of *Streptomyces mobaraensis* transglutaminase functions in cis and in trans to mediate efficient secretion of active enzyme from methylotrophic yeasts. Biosci Biotechnol Biochem 68:2058–2069

Zhang X, Zhang XF, Li HP, Wang LY, Zhang C, Xing XH, Bao CY (2014) Atmospheric and room temperature plasma (ARTP) as a new powerful mutagenesis tool. Appl Microbiol Biotechnol 98:5387–5396

Zhang X, Zhang C, Zhou QQ, Zhang XF, Wang LY, Chang HB, Li HP, Oda Y, Xing XH (2015) Quantitative evaluation of DNA damage and mutation rate by atmospheric and room-temperature plasma (ARTP) and conventional mutagenesis. Appl Microbiol Biotechnol 99:5639–5646

Zhu Y, Rinzema A, Tramper J, Bol J (1995) Microbial transglutaminase—a review of its production and application in food processing. Appl Microbiol Biotechnol 44:277–282

Zotzel J, Keller P, Fuchsbauer HL (2003a) Transglutaminase from *Streptomyces mobaraensis* is activated by an endogenous metalloprotease. Eur J Biochem 270:3214–3222

Zotzel J, Pasternack R, Pelzer C, Ziegert D, Mainusch M, Fuchsbauer HL (2003b) Activated transglutaminase from *Streptomyces mobaraensis* is processed by a tripeptidyl aminopeptidase in the final step. Eur J Biochem 270:4149–4155

6

Enzymatic characterization of a recombinant carbonyl reductase from *Acetobacter* sp. CCTCC M209061

Ping Wei[1,2], Yu-Han Cui[1], Min-Hua Zong[1,2], Pei Xu[1], Jian Zhou[2] and Wen-Yong Lou[1]* ⓘ

Abstract

Background: *Acetobacter* sp. CCTCC M209061 could catalyze carbonyl compounds to chiral alcohols following anti-Prelog rule with excellent enantioselectivity. Therefore, the enzymatic characterization of carbonyl reductase (CR) from *Acetobacter* sp. CCTCC M209061 needs to be investigated.

Results: A CR from *Acetobacter* sp. CCTCC M209061 (AcCR) was cloned and expressed in *E. coli*. AcCR was purified and characterized, finding that AcCR as a dual coenzyme-dependent short-chain dehydrogenase/reductase (SDR) was more preferred to NADH for biocatalytic reactions. The AcCR was activated and stable when the temperature was under 35 °C and the pH range was from 6.0 to 8.0 for the reduction of 4′-chloroacetophenone with NADH as coenzyme, and the optimal temperature and pH were 45 °C and 8.5, respectively, for the oxidation reaction of isopropanol with NAD+. The enzyme showed moderate thermostability with half-lives of 25.75 h at 35 °C and 13.93 h at 45 °C, respectively. Moreover, the AcCR has broad substrate specificity to a range of ketones and ketoesters, and could catalyze to produce chiral alcohol with *e.e.* >99% for the majority of tested substrates following the anti-Prelog rule.

Conclusions: The recombinant AcCR exhibited excellent enantioselectivity, broad substrate spectrum, and highly stereoselective anti-Prelog reduction of prochiral ketones. These results suggest that AcCR is a powerful catalyst for the production of anti-Prelog alcohols.

Keywords: Carbonyl reductase, *Acetobacter* sp., Chiral alcohols, Enzymatic characterization

Background

Chiral alcohols are vitally important building blocks for the synthesis of pharmaceuticals, agricultural chemicals, flavors, and special materials. Currently, both chemical and biological methods can be used to synthesize chiral alcohols. The most prospective biocatalytic method is the asymmetric transfer of hydrogenation to carbonyl groups catalyzed by reductases or microbial cells containing relevant reductases (Li et al. 2014; Zheng et al. 2017). In recent years, the efficient asymmetric reduction of ketones using biocatalysts as its high yield and excellent enantiomeric excess has been extensively used for the production of chiral alcohols in the chemical and pharmaceutical industries (Birolli et al. 2015; Gutierrez et al. 2010).

Carbonyl reductase (CR) or alcohol dehydrogenase (ADH) is NAD(P)H-dependent oxidoreductase which can catalyze a variety of carbonyl compounds to the corresponding alcohols (Kaluzna et al. 2005). Enantioselective CRs or ADHs have been reported from animals (Nakajin et al. 1998; Montfort et al. 2002), plants (Sengupta et al. 2015), and microorganisms (Grosch et al. 2015; Singh et al. 2009). A number of CRs or ADHs have been utilized for asymmetrically reducing carbonyl functionalities, and many excellent biocatalytic processes have been developed (Li et al. 2015). Among all the sources of CRs, microorganisms are the most important providers, and some CRs or ADHs from microorganisms have been reported with high yield and enantioselectivity

*Correspondence: wylou@scut.edu.cn
[1] Lab of Applied Biocatalysis, School of Food Science and Engineering, South China University of Technology, Guangzhou 510640, Guangdong, China
Full list of author information is available at the end of the article

(Musa and Phillips 2011). The carbonyl reductase SCRII from *Candida parapsilosis* could catalyze the asymmetric reduction of 2-hydroxyacetophenone to (S)-1-phenyl-1,2-ethanediol with optical purity of 100% in high yield of 98.1% (Zhang et al. 2011). Xu et al. (2015) found that an NADPH-dependent carbonyl reductase from *Yarrowia lipolytica* ACA-DC 50109 could efficiently convert α-chloroacetophenone to (R)-2-chloro-1-phenylethol with 99% *e.e.* Ni et al. (2011) heterologously overexpressed a β-ketoacyl-ACP reductase from *Bacillus* sp. ECU0013 in *E. coli*, which was used for efficient reduction of ethyl 2-oxo-4-phenylbutyrate at 620 g/L, and the *e.e.* of the product ethyl (S)-2-hydroxy-4-phenylbutyrate was excellent (>99%). Therefore, using carbonyl reductases to catalyze the asymmetric reduction of prochiral carbonyl compounds is an efficient and useful method for the synthesis of chiral alcohols (He et al. 2016; Cui et al. 2017; Qian et al. 2014).

The oxidoreductase from *Acetobacter* is mostly used to produce acetic acid, and rarely used to asymmetric reduction of carbonyl compounds to chiral alcohols. In our previous study, it was found that *Acetobacter* sp. CCTCC M209061 could catalyze the asymmetric reductions of several carbonyl compounds to corresponding chiral alcohols following anti-Prelog rule with excellent enantioselectivity (Cheng et al. 2014; Wei et al. 2015, 2016). In the present work, the AcCR was heterologously expressed and systematically characterized for its substrate spectrum, stereoselectivity, and the capacity of industrial application.

Methods

Chemicals, bacterial strains, and plasmids

The *Acetobacter* sp. CCTCC M209061 strain was previously isolated by our group from Chinese kefir grains and stored at −80 °C. *E. coli* DH5α, BL21(DE3)pLysS, and plasmid pGEX-2T were purchased from Novagen. The restriction enzyme FastDigest AvaI, EcoRI, T4 DNA Ligase, DNA, and protein marker were purchased from Thermo Scientific. KOD FX polymerase for PCR was purchased from Toyobo. The kits used in the construction of the recombinant plasmids were purchased from Generay. The prochiral ketones were purchased from Sigma Aldrich or Aladdin. The primers synthesis and DNA sequencing were completed by Sangon Biotech. All other reagents and solvents were of analytical grade and used without further purification.

Expression of AcCR in *E. coli* BL21(DE3)pLysS

Genomic DNA of *Acetobacter* sp. CCTCC M209061 was extracted and purified using a bacterial genomic DNA Kit. Oligonucleotide primers for *accr* were designed according to the published gene sequence of oxidoreductase from *Acetobacter pasteurianus* 386B (Sequence ID: HF677570.1). The DNA fragment of *accr* was amplified with primer 1 (5′-TCC<u>CCCGGG</u>AATG-GCACGTGTAGCAGGCAAGGTT-3′) and primer 2 (5′-CCG<u>GAATTC</u>CTTATTGCGCGGTGTACCCAC-CATCAAT-3′) and double-digested with AvaI and EcoRI, then the *accr* fragment was inserted into vector pGEX-2T, the resulting plasmid (pGEX-*accr*) was transformed into *E. coli* DH5α for amplification, and then the correct pGEX-*accr* was transformed into *E. coli* BL21(DE3) pLysS to express the recombinant AcCR. The recombinant *E. coli* BL21(DE3)pLysS(AcCR) were cultivated at 37 °C, 180 rpm in 50 mL LB medium (pH 6.5) containing 100 μg/mL of ampicillin. When the optical density at 600 nm (OD_{600}) of the culture reached 1.2, the temperature was changed to 20 °C, and then IPTG was added to a final concentration of 0.4 mM. The cultivation continued at 20 °C for additional 15 h. Then the recombinant cells were harvested by centrifugation (8000 rpm, 5 min) at 4 °C and washed three times with physiological saline (0.85%), and stored at 4 °C for later use.

Measurement of enzymatic activity

The oxidordeuctase activity of the recombinant AcCR was determined by monitoring the change of the absorbance at 340 nm for 3 min on the spectrophotometer (Shinmadzu UV-3010, Japan). The AcCR-catalyzed reduction of 4′-chloroacetophenone was conducted with NADH or NADPH at 35 °C in 2 mL phosphate buffer (pH 6.5 or 5.5). One unit (U) of recombinant AcCR was defined as the amount of enzyme that catalyzed the 1 μmol NADH or NADPH per minute. The NAD^+- and $NADP^+$-linked oxidations catalyzed by AcCR were conducted with isopropanol as substrate and NAD^+ or $NADP^+$ at 45 °C in 2 mL 50 mM citrate–phosphate buffer (pH 8.0). One unit (U) of AcCR was defined as the amount of enzyme that catalyzed the production of 1 μmol NADH or NADPH per minute. The reaction solution was incubated at 35 or 45 °C for 3 min, and then 20 μL recombinant AcCR (about 0.008 mg of the purified enzyme) was added to proceed the reaction.

Purification of recombinant AcCR

The harvested recombinant *E. coli* BL21(DE3) pLysS(AcCR) cells were suspended in 50 mM citrate–phosphate buffer (pH 6.5) at the concentration of 50 mg/mL, ultrasonicated for 3 × 18 min, and then the cells debris was removed by centrifugation at 12,000 rpm at 4 °C for 20 min. The resulting supernatant was filtered with 0.22-μm filter membrane before the purification process. The general step of the purification using NGC Quest™ 10 system was as follows. First, the GST-based affinity chromatography column (5 mL) was pre-equilibrated using buffer A (4.3 mM Na_2HPO_4, 1.47 mM KH_2PO_4, 137 mM NaCl, 2.7 mM KCl, pH7.3), and then the crude enzyme was loaded on the column. After that, the loaded column was first subjected to washing with 10 column volume buffer A to remove the unbound protein fractions, then washed sequentially with buffer B (added 0.5 M NaCl in buffer A) to remove some stubborn protein, and then buffer A was used once again to lower the salinity. At last, the target protein was eluted with buffer C (50 mM Tris–HCl, 2.5 g/L glutathione, pH 8.0). The protein content was measured by the method of Bradford (1976).

Enzymatic characteristic of the recombinant AcCR
Effect of temperature
The effects of temperature on the activity of recombinant AcCR were determined at various temperatures from 20 to 55 °C. For thermal stability determination, the enzyme was pre-incubated at various temperatures ranging from 20 to 55 °C, equivalent was taken at a certain time, and the residual activity was determined as described in "Measurement of enzymatic activity." The coenzyme NAD^+ and NAD^+ was used for the reduction of 4′-chloroacetophenone (5 mM) and the oxidation of the isopropanol (150 mM), respectively. The enzyme activity of the first measurement was defined as 100%.

Effect of pH
The optimum pH of recombinant AcCR was investigated within a pH range of 4.5–9.5 using various buffer systems at 50 mM. The buffers were as follows: citrate–phosphate (pH 4.5–8.0), Tris–HCl (pH 8.0–8.5) and glycine–NaOH (pH 8.6-9.5). The pH stability was evaluated by pre-incubating the recombinant AcCR in different pH buffers (4.5–9.5) at 4 °C for 4 days. Samples were taken at a certain time, and the residual activity was determined as described above with the non-incubated recombinant AcCR as the control.

Effect of metal ions and chemical agents
The influence of metal ions and chemical agents on the catalytic activity of recombinant AcCR was investigated by pre-incubating the enzyme with various additives in citrate–phosphate buffer (50 mM, pH 6.5) at 35 °C for 30 min. The enzyme activity was determined under the condition described above using NADH as coenzyme. The enzyme activity of the recombinant AcCR in the absence of additives was recorded as 100%.

Substrate specificity and bioconversion of various carbonyl compounds
The activity of recombinant AcCR for each specific substrate was measured by the method described above using NADH as the coenzyme. The relative activity of recombinant AcCR to 4′-chloroacetophenone was defined as 100%.

The reduction of various carbonyl compounds were performed in a 10-mL conical flask with 4 mL citrate–phosphate buffer (50 mM, pH 6.5) containing 50 mM each substrate, 0.1 mM NADH, and 150 mM isopropanol. The reactant was pre-incubated in a shaker at 35 °C for 10 min, and then 8 U recombinant AcCR was added to initiate the reaction. Samples (25 µL) were withdrawn after reaction for a certain time, and then the product and residual substrate were extracted with ethyl acetate (2 × 25 µL, 5 mM n-dodecane as internal standard) before GC analysis.

Kinetic parameters assays
For the kinetic analysis, the initial reaction rates of the recombinant AcCR were determined under the optimum conditions. For the reduction of 4′-chloroacetophenone, the coenzyme NADH or NADPH varied from 0.05 to 0.6 mM or from 0.05 to 1.0 mM, and the concentrations of 4′-chloroacetophenone were from 0.5 to 10 mM. For the oxidation of isopropanol, the concentrations of NAD^+ or $NADP^+$ varied from 0.05 to 2 mM or from 0.5 to 20 mM. The isopropanol concentrations were from 30 to 275 mM. All the measurements were carried out in duplicate. Michaelis–Menten was used to fit the data, and the kinetic parameters of recombinant AcCR-catalyzed reduction and oxidation reactions, K_m and V_{max} values, were obtained in the fit before computing the K_{cat}.

GC methods
The organic extracts of reaction mixtures were analyzed by a GC analysis (Shimadzu Corp GC 2010). GC was performed using chiral columns HP Chiral 10B and CP-Chiralsil-Dex-CB. The initial reaction rate, the yield, and the product $e.e.$ of those reactions were calculated as described in our previous report (Wang et al. 2013).

Results and discussion
Sequence analysis of AcCR
A 762-bp polynucleotide sequence was amplified by PCR from the genome DNA of Acetobacter sp. CCTCC M209061. The sequence was an intact open reading frame and encoded a predicted protein of 253 amino acid residues with a molecular weight of about 26.4 kDa. The GenBank accession number of the nucleotide sequence of AcCR gene is MF419650. The result of multiple sequence alignment with NCBI protein blast is shown in Fig. 1. The AcCR amino acid sequence displayed a high level of similarity to the identical proteins from other Acetobacter and related bacteria. The identities of AcCR with other proteins were as follows: 84% with 3-beta

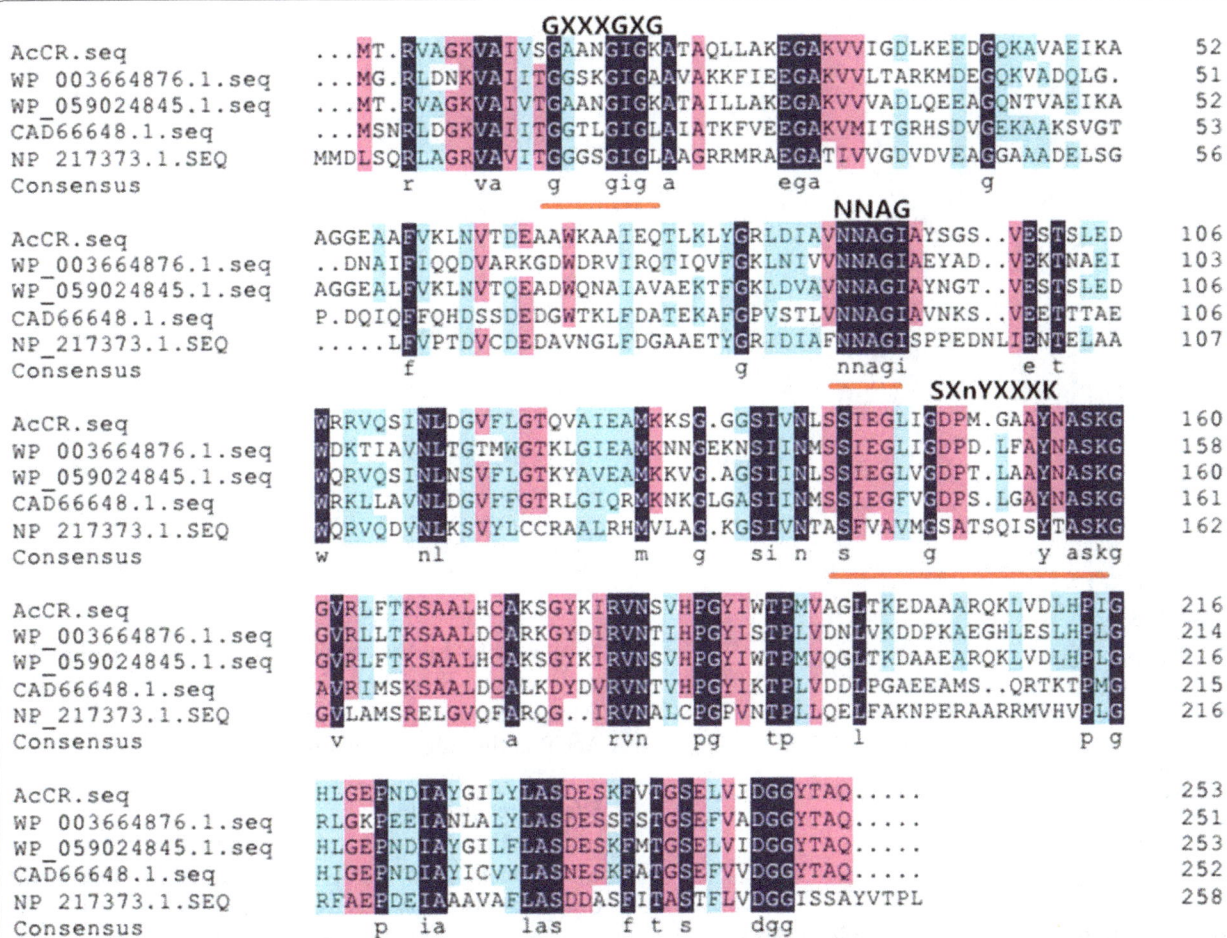

Fig. 1 Alignment of multiple deduced amino acid sequences of AcCR and carbonyl reductases from other sources

hydroxysteroid dehydrogenase from *Acetobacter ghanensis* (WP_059024845.1), 56% with Cyclopentanol dehydrogenase from *Mesorhizobium plurifarium* (CDX57396.1), 51% with *R*-specific alcohol dehydrogenase from *Lactobacillus brevis* (CAD66648.1), and 37% with short-chain type dehydrogenase/reductase from *Mycobacterium tuberculosis* H37Rv (NP_217373.1), respectively. The amino acid sequence alignments of the deduced polypeptides of AcCR with the short-chain dehydrogenase/reductase (SDR) proteins from GenBank database were performed. The conserved sequence (GXXXGXG) (Rossman fold motif) (Masud et al. 2011; Jörnvall et al. 2010) reported importantly of the coenzyme binding site of SDR family enzyme was found at the N-terminal 13–19 position of the AcCR. And the common active site conserved amino acid pattern SXnYXXXK was emerged at the mid-chain pattern 142–159, as well as the NNAG (89–92) sequence which has a function in stabilizing

β-strands of classical SDR (Persson et al. 2003; Filling et al. 2002). Therefore, the AcCR could be classified as the SDR which belongs to a bulky dehydrogenase/reductase family.

Expression and purification of recombinant AcCR

The *accr* was constructed into plasmid pGEX-2T and transformed into *E. coli.* BL21(DE3)pLysS. SDS-PAGE analysis of the crude extracts indicated that the recombinant AcCR was successfully expressed in the soluble form with the GST tag (shown in Fig. 2). As a GST-tagged fusion protein, the GAcCR (recombinant AcCR) was purified by a single-step affinity chromatography using a GST-tagged column. As expected, the purified GAcCR was observed at the position of approximately 53 kDa (the GST tag is about 26 kDa) which was consistent with the estimated value (Fig. 2). The purified recombinant AcCR activity was 5.17 U/mg and stored at 4 °C for later use.

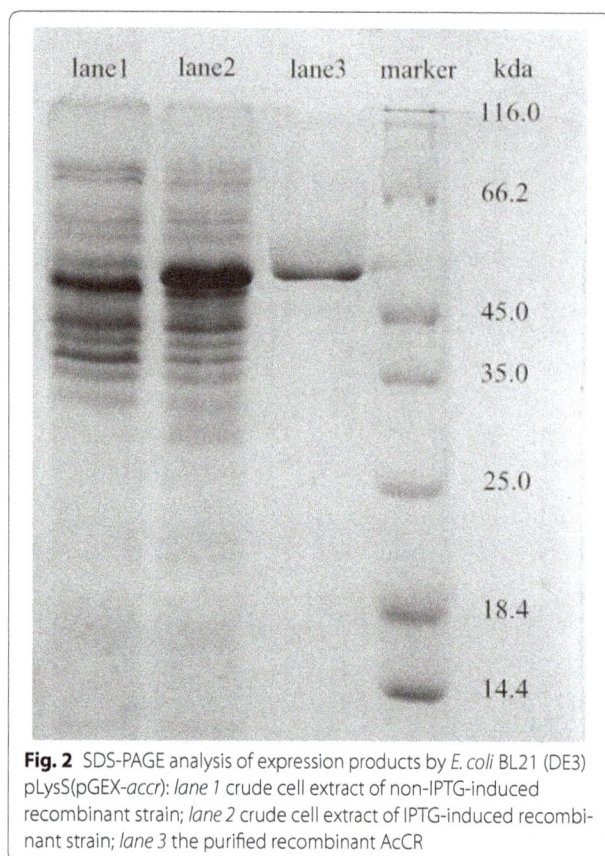

Fig. 2 SDS-PAGE analysis of expression products by *E. coli* BL21 (DE3) pLysS(pGEX-*accr*): *lane 1* crude cell extract of non-IPTG-induced recombinant strain; *lane 2* crude cell extract of IPTG-induced recombinant strain; *lane 3* the purified recombinant AcCR

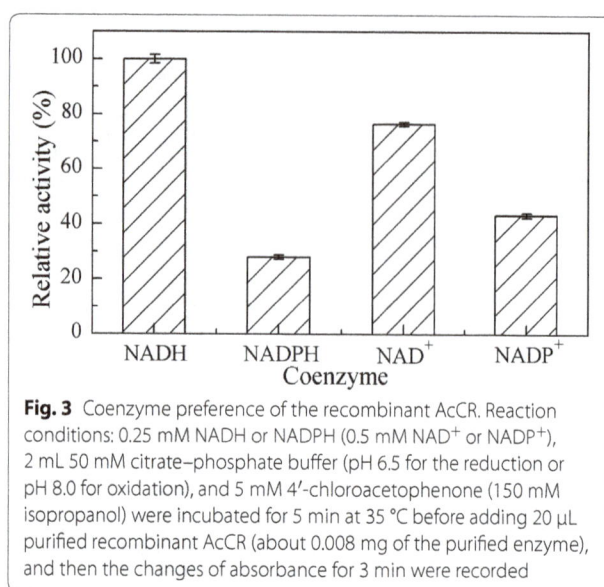

Fig. 3 Coenzyme preference of the recombinant AcCR. Reaction conditions: 0.25 mM NADH or NADPH (0.5 mM NAD$^+$ or NADP$^+$), 2 mL 50 mM citrate–phosphate buffer (pH 6.5 for the reduction or pH 8.0 for oxidation), and 5 mM 4′-chloroacetophenone (150 mM isopropanol) were incubated for 5 min at 35 °C before adding 20 μL purified recombinant AcCR (about 0.008 mg of the purified enzyme), and then the changes of absorbance for 3 min were recorded

Enzymatic characteristic of the recombinant AcCR
Coenzyme preference of recombinant AcCR
SDRs act on the substrate by transferring electrons from or to coenzyme. The recombinant AcCR was found to be a kind of carbonyl reductase which could use both NAD(H) and NADP(H) as coenzyme to execute the redox reactions. The coenzyme preference of the recombinant AcCR was examined by measuring enzyme activity using 4′-chloroacetophenone or isopropanol as substrate. As shown in Fig. 3, the recombinant AcCR showed redox activity with both NAD(H) and NADP(H) as coenzymes. Meanwhile, the relative activity of the recombinant AcCR with NADPH as coenzyme was only about 40% of that with NADH, which confirmed that AcCR exhibited coenzyme specificity for NADH over NADPH. In the recent reports, many biocatalysts catalyzing asymmetric reductions utilize NADPH as hydrogen donor (Leuchs and Greiner 2011; Richter and Hummel 2011; Ma et al. 2013; Zhang et al. 2015). ADHs capable of utilizing NADH as the coenzyme outperformed NADPH-dependent ones, since NADH was more stable than NADPH under operational conditions (Li et al. 2015) and economic than NADPH. The NAD(H)-preferred CRs could be advantageous for their application in industrial bioreduction system. Therefore, NAD(H) was used as the coenzyme of the enzymatic characteristic of the recombinant AcCR in the subsequent study.

Effect of temperature
Initially, the effect of varying temperatures on the activity of recombinant AcCR was explored on both reduction and oxidation reactions. The optimal temperature was found to be 35 °C for the reduction of 4′-chloroacetophenone when NADH was used as the coenzyme (Fig. 4a). For the case of the oxidation of isopropanol, the maximum activity was observed at 45 °C with NAD$^+$ as coenzyme (Fig. 4a). The thermal stability of the recombinant AcCR was investigated at various temperatures ranging from 20 to 55 °C. As illustrated in Fig. 4b, the enzyme activity decreased slowly when the incubation temperature was less than 35 °C. The relative activity was over 50% after 36 h incubation at 30 °C, and even more than 62% after incubating for 24 h at 35 °C and over 5 h at 50 °C. Thus, the thermal stability of recombinant AcCR was excellent and much better than many reported carbonyl reductases from other sources (Xu et al. 2015; Singh et al. 2009; Luo et al. 2015; Wang et al. 2015). The enzyme exhibited half-lives of 25.75 h at 35 °C and 13.93 h at 45 °C, respectively, for the reduction and oxidation reactions.

Effect of pH
The effect of pH on recombinant AcCR was detected in different buffer systems (50 mM) at the range from 4.5 to 9.5 for both reduction and oxidation. The maximum

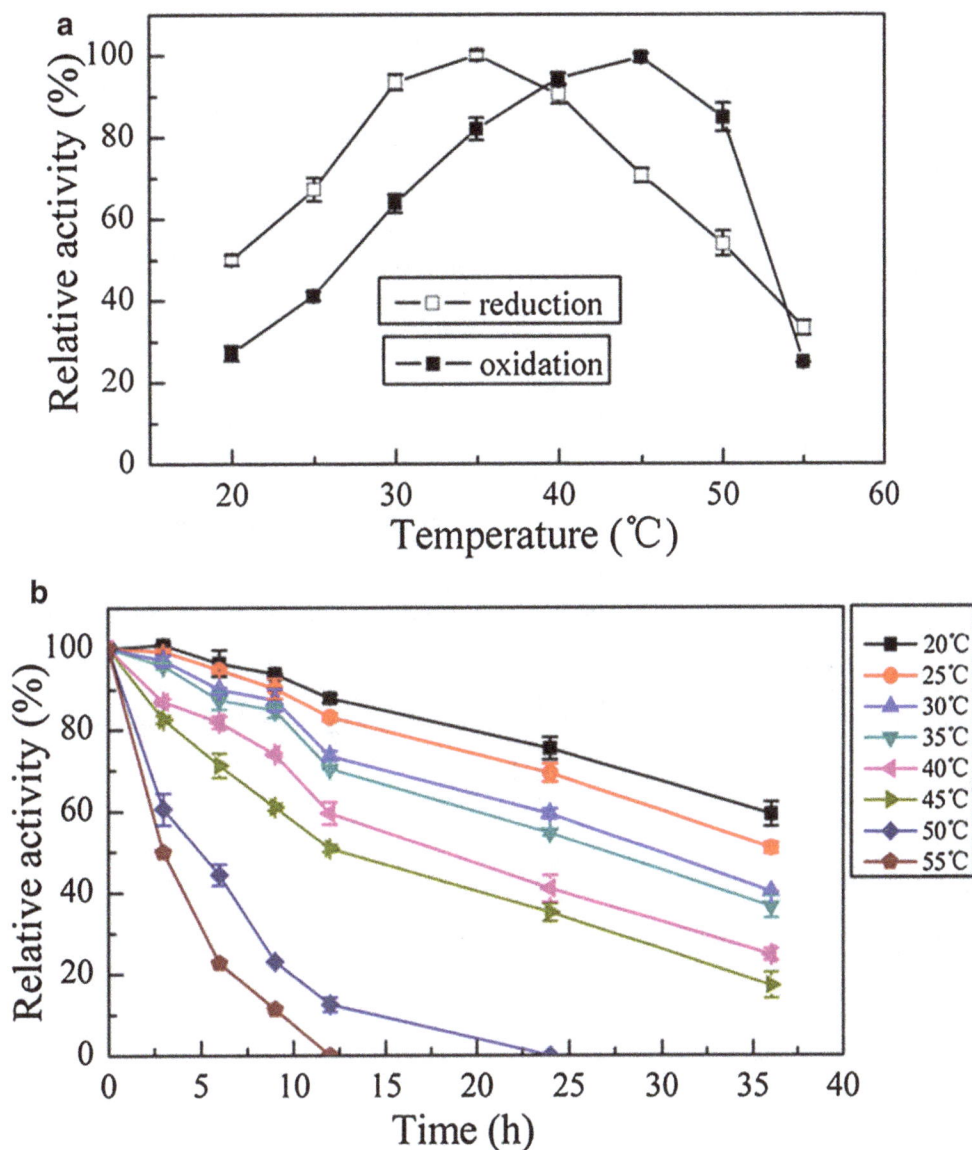

Fig. 4 Effects of temperature on the activity and stability of the recombinant AcCR. Reaction conditions: **a** 0.25 mM NADH or 0.5 mM NAD$^+$, 2 mL 50 mM citrate–phosphate buffer (pH 6.5 or 8.0), and 5 mM 4′-chloroacetophenone or 150 mM isopropanol were incubated at 20–55 °C for 5 min before adding 20 μL purified recombinant AcCR (about 0.008 mg of the purified enzyme), and the changes of absorbance for 3 min at 20–55 °C, respectively, were recorded; **b** 0.25 mM NADH, 2 mL 50 mM citrate–phosphate buffer (pH 6.5), and 5 mM 4′-chloroacetophenone were incubated for 5 min at 35 °C, then added 20 μL purified recombinant AcCR (about 0.008 mg of the purified enzyme, incubated at 20–55 °C), and the changes of absorbance for 3 min were recorded

activity of the recombinant AcCR for reduction of 4′-chloroacetophenone was obtained at pH 6.5. As shown in Fig. 5a, the relative activity was over 90% in the reduction reaction at the pHs between 6.0 and 8.0. The optimal pH for oxidation of isopropanol was found to be pH 8.5 when NAD$^+$ was acted as coenzyme (Fig. 5b). The optimal pH of the recombinant AcCR for catalyzing reduction and oxidation is close to that of the oxidoreductase from *Gluconobacter oxydans* (Liu et al. 2014).

To investigate the pH stability, the recombinant AcCR was incubated at 4 °C in varying pHs from 4.5 to 9.5 and tested the reduction activity using NADH as coenzyme. As shown in Fig. 5c, the optimal storage pH was 6.5, at which pH almost 80% relative activity remained after 96 h. Almost 70% relative activities remained at both pH 7.0 and 7.5 after 96 h, as well as at pH 6.0 after 75 h. The results showed that the recombinant AcCR had broad pH tolerant scope and excellent pH stability.

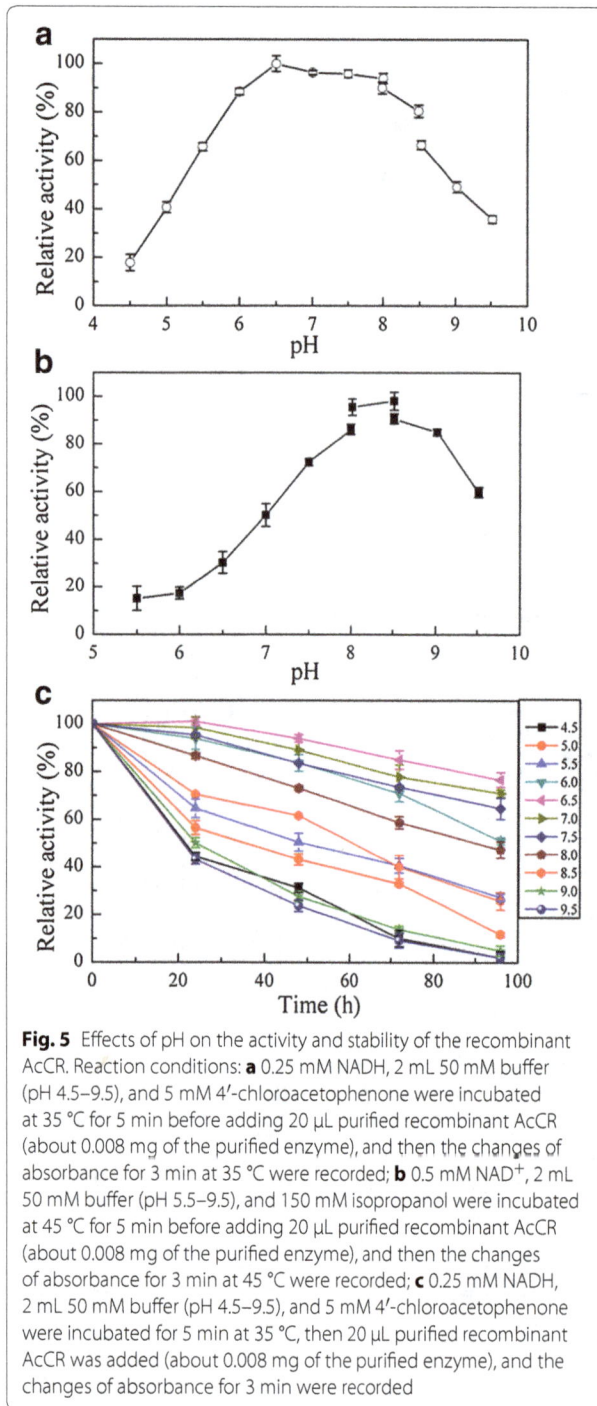

Fig. 5 Effects of pH on the activity and stability of the recombinant AcCR. Reaction conditions: **a** 0.25 mM NADH, 2 mL 50 mM buffer (pH 4.5–9.5), and 5 mM 4′-chloroacetophenone were incubated at 35 °C for 5 min before adding 20 μL purified recombinant AcCR (about 0.008 mg of the purified enzyme), and then the changes of absorbance for 3 min at 35 °C were recorded; **b** 0.5 mM NAD⁺, 2 mL 50 mM buffer (pH 5.5–9.5), and 150 mM isopropanol were incubated at 45 °C for 5 min before adding 20 μL purified recombinant AcCR (about 0.008 mg of the purified enzyme), and then the changes of absorbance for 3 min at 45 °C were recorded; **c** 0.25 mM NADH, 2 mL 50 mM buffer (pH 4.5–9.5), and 5 mM 4′-chloroacetophenone were incubated for 5 min at 35 °C, then 20 μL purified recombinant AcCR was added (about 0.008 mg of the purified enzyme), and the changes of absorbance for 3 min were recorded

Effect of metal ions and chemical agents

It is well known that metal ions have remarkable effects on the activity of carbonyl reductase. For example, Zn^{2+} is necessary for many medium-chain dehydrogenases/reductases (Zn-MDR). The Zn^{2+} appears in those enzymes to give the protein an additional strength, as if compensating for domain variability and evolutionary

changes in the protein scaffold (Jörnvall et al. 2010). Besides, some metal ions also have negative effect on the activity of a number of enzymes. For example, Co^{2+}, Cu^{2+}, and Zn^{2+} can severely inhibit the enzyme activity of carbonyl reductase SCRII from *Candida parapsilosis* (Zhang et al. 2011). So it is of significantly important to explore the effects of various additives on the recombinant AcCR. As shown in Table 1, among the tested metal ions, Mn^{2+}, K^+, Fe^{2+}, Mg^{2+}, Ca^{2+}, and Co^{2+} had slight stimulation on the recombinant AcCR, while the other metal ions had inhibitory effects to some extent. Enzyme activity was inhibited about 5 and 15% when 2 and 5 mM Zn^{2+} was added, and the residue enzyme activity was only 43.0 and 24.9% when 2 and 5 mM Cu^{2+} was added, respectively. Ag^+ and Hg^+ inhibited the enzyme activity completely.

The effect of chemical agents including metal-chelator, surfactants, and sulfhydryl reagents on the recombinant AcCR was presented in Table 2. EDTA had almost no effect on the enzyme activity of the recombinant

Table 1 Effects of metal ions on the enzymatic activity of the recombinant AcCR

Reagent	Concentration (mM)	Relative activity (%)
Control		100.0 ± 0.8
Mn^{2+}	2	122.3 ± 0.4
	5	119.9 ± 0.7
K^+	2	110.6 ± 1.2
	5	103.4 ± 0.8
Fe^{2+}	2	106.7 ± 1.0
	5	106.0 ± 2.9
Zn^{2+}	2	94.8 ± 1.2
	5	85.8 ± 1.3
Mg^{2+}	?	115.7 ± 0.7
	5	123.5 ± 2.4
Ba^{2+}	2	100.8 ± 0.5
	5	97.9 ± 1.7
Ca^{2+}	2	116.7 ± 0.6
	5	113.9 ± 1.0
Cu^{2+}	2	43.0 ± 0.7
	5	24.9 ± 0.7
Co^{2+}	2	99.6 ± 2.1
	5	113.8 ± 0.2
Hg^{2+}	2	Nd
	5	Nd
Ag^+	2	Nd
	5	Nd

Reaction conditions: 0.5 mM NADH, 2 mL 50 mM citrate–phosphate buffer (pH 6.5) with different metal ions, and 5 mM 4′-chloroacetophenone were incubated for 10 min at 35 °C before adding 20 μL purified recombinant AcCR (about 0.008 mg of the purified enzyme), and recorded the changes of absorbance for 3 min

Table 2 Effects of chemical agents on the recombinant AcCR

Reagent	Concentration (mM)	Relative activity (%)
Control		100.0 ± 0.8
EDTA	2	95.7 ± 2.3
	5	92.4 ± 2.9
SDS	2	68.4 ± 1.2
	5	53.4 ± 1.4
β-Mercaptoethanol	20	94.2 ± 2.0
	40	88.1 ± 2.0
Iodoacetamide	20	97.1 ± 1.4
	40	85.3 ± 1.2
Tween-80	1% (v/v)	74.1 ± 1.9
	2% (v/v)	66.9 ± 1.8
Triton X-100	1% (v/v)	86.8 ± 1.9
	2% (v/v)	71.0 ± 2.0

Reaction conditions: 0.25 mM NADH, 2 mL 50 mM citrate–phosphate buffer (pH 6.5) with different chemical agents, 5 mM 4′-chloroacetophenone were incubated for 10 min at 35 °C before adding 20 µL purified recombinant AcCR (about 0.008 mg of the purified enzyme), and the changes of absorbance for 3 min were recorded

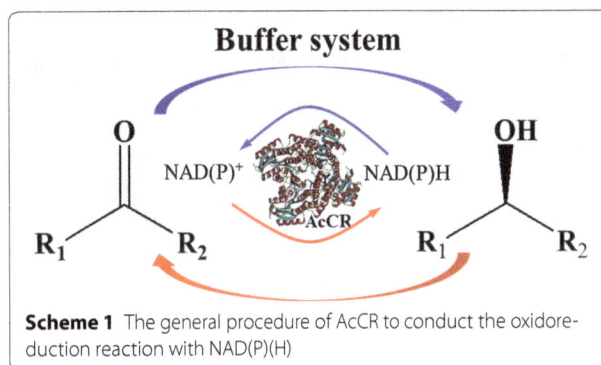

Scheme 1 The general procedure of AcCR to conduct the oxidoreduction reaction with NAD(P)(H)

AcCR, revealing that the recombinant AcCR was a metal-ion-independent enzyme. The metal-ion-independent enzyme was conformed to the characteristic of the vast majority of SDRs (Kavanagh et al. 2008; Persson and Kallberg 2013). The sulfhydryl reagents, β-mercaptoethanol and iodoacetamide, had no significant effect on the recombinant AcCR activity, suggesting that there were no essential disulfide linkages and -SH groups at the catalytic sites of the recombinant AcCR (Dako et al. 2008). Surfactants were reported to play a key role in the catalytic activity of several enzymes (Ye et al. 2009). The surfactants such as SDS, Tween-80, and Triton X-100 had obviously inhibitory effect on the activity of the recombinant AcCR, which could be due to fact that the surfactants partially disrupted the hydrophobic interactions and increased the internal repulsive forces (Adler et al. 2000).

Substrate specificity and bioconversion of various carbonyl compounds

Most carbonyl reductases can catalyze reversible reaction mediated by different coenzymes, and the substrate specificity and stereoselectivity are different towards certain carbonyl reductase (He et al. 2014; Zhang et al. 2015; Takeuchi et al. 2015; Tang et al. 2014). Therefore, it was of great interest to investigate the capability of the recombinant AcCR for catalyzing redox reaction, substrate specificity, and stereoselectivity. Seventeen of prochiral carbonyl compounds and their corresponding alcohols were explored using NADH or NAD⁺ as coenzyme (Scheme 1), and the results are shown in Table 3. Glancing

at the results, we found that the AcCR could not only catalyze the reduction of carbonyl groups but also could complete the dehydrogenization of the partial tested alcohols. The activities of ketones reduction were higher than those of oxidation of alcohols obviously, which was beneficial to proceed ketones reduction. The oxidation of ethyl (R)-(+)-4-chloro-3-hydroxybutyrate or ethyl (S)-(-)-4-chloro-3-hydroxybutyrate was investigated, respectively, and no oxidative activity was tested. But when ethyl-4-chloroacetoacetate was used as substrate, the relative activity was tested to be 120.6%, indicating that the reaction was not reversible towards ethyl-4-chloroacetoacetate and ethyl (4-chloro-3-hydroxybutyrate. In addition, using 2-(R)-octanol or 2-(S)-octanol as substrate, the relative activity of AcCR was far from each other (107.7% vs 2.9%), which attested that the AcCR also had stereoselectivity on the oxidation of enantiomer alcohols substrates.

Chiral alcohols are one of the most useful building blocks for the preparation of pharmaceuticals. As the excellent relative activity of the recombinant AcCR towards those prochiral carbonyl-group substrates, it is of great interest to investigate the product yields and enantiomeric excess (e.e.) of the asymmetric reduction of the carbonyl compounds catalyzed by the recombinant AcCR. As shown in Table 4, AcCR exhibits a broad substrate spectrum and the enantioselectivity of this enzyme follows anti-Prelog rule which are relatively rare in nature (Tang et al. 2014; Li et al. 2015). The results shown in Tables 3 and 4 indicated that the relative activity was affected by the position of the substituents, such as the 2′-, 3′- and 4′-positions substituted acetophenone, 2′-methoxyacetophenone (16.5%), 3′-methoxyacetophenone (3.1%), and 4′-methoxyacetophenone (10.7%). High product yield and excellent enantioselectivity (>99%) were achieved when the 4′-position of acetophenone was substituted by electron-withdrawing groups such as $-F$, $-Br$, and $-NO_3$, whereas the electron-releasing groups like -methyl had a negative effect on the activity of the recombinant AcCR. Moreover, it is noteworthy that the

Table 3 The relative activity of the recombinant AcCR to catalyze the oxidation of alcohols and reduction of ketones

Substrate	Structure	Relative activity (%)	Substrate	Structure	Relative activity (%)
Isopropanol		45.4			
1-Phenethyl alcohol		Nd	Acetophenone		9.2
1-(4-Methylphenyl)ethanol		16.6	4-Methylacetophenone		16.5
2-Methoxyphenethyl alcohol		1.9	2-Methoxyacetophenone		3.1
3-Methoxyphenethyl alcohol		7.7	3-Methoxyacetophenone		10.7
4-Methoxyphenethyl alcohol		11.6	4-Methoxyacetophenone		5.1
3,3-Dimethyl-2-butanol		36.5	3,3-dimethyl-2-butanone		47.5
1-(Trimethylsilyl)-ethanol		Nd	1-(Trimethylsilyl)ethanone		156.5
4-(Trimethylsilyl)-3-butyn-2-ol		2.9	4-(Trimethylsilyl)-3-butyn-2-one		114.4
Methyl 3-hydroxybutyrate		13.1	Methyl acetoacetate		81.1
Ethyl 3-hydroxybutyrate		6.2	Ethyl acetoacetate		52.1
Ethyl (R)-4-chloro-3-hydroxybutyrate		Nd	Ethyl-4-chloroacetoacetate		120.6
Ethyl (S)-4-chloro-3-hydroxybutyrate		Nd			
1-(4-Fluorophenyl)ethanol		55.1	4-Fluoroacetophenone		85.1
1-(4-Chlorophenyl)ethanol		52.3	4-Chloroacetophenone		100.0
1-(4-Bromophenzyl) alcohol		44.2	4-Bromoacetophenone		117.1
1-(4-Nitrophenzyl) alcohol		49.6	4-Nitroacetophenone		98.7
2-Pentanol		98.6	2-pentanone		158.6

Table 3 continued

Substrate	Structure	Relative activity (%)	Substrate	Structure	Relative activity (%)
2- (R)-Octanol		107.7	2-Octanone		87.58
2- (S)-Octanol		2.9			
Ethyl 2-hydroxy-4-phenylb-utyrate		2.3	Ethyl 2- oxo-4-phenylbutyrate		35.8

Reaction conditions: for oxidation reaction (a) 0.5 mM NAD$^+$, 2 mL 50 mM Tris–HCl buffer (pH 8.5), 50 mM alcohol compound incubated at 45 °C for 5 min before adding 20 μL purified recombinant AcCR (about 0.008 mg the purified enzyme), recorded the changes of absorbance for 3 min at 45 °C; for reduction reaction (b) 0.25 mM NADH, 2 mL 50 mM citrate–phosphate buffer (pH 6.5), 50 mM carbonyl compound at 35 °C for 5 min before adding 20 μL purified recombinant AcCR (about 0.008 mg the purified enzyme) recorded the changes of absorbance for 3 min at 35 °C

Table 4 Asymmetric reduction of prochiral carbonyl compounds catalyzed by the recombinant AcCR

Substrates	Products	Reaction time (h)	Con.[a] Yield (%)	Yield (%)	e.e. (%)	Config.[b]
Acetophenone	Phenethyl alcohol	3	48.2	45	>99	R
2'-Methoxyacetophenone	2-Methoxyphenethyl ethanol	12	17.1	10.8	>99	R
3'-Methoxyacetophenone	3-Methoxyphenethyl ethanol	7	67.5	61.6	>99	R
4'-Methoxyacetophenone	4-Methoxyphenethyl ethanol	12	56.1	50.8	>99	R
4'-Methylacetophenone	1-(4-Methylphenyl) ethanol	3	62.8	57.7	>99	R
4'-Fluoroacetophenone	1-(4-Fluorophenyl) ethanol	7	99.1	95.6	>99	R
4'-Chloroacetophenone	1-(4-Chlorophenyl) ethanol	7	99.7	99.4	>99	R
4'-Bromoacetophenone	1-(4-Bromophenyl) ethanol	7	99.5	99.1	>99	R
4'-Nitroacetophenone	1-(4-Nitrophenyl) ethanol	7	99.8	98.6	>99	R
Methyl acetoacetate	Methyl 3-hydroxybutyrate	3	98.7	94.3	>99	R
Ethyl acetoacetate	Ethyl 3-hydroxybutyrate	3	91.7	88.5	97.6	R
Ethyl 4'-chloroacetoacetate	Ethyl 4-chloro-3-hydroxybutanoate	5	95.3	92.1	>99	S
Ethyl 2-oxo-4-phenylbutyrate	Ethyl 2-hydroxy-4-phenylbutyrate	3	72.9	68.4	83.1	S
4-(Trimethylsilyl)-3-butyn-2-one	4-(Trimethylsilyl)-3-butyn-2-ol	7	89.2	62.5	>99	R
3,3-dimethyl-2-butanone	3,3-dimethyl-2-butanol	7	76.5	61.7	80.7	R
2-Octanone	2-Octanol	3	71.6	66.9	>99	R

[a] Conversion

[b] Configuration

Reaction conditions: 4 mL citrate–phosphate buffer (50 mM, pH 6.5) containing 50 mM substrate, 0.1 mM NADH, 150 mM isopropanol, 8 U recombinant AcCR, 35 °C, 180 rpm

AcCR was more active to ethyl 4'-chloroacetoacetate among the investigated α-ketoesters, and the relative activity and yield were over 120.6 and 92.1%, respectively. The recombinant AcCR exhibited 114.4% relative activity to the 4-(trimethylsilyl)-3-butyn-2-one, and the product e.e. was over 99%, but the yield was only 62.5%, possibly because the substrate could be easily decomposed into a carbonyl alkyne and trimethyl hydroxysilane in buffer system with pH over 6.0 (Zhang et al. 2008). The product yield of reducing 3,3-dimethyl-2-butanone to 3,3-dimethyl-2-butanol was 61.7% and the product e.e was 80.7%, and if further reaction continued, the product e.e decreased markedly. What is more, unlike the aceto-phenone reduced to corresponding (R)-products, ethyl 4'-chloroacetoacetate and ethyl 2-oxo-4-phenylbutyrate were catalyzed to ethyl (S)-4-chloro-3-hydroxybutyrate and (S)-2-hydroxy-4-phenylbutyrate, respectively, which accorded with the anti-Prelog rule that the Cahn-Ingold-Prelog priority system was conversely exhibited when the substituted group contained higher priority atoms at its chiral center (Itoh 2014).

Kinetic parameters assays

The kinetic constants of the purified AcCR were calculated by fitting data by linear regression to a Lineweaver–Burk double reciprocal plot. All the kinetic constants for 4'-chloroacetophenone, isopropanol, NAD(P)H, and NAD(P)$^+$ are presented in Table 5. The results for reduction

Table 5 Kinetic parameters of different substrates for the recombinant AcCR

Substrate	K_m (mM)	V_{max} (μmol min^{-1} mg^{-1})	k_{cat} (min^{-1})
NADH	0.16	5.19	145.7
4'-Chloroacetophenone	0.26		
NADPH	0.44	0.66	17.71
4'-Chloroacetophenone	2.75		
NAD$^+$	0.38	11.8	265.8
Isopropanol	23.8		
NADP$^+$	1.39	1.32	35.94
Isopropanol	54.1		

of 4'-chloroacetophenone showed that the enzyme had a much higher affinity for NADH than NADPH. Many carbonyl reductases from microorganism preferred NADPH much more, such as ADHs from *Lactobacillus* (Weckbecker and Hummel 2009; Leuchs and Greiner 2011), carbonyl reductase from *Gluconobacter oxydans* (Chen et al. 2015), and aldo–keto reductases from *Candida parapsilosis* (Guo et al. 2014). The NADH as coenzyme for carbonyl reductase was heavily favored for its economy and stability (Leuchs and Greiner 2011). The results for the oxidation of isopropanol (*Km* value 0.38 and 23.8 Vs 1.39 and 54.1) gave a proof that the AcCR was more affiliative to NAD$^+$ than NADP$^+$ when conducting the oxidation. Above all, the AcCR was an NAD(H) and NADP(H) dependent oxidordeuctase and more preferred NAD(H) as coenzyme in the redox reaction.

Conclusions

The anti-Prelog carbonyl reductase AcCR from *Acetobacter* sp. CCTCC M209061 was cloned and its polypeptide sequence had been predicted and analyzed to confirm that the AcCR belongs to SDRs superfamily. Then AcCR was expressed heterologously, purified, and characterized. The purified enzyme preferred the inexpensive coenzyme NADH as specific electron donor. The recombinant AcCR exhibited excellent enantioselectivity, broad substrate spectrum, and highly stereoselective anti-Prelog reduction of prochiral ketones. These results suggest that AcCR is a powerful chiral tool for the production of anti-Prelog alcohols.

Abbreviations
CR: carbonyl reductase; ADH: alcohol dehydrogenase; SDR: short-chain dehydrogenase/reductase; AcCR: carbonyl reductase from *Acetobacter* sp. CCTCC M209061; GC: gas chromatography; IPTG: isopropyl-β-d-thiogalactoside.

Authors' contributions
Conceived and designed the experiments: WYL MHZ. Performed the experiments: PW YHC PX. Analyzed the data: WYL PW. Contributed reagents/materials/analysis tools: MHZ WYL JZ. Wrote the paper: PW WYL. All authors read and approved the final manuscript.

Author details
[1] Lab of Applied Biocatalysis, School of Food Science and Engineering, South China University of Technology, Guangzhou 510640, Guangdong, China. [2] School of Chemistry and Chemical Engineering, South China University of Technology, Guangzhou 510640, Guangdong, China.

Acknowledgements
Not applicable.

Competing interests
The authors declare that they have no competing interests.

Funding
The National Natural Science Foundation of China (21336002; 21676104; 21376096), the Fundamental Research Funds for the Chinese Universities (2015PT002; 2015ZP009), the Program of State Key Laboratory of Pulp and Paper Engineering (2017ZD05), and the Open Funding Project of the State Key Laboratory of Bioreactor Engineering.

References
Adler JJ, Singh PK, Patist A, Rabinovich YI, Shah DO, Moudgil BM (2000) Correlation of particulate dispersion stability with the strength of self-assembled surfactant films. Langmuir 16:7255–7262

Birolli WG, Ferreira IM, Alvarenga N, Santos DDA, de Matos IL, Comasseto JV, Porto ALM (2015) Biocatalysis and biotransformation in Brazil: an overview. Biotechnol Adv 33:481–510

Bradford MM (1976) A rapid and sensitive method for the quantitation of microgram quantities of protein utilizing the principle of protein-dye binding. Anal Biochem 72:248–254

Chen R, Liu X, Wang JL, Lin JP, Wei DZ (2015) Cloning, expression, and characterization of an anti-Prelog stereospecific carbonyl reductase from *Gluconobacter oxydans* DSM2343. Enzyme Microb Technol 70:18–27

Cheng J, Lou W, Zong M (2014) Biocatalytic asymmetric oxidation of racemic 1-(4-methoxyphenyl) ethanol using immobilized *Acetobacter* sp CCTCC M209061 cells in organic solvent-containing biphasic system. Chem J Chin Univ Chin 35:1529–1535

Cui ZM, Zhang JD, Fan XJ, Zheng GW, Chang HH, Wei WL (2017) Highly efficient bioreduction of 2-hydroxyacetophenone to (S)- and (R)-1-phenyl-1,2-ethanediol by two substrate tolerance carbonyl reductases with cofactor regeneration. J Biotechnol 243:1–9

Dako E, Jankowski CK, Bernier A-M, Asselin A, Simard RE (2008) A new approach for the purification and characterisation of PA49.5, the main prebiotic of *Lactococcus lactis* subsp. cremoris. Int J Food Microbiol 126:186–194

Filling C, Berndt KD, Benach J, Knapp S, Prozorovski T, Nordling E, Ladenstein R, Jornvall H, Oppermann U (2002) Critical residues for structure and catalysis in short-chain dehydrogenases/reductases. J Biol Chem 277:25677–25684

Grosch J-H, Loderer C, Jestel T, Ansorge-Schumacher M, Spieß AC (2015) Carbonyl reductase of *Candida parapsilosis*—stability analysis and stabilization strategy. J Mol Catal B Enzym 112:45–53

Guo R, Nie Y, Mu XQ, Xu Y, Xiao R (2014) Genomic mining-based identification of novel stereospecific aldo-keto reductases toolbox from *Candida parapsilosis* for highly enantioselective reduction of carbonyl compounds. J Mol Catal B Enzym 105:66–73

Gutierrez MC, Ferrer ML, Yuste L, Rojo F, del Monte F (2010) Bacteria incorporation in deep-eutectic solvents through freeze-drying. Angew Chem Int Ed 49:2158–2162

He S, Wang Z, Zou Y, Chen S, Xu X (2014) Purification and characterization of a novel carbonyl reductase involved in oxidoreduction of aromatic β-amino ketones/alcohols. Process Biochem 49:1107–1112

He YC, Zhang DP, Di JH, Wu YQ, Tao ZC, Liu F, Zhang ZJ, Chong GG, Ding Y, Ma CL (2016) Effective pretreatment of sugarcane bagasse with combination pretreatment and its hydrolyzates as reaction media for the biosynthesis of ethyl (S)-4-chloro-3-hydroxybutanoate by whole cells of *E. coli* CCZU-K14. Bioresour Technol 211:720–726

Itoh N (2014) Use of the anti-Prelog stereospecific alcohol dehydrogenase from *Leifsonia* and *Pseudomonas* for producing chiral alcohols. Appl Microbiol Biotechnol 98:3889–3904

Enzymatic characterization of a recombinant carbonyl reductase from Acetobacter sp. CCTCC M209061

65

Jörnvall H, Hedlund J, Bergman T, Oppermann U, Persson B (2010) Super-families SDR and MDR: from early ancestry to present forms. Emergence of three lines, a Zn-metalloenzyme, and distinct variabilities. Biochem Biophys Res Commun 396:125–130

Kaluzna IA, Rozzell JD, Kambourakis S (2005) Ketoreductases: stereoselective catalysts for the facile synthesis of chiral alcohols. Tetrahedron Asymmetry 16:3682–3689

Kavanagh KL, Jornvall H, Persson B, Oppermann U (2008) Medium- and short-chain dehydrogenase/reductase gene and protein families: the SDR superfamily: functional and structural diversity within a family of metabolic and regulatory enzymes. Cell Mol Life Sci 65:3895–3906

Leuchs S, Greiner L (2011) Alcohol dehydrogenase from Lactobacillus brevis: a versatile robust catalyst for enantioselective transformations. Chem Biochem Eng Q 25:267–281

Li BJ, Li YX, Bai DM, Zhang X, Yang HY, Wang J, Liu G, Yue JJ, Ling Y, Zhou DS, Chen HP (2014) Whole-cell biotransformation systems for reduction of prochiral carbonyl compounds to chiral alcohol in Escherichia coli. Sci Rep 4:6750. doi:10.1038/srep06750

Li A, Ye L, Wu H, Yang X, Yu H (2015) Characterization of an excellent anti-Prelog short-chain dehydrogenase/reductase EbSDR8 from Empedobacter brevis ZJUY-1401. J Mol Catal B Enzym 122:179–187

Liu X, Chen R, Yang Z, Wang J, Lin J, Wei D (2014) Characterization of a putative stereoselective oxidoreductase from Gluconobacter oxydans and its application in producing ethyl (R)-4-chloro-3-hydroxybutanoate ester. Mol Biotechnol 56:285–295

Luo X, Wang YJ, Zheng YG (2015) Cloning and characterization of a NADH-dependent aldo-keto reductase from a newly isolated Kluyveromyces lactis XP1461. Enzym Microb Technol 77:68–77

Ma CW, Zhang L, Dai JY, Xiu ZL (2013) Characterization and cofactor binding mechanism of a novel NAD(P)H-dependent aldehyde reductase from Klebsiella pneumoniae DSM2026. J Microbiol Biotechnol 23:1699–1707

Masud U, Matsushita K, Theeragool G (2011) Molecular cloning and characterization of two inducible NAD(+)-adh genes encoding NAD(+)-dependent alcohol dehydrogenases from Acetobacter pasteurianus SKU1108. J Biosci Bioeng 112:422–431

Montfort L, Frenette G, Sullivan R (2002) Sperm-zona pellucida interaction involves a carbonyl reductase activity in the hamster. Mol Reprod Dev 61:113–119

Musa MM, Phillips RS (2011) Recent advances in alcohol dehydrogenase-catalyzed asymmetric production of hydrophobic alcohols. Catal Sci Technol 1:1311

Nakajin S, Minamikawa N, Baker ME, Toyoshima S (1998) An NADPH-dependent reductase in neonatal pig testes that metabolizes androgens and xenobiotics. Biol Pharm Bull 21:1356–1360

Ni Y, Li CX, Zhang J, Shen ND, Bornscheuer UT, Xu JH (2011) Efficient reduction of ethyl 2-oxo-4-phenylbutyrate at 620 g/L⁻¹ by a bacterial reductase with broad substrate spectrum. Adv Synth Catal 353:1213–1217

Persson B, Kallberg Y (2013) Classification and nomenclature of the superfamily of short-chain dehydrogenases/reductases (SDRs). Chem Biol Interact 202:111–115

Persson B, Kallberg Y, Oppermann U, Jörnvall H (2003) Coenzyme-based functional assignments of short-chain dehydrogenases/reductases (SDRs). Chem Biol Interact 143–144:271–278

Qian XL, Pan J, Shen ND, Ju X, Zhang J, Xu JH (2014) Efficient production of ethyl (R)-2-hydroxy-4-phenylbutyrate using a cost-effective reductase expressed in Pichia pastoris. Biochem Eng J 91:72–77

Richter N, Hummel W (2011) Biochemical characterisation of a NADPH-dependent carbonyl reductase from Neurospora crassa reducing alpha- and beta-keto esters. Enzyme Microb Technol 48:472–479

Sengupta D, Naik D, Reddy AR (2015) Plant aldo-keto reductases (AKRs) as multi-tasking soldiers involved in diverse plant metabolic processes and stress defense: a structure-function update. J Plant Physiol 179:40–55

Singh A, Bhattacharyya MS, Banerjee UC (2009) Purification and characterization of carbonyl reductase from Geotrichum candidum. Process Biochem 44:986–991

Takeuchi M, Kishino S, Park S-B, Kitamura N, Ogawa J (2015) Characterization of hydroxy fatty acid dehydrogenase involved in polyunsaturated fatty acid saturation metabolism in Lactobacillus plantarum AKU 1009a. J Mol Catal B Enzym 117:7–12

Tang TX, Liu Y, Wu ZL (2014) Characterization of a robust anti-Prelog short-chain dehydrogenase/reductase ChKRED20 from Chryseobacterium sp. CA49. J Mol Catal B Enzym 105:82–88

Wang XT, Yue D-M, Zong MH, Lou WY (2013) Use of ionic liquid to significantly improve asymmetric reduction of ethyl acetoacetate catalyzed by Acetobacter sp. CCTCC M209061 cells. Ind Eng Chem Res 52:12550–12558

Wang YJ, Liu XQ, Luo X, Liu ZQ, Zheng YG (2015) Cloning, expression and enzymatic characterization of an aldo-keto reductase from Candida albicans XP1463. J Mol Catal B Enzym 122:44–50

Weckbecker A, Hummel W (2009) Cloning, expression, and characterization of an (R)-specific alcohol dehydrogenase from Lactobacillus kefir. Biocatal Biotransform 24:380–389

Wei P, Xu P, Wang XT, Lou WY, Zong MH (2015) Asymmetric reduction of ethyl acetoacetate catalyzed by immobilized Acetobacter sp CCTCC M209061 cells in hydrophilic ionic liquid hybrid system. Biotechnol Bioprocess Eng 20:324–332

Wei P, Liang J, Cheng J, Zong M-H, Lou W-Y (2016) Markedly improving asymmetric oxidation of 1-(4-methoxyphenyl) ethanol with Acetobacter sp CCTCC M209061 cells by adding deep eutectic solvent in a two-phase system. Microb Cell Fact 15:5

Xu Q, Xu X, Huang H, Li S (2015) Efficient synthesis of (R)-2-chloro-1-phenylethol using a yeast carbonyl reductase with broad substrate spectrum and 2-propanol as cosubstrate. Biochem Eng J 103:277–285

Ye Q, Yan M, Yao Z, Xu L, Cao H, Li Z, Chen Y, Li S, Bai J, Xiong J, Ying H, Ouyang P (2009) A new member of the short-chain dehydrogenases/reductases superfamily: purification, characterization and substrate specificity of a recombinant carbonyl reductase from Pichia stipitis. Biores Technol 100:6022–6027

Zhang BB, Lou WY, Zong MH, Wu H (2008) Efficient synthesis of enantio-pure (S)-4-(trimethylsilyl)-3-butyn-2-ol via asymmetric reduction of 4-(trimethylsilyl)-3-butyn-2-one with immobilized Candida parapsilosis CCTCC M203011 cells. J Mol Catal B Enzym 54:122–129

Zhang R, Geng Y, Xu Y, Zhang W, Wang S, Xiao R (2011) Carbonyl reductase SCRII from Candida parapsilosis catalyzes anti-Prelog reaction to (S)-1-phenyl-1,2-ethanediol with absolute stereochemical selectivity. Biores Technol 102:483–489

Zhang Y, Ujor V, Wick M, Ezeji TC (2015) Identification, purification and characterization of furfural transforming enzymes from Clostridium beijerinckii NCIMB 8052. Anaerobe 33:124–131

Zheng YG, Yin HH, Yu DF, Chen X, Tang XL, Zhang XJ, Xue YP, Wang YJ, Liu ZQ (2017) Recent advances in biotechnological applications of alcohol dehydrogenases. Appl Microbiol Biotechnol 101:987–1001

7

Germ soak water as nutrient source to improve fermentation of corn grits from modified corn dry grind process

Ankita Juneja[†], Deepak Kumar[†] and Vijay Singh[*]

Abstract

Corn fractionation in modified dry grind processes results in low fermentation efficiency of corn grits because of nutrient deficiency. This study investigated the use of nutrient-rich water from germ soaking to improve grits fermentation in the conventional dry grind and granular starch hydrolysis (GSH) processes. Comparison of germ soak water with the use of protease and external B-vitamin addition in improving grits fermentation was conducted. Use of water from optimum soaking conditions (12 h at 30 °C) resulted in complete fermentation with 29 and 8% higher final ethanol yields compared to that of control in conventional and GSH process, respectively. Fermentation rate (4–24 h) of corn grits with germ soak water (0.492 v/v-h) was more than double than that of control (0.208 v/v-h) in case of conventional dry grind process. The soaking process also increased the oil concentration in the germ by about 36%, which would enhance its economic value.

Keywords: Ethanol, Dry grind, Granular starch hydrolysis, Yeast nutrition, Fermentation

Background

Bioethanol is considered as one of the most promising renewable alternatives to petroleum-based transportation fuel. In the conventional dry grind process, corn starch is liquefied to dextrins at high temperature and pressure, which are further converted to glucose during the saccharification process. Glucose is simultaneously fermented to ethanol by yeast, and this combined process is known as simultaneous saccharification and fermentation (SSF). In an alternate approach, granular starch hydrolyzing enzymes (GSHE) can directly hydrolyze the raw granular starch into glucose at low temperatures, without the need of liquefaction step. At the end of both processes, remaining non-fermentable components (germ, fiber, protein, and residual starch) are recovered as DDGS (distillers dried grains with solubles), a coproduct primarily used as ruminant animal food. Fractionation of corn to recover germ and pericarp, prior to hydrolysis, is

one way to generate valued coproducts and simultaneously improve nutritional value of DDGS (low fiber due to removal of pericarp) (Murthy et al. 2006a, b). Germ and pericarp obtained from the modified process can be refined to obtain valuable products including corn oil from corn germ and corn fiber oil from pericarp fiber. Corn fiber oil has very high economic value because its constituents have nutraceutical properties (Moreau et al. 1996; Murthy et al. 2006b). Grits obtained after germ and pericarp removal contain relatively high amount of starch, which would produce higher ethanol concentrations compared to whole corn at same solid loadings.

However, removal of germ during corn fractionation also removes the soluble proteins and micronutrients present in the germ that are essential for yeast during the fermentation process. Also the lipids, present in germ and the aleurone layer below the pericarp, are essential to maintain membrane integrity and yeast activity, especially during high glucose and ethanol concentrations. Murthy et al. (2006a) reported that both initial rate of fermentation and final ethanol concentrations were low for endosperm obtained from 3D process compared to those from wet fractionation (E-milling). One way to

*Correspondence: vsingh@illinois.edu
[†]Ankita Juneja and Deepak Kumar are first authors
Department of Agricultural and Biological Engineering, University of Illinois at Urbana-Champaign, Urbana, IL 61801, USA

address this problem to some extent is addition of protease enzymes. Addition of proteases causes hydrolysis of the protein matrix surrounding the starch granules, which produces free amino nitrogen (FAN) as well as improve accessibility of starch to enzymes. Fermentation efficiency can also be improved by adding external nutrition, such as yeast extract, lipid supplementation, and B-vitamin complex. However, both protease enzymes and external nutrient add up to the cost of the process and counter the benefits of fractionation. One potential cost-effective approach could be the extraction of these nutrients from the recovered germ, as suggested by Murthy et al. (2006a). The study reported that the water obtained after soaking of fractionated germ (2 h soaking) resulted in increase of final ethanol concentrations from 12.3 to 14.7% (v/v) during conventional dry grind processing of corn grits.

This study aims to investigate this approach in detail and optimize the process conditions (germ soaking time and amount) to maximize the fermentation rate and final ethanol concentrations of corn grits during conventional dry grind as well as GSH process. Germ water was obtained from two soaking conditions and its effect on fermentation performance of corn grits was compared to those from control, protease addition, and B-vitamin addition. Combination of germ water and B-vitamins was also investigated to determine the maximum achievable ethanol efficiency. Composition of raw germ and germ after soaking was also evaluated to determine the changes in oil concentrations.

Methods
Materials
Flaking grits and germ samples were obtained from a commercial corn dry-milling plant (Bunge, Danville, IL, USA). Samples were stored in refrigerator at 4 °C till analysis. All enzymes including conventional α-amylase (Spezyme RSL with reported activity of 20,100 NLC/g), conventional glucoamylase [distillase SSF, with reported activity of 380 GAU/g (GAU: glucoamylase unit)], GSHE (Stargen 002), and Protease (Fermgen) are commonly used commercial enzymes and were generously donated by DuPont Industrial Biosciences (Palo Alto). GSHE contained α-amylase from *A. kawachi* expressed in *T. reesei* and glucoamylase from T. *reesei*, and had an activity of >570 GAU/g. Protease enzyme contained fungal protease obtained from genetically modified selected strain of *T. reesei*, with an activity of >1000 SAPU/g (SAPU is spectrophotometric acid protease units). Conventional active dry yeast (ethanol red) was obtained from the Fermentis-Lesaffre Yeast Corporation (Milwaukee, Wisconsin).

Corn grits and germ composition
Composition analysis of corn grits, raw germ, and soaked germ was performed as per American Association of Cereal Chemists International (AACCI) standard procedures. The moisture content of corn grits was determined by drying the samples in hot air oven at 135 °C for 2 h (AACC international approved method 44-19.01). Corn grits and germ (before and after soaking) were analyzed for crude protein content (method 990.03), oil (method 920.39), and neutral detergent fiber (method 2002.04) in a commercial analytical laboratory (Illinois crop improvement association, Champaign, IL, USA). All analyses were conducted in duplicates. Starch content in the ground corn grits was determined using acid hydrolysis method (Vidal et al. 2009). Briefly, about 1 g of ground corn samples (~1 g) were mixed with 50 mL dilute HCl (0.4 N) in 100 mL autoclavable bottles, and the slurry was autoclaved at 126 °C for 1 h (Napco Model 9000D, Thermo Fisher 157 Scientific, Waltham, MA). Pure glucose and starch samples were used to determine glucose recovery factors. After cooling, 1 mL aliquot samples was withdrawn and centrifuged at $1500 \times g$ for 5 min (Model 5415 D, Brinkmann–Eppendorf, Hamburg, Germany). The supernatants were analyzed in the HPLC for glucose determination.

Dry grind process
A simple schematic of lab-scale dry grind and GSH process is shown in Fig. 1. Corn grits were ground in a laboratory-scale mill (model MHM4, Glen Mills, Clifton, NJ) at 500 rpm and using a 0.5-mm screen. All experiments were performed at 250 mL scale in 500 mL stainless steel reactors in duplicate. Ground grits were mixed with water or germ-soaked water (Table 1) to make a slurry with 25% solids on dry basis. For liquefaction, the pH of the slurry was adjusted to 5.1 using 10 N sulfuric acid and 16 μL α-amylase was added, as per manufacturer's recommendations. The liquefaction was performed in Labomat Incubator (Labomat BFA-12, Werner Mathis AG, Switzerland) at 85 °C for 120 min with continuous shaking. Heating and cooling of the samples in the incubator were at the rate of 3 °C/min (this time was additional to 90 min of liquefaction time). The liquefied slurry was then prepared for simultaneous saccharification and fermentation (SSF). The pH of the slurry was adjusted to 4.3 using 10 N sulfuric acid and GA (54.7 mL), urea (0.4 mL of 50% w/v solution), and yeast inoculum (2 mL) were added. Yeast inoculum was prepared by dispersing 5 g of active dry yeast in 25 mL of distilled water and agitated at 100 rpm and 32 °C for 20 min. The broth was fermented at 32 °C for 72 h in an automatic incubator with continuous agitation (150 rpm). Samples (2 mL) were drawn at 4, 8, 24, 48, and 72 h to monitor the fermentation.

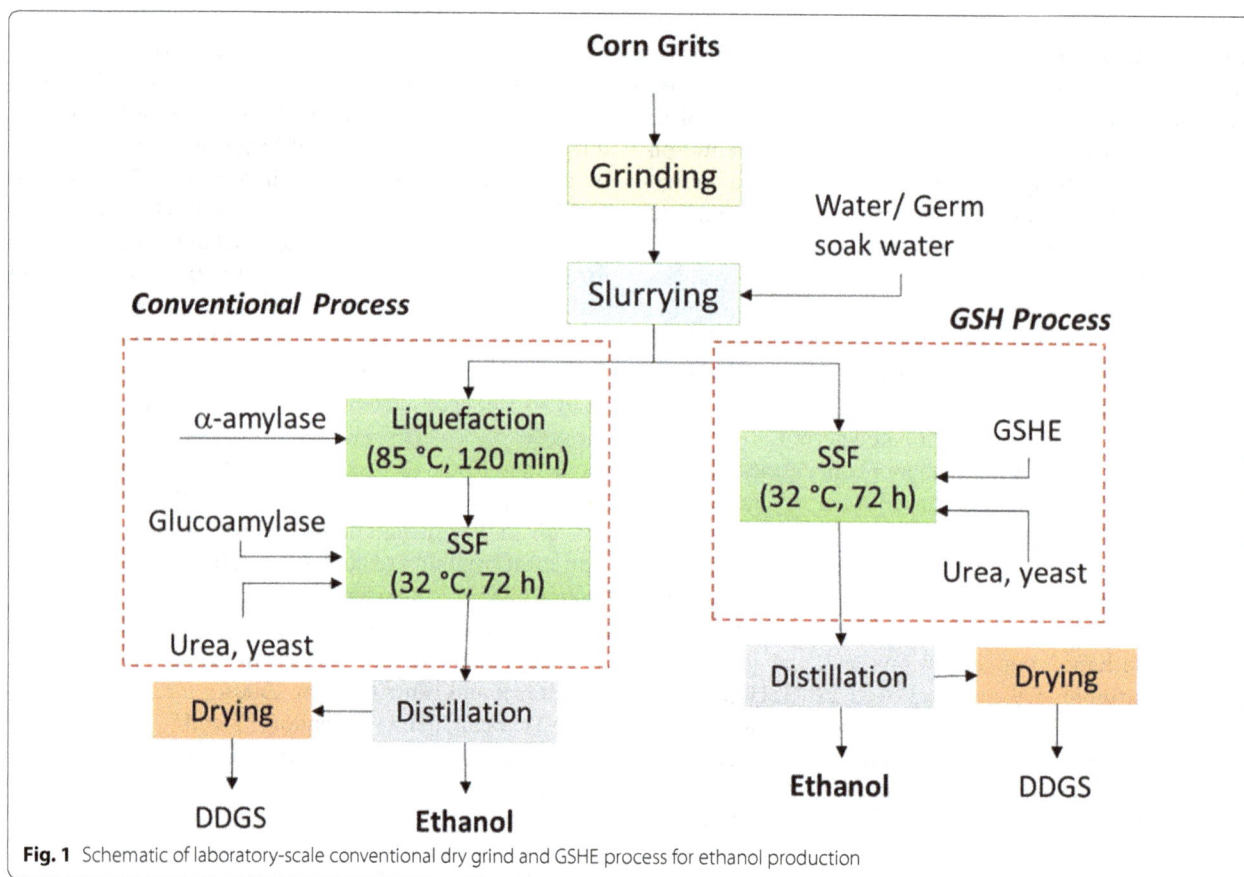

Fig. 1 Schematic of laboratory-scale conventional dry grind and GSHE process for ethanol production

Table 1 Description of treatments investigated in processing corn grits using conventional dry grind and GSH process

Treatment	Description	DI water (% of liquid in slurry)	Germ soak water (% of liquid in slurry)	Germ soaking conditions		Protease	B-vitamin
				Temp (°C)	Time (h)		
T1	Control	100	0	–	–	No	No
T2	Control with protease	100	0	–	–	Yes	No
T3	Partial germ water	66.66	33.33	30	2	No	No
T4	Partial germ water—long time	66.66	33.33	30	12	No	No
T5	Full germ water	0	100	30	12	No	No
T6	B-vitamin	100	0	–	–	No	Yes
T7	Partial germ water and B-vitamin	66.66	33.33	30	12	No	Yes

Germ soak water was obtained by soaking the germ under two conditions: (i) 30 °C for 2 h and (ii) 30 °C for 12 h (Table 1). In each case, 25 g of germ was mixed in 250 mL of deionized (DI) water in 500 mL flasks and was incubated as per conditions mentioned in Table 1, with continuous shaking at 125 rpm. After soaking, the liquid was vacuum-filtered through Whatman No. 4 filter paper and used to make slurry as described in Table 1. Two dosages of germ water were investigated: (1) one-third (33.33%) of total liquid in slurry, referred as partial germ water (treatments T3, T4, T7 in

Table 1), (2) 100% of liquid in slurry, referred as full germ water (treatment T5 in Table 1) in the article. The first case (partial germ water) represents the water obtained from the soaking of germ proportional (10%) to corn grits used in the experiment. In the current study, 62.5 mL of germ water was added in total 250 mL slurry (62.5 g corn grits and 187.5 mL liquid). Full germ water case was investigated to determine the effect of adding excess nutrients on the fermentation efficiency. Other than germ soak water, two additional set of treatments (T6 and T7) were performed by addition of

B-vitamins. In treatment T6, conditions were similar to that of control, except excess of vitamin B12 and B-complex were added at the start of SSF process. In the case of treatment T7, combined effect of germ soak water and B-vitamins was investigated and excess of vitamin B12 and B-complex was added in addition to germ soak water (Table 1).

GSH process

The front-end operations (cleaning, grinding, and slurry making) were similar to that of conventional dry grind process described above (Fig. 1). Liquefaction step is not required in this process. The slurry prepared was adjusted to a pH of 4.1 using 10 N sulfuric acid, and GSHE (0.234 mL), urea (0.4 μL of 50% w/v solution) and yeast inoculum (2 mL) were added. Yeast inoculum was prepared as described in the previous section. The slurry was incubated at 32 °C for 72 h in an automatic incubator with continuous agitation (150 rpm), and 2 mL of samples were drawn at 4, 8, 12, 24, 48, and 72 h to monitor the fermentation. This process was also investigated for all conditions presented in Table 1.

HPLC analysis

Samples collected were centrifuged at 9729 g (5415 D, Brinkmann Eppendorf, Hamburg, Germany) for 10 min, and clear liquid was passed through 0.2 μm syringe filters (nylon Acrodisc WAT200834, Pall Life Sciences, Port Washington, NY) into 150 μL HPLC vials. The vials were immediately stored at −20 °C until analysis. The filtrate was then analyzed using HPLC with an ion-exclusion column (Aminex HPX-87H, Bio-Rad, Hercules, CA, USA). The mobile phase was 0.005 M sulfuric acid at 50 °C at a flow rate of 0.6 mL/min. The amounts of sugars, alcohols, and organic acids were quantified using a refractive index detector and multiple standards.

Fermentation rate and ethanol yield

For each treatment, ethanol and glucose concentration was measured at every time point as described above and were plotted against time. Fermentation rates (ethanol production rates) between different time points were calculated using Eq. 1.

$$\text{Fermentation rate} = \frac{E_{t_2} - E_{t_1}}{t_2 - t_1}, \tag{1}$$

where E_{t_2} and E_{t_1} are ethanol concentrations (% v/v) at fermentation times t_2 and t_1, respectively.

Starch-to-ethanol conversion efficiencies were calculated as the ratio of actual ethanol yields with the theoretical ethanol yield (Eq. 2).

$$\eta_{\text{EtOH}} = \frac{E_{\text{EtOH}}}{E_{\text{Th\_EtOH}}} * 100, \tag{2}$$

where $E_{\text{Th\_EtOH}}$ is theoretical ethanol yield, L/kg dry corn grits; E_{EtOH} is the actual ethanol yield, L/kg dry corn grits.

Theoretical yields were estimated based on the starch content, assuming complete starch conversion and 100% fermentation efficiency. Actual ethanol yields were determined by calculating liquid volume in final slurry at end of fermentation (Kumar and Singh 2016). Final slurry was weighed and a sample of the slurry was dried in hot air oven at 105 °C till constant weight was achieved (~24 h) to estimate the solid percent in the slurry. The actual ethanol yields were calculated using following Eqs. 3–5.

$$W_{\text{L}} = W_{\text{slurry}} * (1 - S_{\text{slurry}}) \tag{3}$$

$$V_{\text{EtOH}} = \frac{W_{\text{L}}}{\rho_{\text{H}_2\text{O/EtOH}}} * C_{\text{EtOH}} \tag{4}$$

$$E_{\text{EtOH}} = \frac{V_{\text{EtOH}}}{W_{\text{C}} * (1 - \text{MC}_{\text{C}}/100)}, \tag{5}$$

where W_{L} is the weight of liquid in the fermented slurry, g; W_{slurry} is the weight of fermented slurry, g; S_{slurry} is the solid fraction in the slurry; V_{EtOH} is the volume of ethanol produced, mL; $\rho_{\text{H}_2\text{O/EtOH}}$ is the density of water–ethanol mixture (g/L) at final ethanol concentration; C_{EtOH} is the final ethanol concentration, mL/L; E_{EtOH} is the actual ethanol yield, L/kg; MC_{C} is the moisture content in grits, %; and W_{L} is the weight of the grits.

Statistical analysis

Analysis of variance (1-way ANOVA) and Fisher's least significant difference (LSD) tests were used to compare the glucose and ethanol concentrations (SAS version 9.3). The level selected to show the statistical significance in all cases was 5% ($P < 0.05$).

Results and discussion
Composition of corn grits

Starch content in the corn grits was estimated as 86.5% on dry basis. Crude protein, oil, and neutral detergent fiber (NDF) were 6.1, 0.6, and 0.9% (dry basis), respectively. Based on this composition, the theoretical ethanol yield was calculated 0.62 L/kg dry corn grits (4.17 gal/bu).

Conventional dry grind process
Effect of germ soak water

Figure 2 illustrates the comparison of ethanol and glucose concentration during SSF of corn grits, for control and germ water addition from two soaking conditions. As expected, addition of germ water improved the fermentation profile compared to that of control. The addition of germ water from soaking at 30 °C for 2 h resulted in an increase in the final ethanol concentration from 12.79 to 14.52% (v/v), which was similar to as observed by Murthy

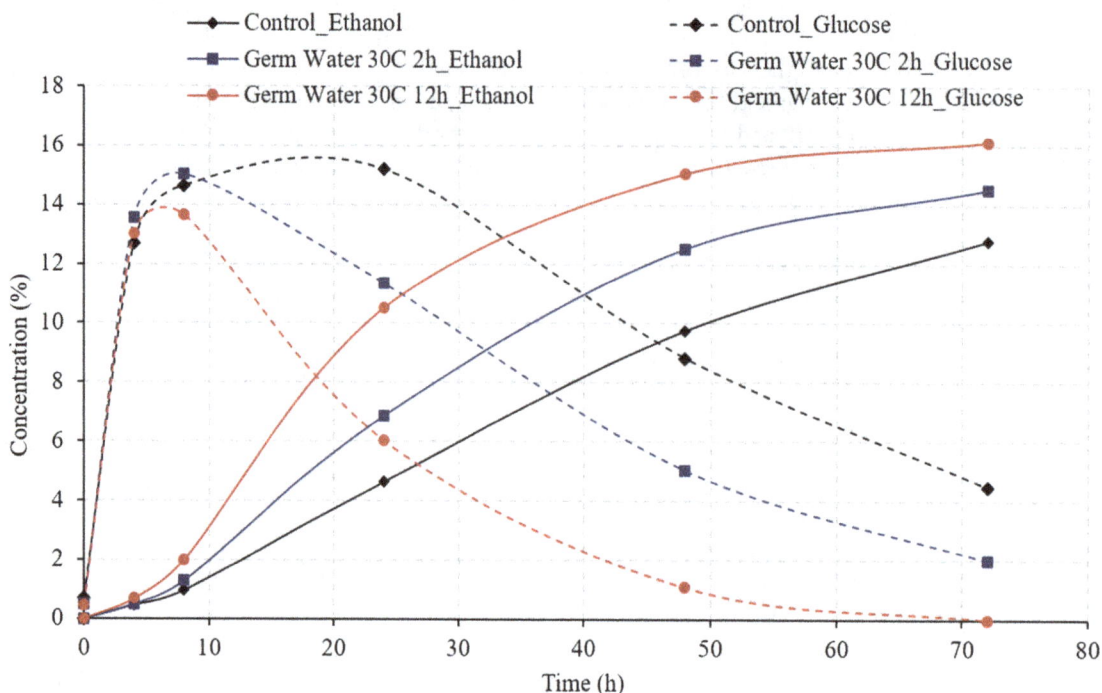

Fig. 2 Effect of germ water addition on fermentation of corn grits in conventional dry grind process. (Ethanol concentrations in % v/v and glucose concentration % w/v)

et al. (2006a). Likewise, there was unconverted glucose (2% w/v) observed at the end of the fermentation. However, the addition of germ water from longer soaking conditions (30 °C for 12 h) resulted in complete fermentation with a final ethanol concentration of 16.14% and no residual glucose. The average final ethanol concentration of the 12-h germ-soaked water was 28.3 and 12.9% higher than the control and 2-h germ-soaked water, respectively. Final glycerol production in case of germ water (30 °C for 12 h)-supplemented samples was about 23% less than that of control (1.16 vs. 1.51%).

About 4.5% glucose remained unconverted in the case of control, which along with high glycerol production resulted in very low starch-to-ethanol conversion efficiency (62.4%). Efficient fermentation with germ water (30 °C and 12 h) addition led to an increase in conversion efficiency to 82.7%, which was 12% (in relative terms) higher than that in the case of addition of 2-h germ-soaked water (73.8%). Since the glucose released in the first 8 h is similar for all three conditions (Fig. 2), it can be stated that the increased rate of fermentation in the 12-h germ-soaked water is due to the better functioning of the yeast in the presence of micronutrients and free amino acids present in germ soak water. These results indicate that longer soaking resulted in leaching out more nutrients that improved the yeast performance and led to

high ethanol yields and fermentation rates. Due to the lack of these micronutrients in the control, fermentation was observed to be slowest among all treatments. Ethanol yields from control, treatment with 2-h germ water and treatment with 12-h germ water were estimated as 0.39, 0.46, and 0.51 L/kg dry grits (2.6, 3.1, and 3.5 gal/bu) respectively.

Germ water vs. protease addition

Earlier studies have shown the addition of protease enzymes increases the fermentation rate and ethanol yield in the dry grind process (Johnston and McAloon 2014; Vidal et al. 2009). However, protease are relatively expensive enzymes and add up significant cost in the process (Wang 2008). Figure 3 illustrates the fermentation profile (glucose and ethanol concentrations) of control, treatment with protease addition, and germ soak water (30 °C, 12 h) addition (treatment T1, T2, and T4 in Table 1) during SSF of corn grits. As expected, the addition of protease enzymes improved the fermentation efficiency compared to that of control and resulted in final ethanol concentration 16.2% (v/v) compared to only 12.79% for control. As discussed in earlier section, germ water from new soaking conditions (30 °C, 12 h) resulted in complete fermentation with 16.14% ethanol (same as that of protease); however, the initial fermentation rates in case of germ soak water

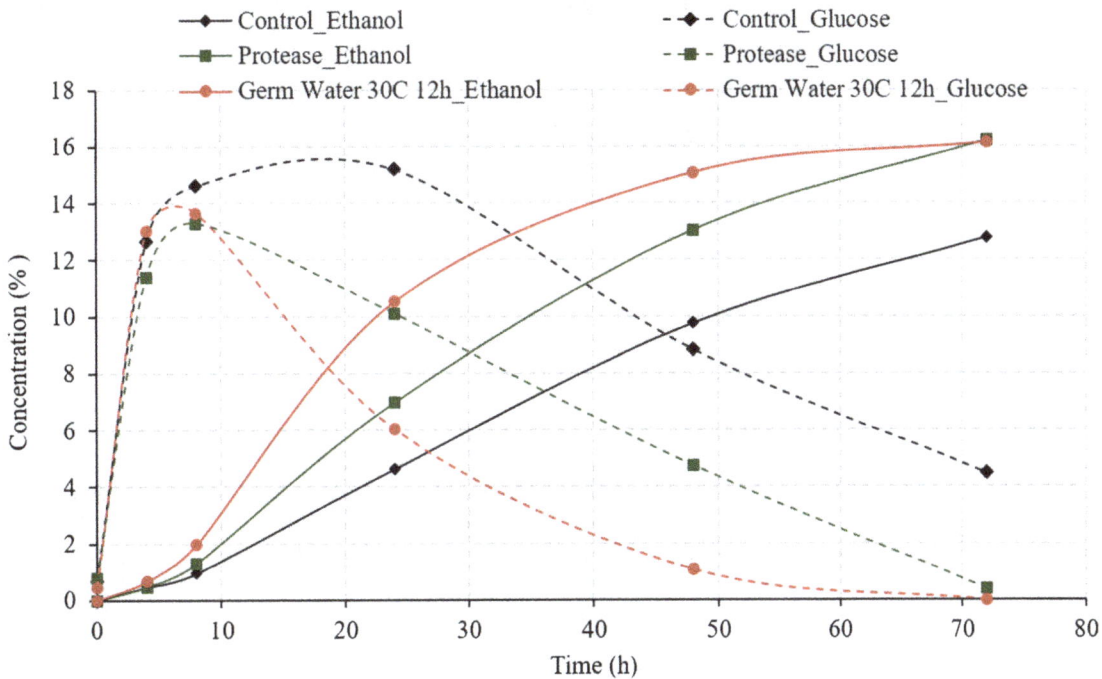

Fig. 3 Comparison of germ soak water and protease enzyme on fermentation profile of corn grits during conventional dry grind process. (Ethanol concentrations in % v/v and glucose concentrations % w/v)

were observed to be higher (0.49 vs. 0.32 v/v-h in 4–24 h). Ethanol yields of 0.51 and 0.52 L/kg dry grits (3.45 and 3.46 gal/bu) were similar for treatments with germ water supplementation and protease addition. The results indicated that the addition of 12-h germ-soaked water could potentially replace the protease enzyme, with even higher ethanol production rate.

Effect of water amount

To further investigate the process, the amount of germ water addition was also varied. Instead of adding one-third of total liquid during slurry formation, the slurry was prepared with 100% germ soak water (30 °C, 12 h) in this case. Ethanol and glucose concentrations during SSF of corn grits without and with addition of two dosages of germ soak water (partial and full germ soak water as explained in Table 1) are given in Table 2. Final ethanol concentration was observed similar for both treatments (partial germ water and full 100% germ water). However, the ethanol production rate in case of 100% germ soak water was higher (0.54 vs. 0.49% v/v/h) than of partial germ water case. No residual sugars in the broth indicated that maximum ethanol potential had been reached. Ethanol yields of 0.51 L/kg dry biomass were similar in both treatments. Considering the similar final ethanol concentrations and yields, it can be interpreted that micronutrients in partial germ water slurry were

sufficient for the yeast and there would not be a huge advantage of making slurry with only germ soak water.

Effect of B-vitamins

Vitamins are essential for yeast metabolism and functioning, however, yeast cannot synthesize many of these vitamins. B-vitamin complex consists of essential coenzymes involved in carbohydrate metabolism and provides necessary metabolic precursors for yeast growth. Other vitamins such as nicotinic acid and pantothenic acid are also helpful in improving yeast performance (White 2012). Riboflavin (vitamin B2) is essential for lipid synthesis, and vitamin B6 is essential for nitrogen metabolism in yeast (Murthy et al. 2006a). Supplementation of vitamin B1, B12, and B-complex have previously shown to increase the ethanol concentrations during fermentation (Laser 1941; Murthy et al. 2006a). Effect of addition of B-vitamins on the fermentation profile of the corn grits was investigated under two conditions: (i) control with addition of excess of B12 and B-complex vitamins, and (ii) addition of both germ soak water (30 °C, 12 h) and excess B-vitamins (treatments T6 and T7 in Table 1). These conditions would answer two questions: (i) can germ water addition improve fermentation performance similar to that of B-vitamins, and (ii) what is the maximum achievable fermentation improvement. Results from these conditions are illustrated in Figs. 4 and 5.

Table 2 Effect of using partial vs. full germ water on the glucose and ethanol concentrations of corn grits during conventional dry grind process

	Treatment	Time (h)					
		0	4	8	24	48	72
Ethanol (% v/v)	Control	0.00	0.48	1.00	4.72	9.61	12.58
	Partial GW	0.00	0.68	1.99	10.52	15.07	16.14
	Full GW	0.00	0.69	2.29	11.41	16.03	16.32
Glucose (% w/v)	Control	0.64	13.84	15.45	15.37	9.17	4.88
	Partial GW	0.47	13.01	13.64	6.04	1.08	0.00
	Full GW	0.67	7.02	8.79	1.77	0.10	0.00

Final ethanol concentrations and yields with the use of only germ soak water were similar to those of treatments with addition of B-vitamins and both germ water and B-vitamins ($P > 0.05$) (Fig. 4; Table 3). These results indicate that germ soak water has sufficient nutrients required to achieve similar ethanol profiles as with B-vitamins. The initial ethanol production rate was higher in the treatment using both germ water and B-vitamins. In all the three cases (germ water, B-vitamin, and germ water plus B-vitamin), the glucose concentration at the end of the fermentation is negligible, which indicates complete fermentation at 72 h. These results suggest that addition of germ water can replace the need for the addition of expensive vitamins, and maximum fermentation rate can be achieved by adding both germ water and B-vitamins.

Final ethanol concentrations, fermentation rates (4–24 h), and ethanol conversion efficiencies for all treatment have been compiled in Table 3. Except for control and germ water from soaking at 30 °C and 2 h, the ethanol conversion efficiency was more than 80% in all treatments. Although conversion efficiency is similar in all other cases, the fermentation rate (4–24 h) was maximum for full germ slurry and treatment using both germ water and B-vitamins.

Granular starch hydrolysis process

Considering the advantages (low energy use and low glucose inhibition) and increasing trend of granular starch hydrolysis process in corn ethanol industry, it was important to investigate the effect of germ soak water on yeast performance in GSH process also. Performance of germ

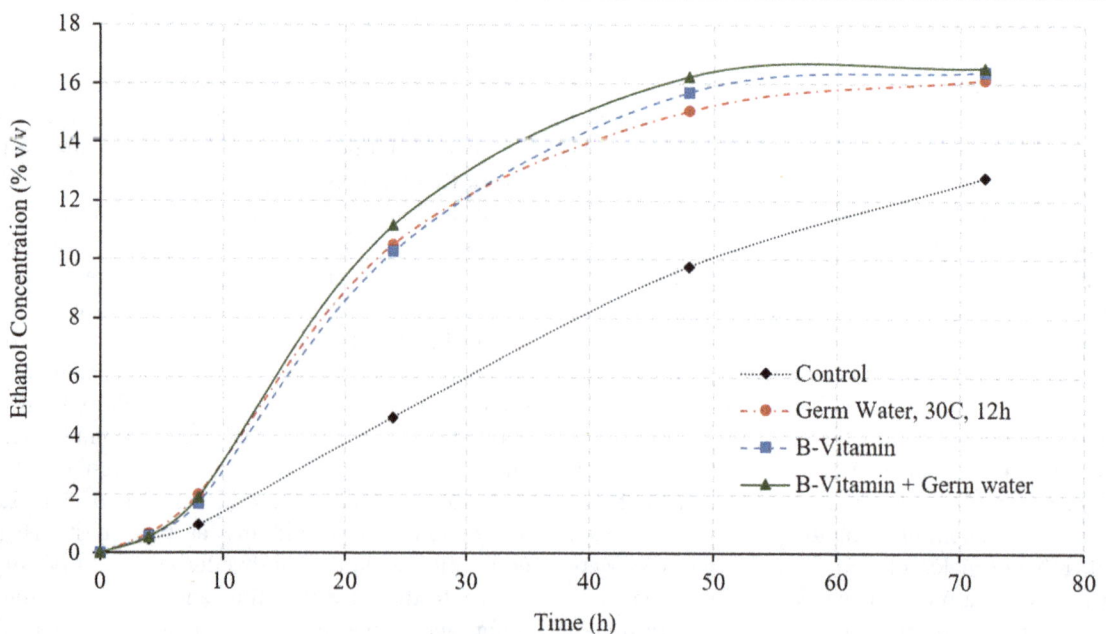

Fig. 4 Ethanol concentrations during fermentation of corn grits in conventional dry grind process with various treatments

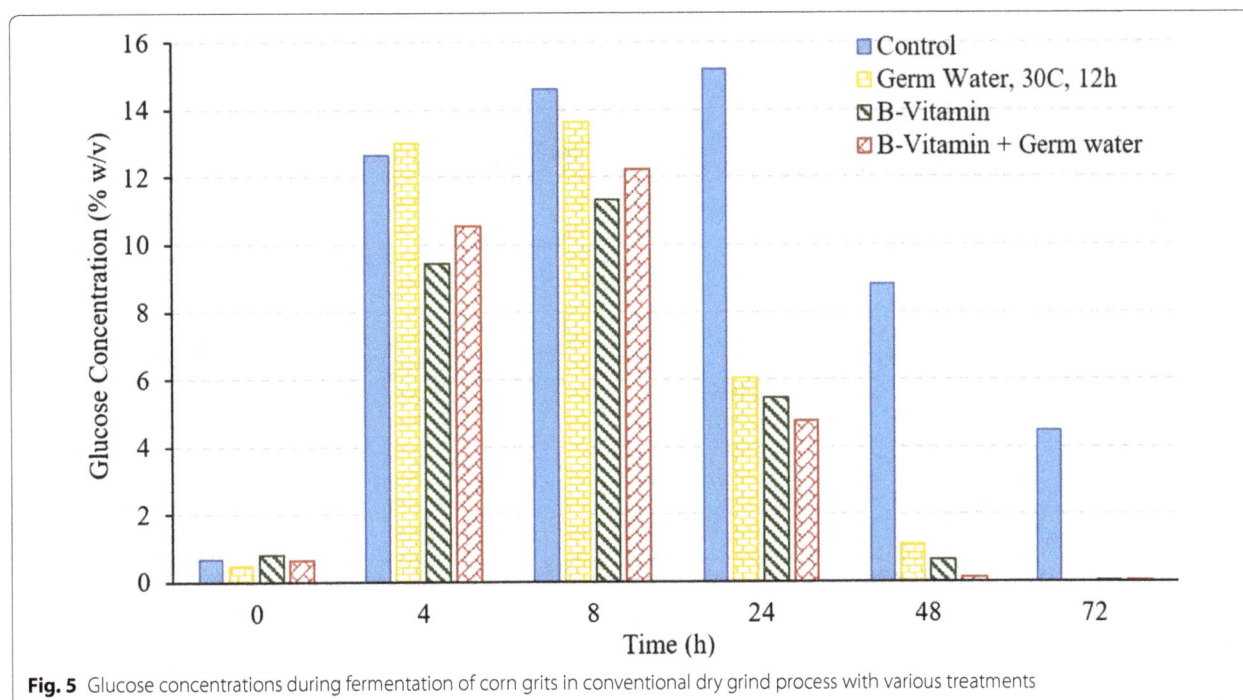

Fig. 5 Glucose concentrations during fermentation of corn grits in conventional dry grind process with various treatments

Table 3 Ethanol yields and conversion efficiencies for all treatments in dry grind process

Treatments	Final ethanol concentration (%)	Ethanol conversion efficiency (%)	Fermentation rates (% v/v/h)
Control	12.79 d	62.44 e	0.208 d
Protease	16.20 a b	82.83 a b c	0.324 c
Germ soak water 2 h 30 °C	14.52 c	74.10 d	0.318 c
Germ soak water 12 h 30 °C-PS	16.14 a b	82.67 a b c	0.492 b
Germ soak water 12 h 30 °C-FS	16.32 a	82.14 b c	0.536 a
B-vitamins	16.39 a	84.02 a	0.484 b
B-vitamins + Germ soak water[a]	16.54 a	83.43 ab	0.531 a

Means followed by the same letter in one column are statistically not different (at $P < 0.05$)

PS partial slurry, *FS* full slurry

[a] Partial slurry of 12 h 30 °C germ soak water

water for all conditions listed in Table 1 was studied and compared with control, protease addition, and B-vitamin addition.

Effect of germ soak water

Similar to conventional dry grind process, germ water addition improved the fermentation performance compared to that of control (Figs. 6, 7). The improvement was better with addition of germ water obtained from soaking at 30 °C for 12 h. With addition of germ water from soaking at 30 °C for 2 h, the ethanol production rate was higher (0.43 vs. 0.25% v/v-h), however, the final ethanol concentration was similar as that of control (Table 4). The average final ethanol concentration in treatment

supplemented with germ water obtained from longer soaking (12 h) was 8.3% higher than the control (16.35 vs. 15.10% v/v), with negligible unconverted glucose at the end of fermentation. About 0.72% (w/v) and 0.82% (w/v) glucose remained unconverted in cases of control and germ water from soaking for 2 h (Fig. 7). The increase in final ethanol concentration in this process (8.3%) was less than that observed in case of conventional process (28.3%). This was attributed to the higher ethanol concentrations obtained with GSH enzymes in control due to relatively low glucose inhibition. The peak glucose concentration for the conventional process was almost double (16.12%) compared to that from GSH process (8.77%). The conversion was higher because of different

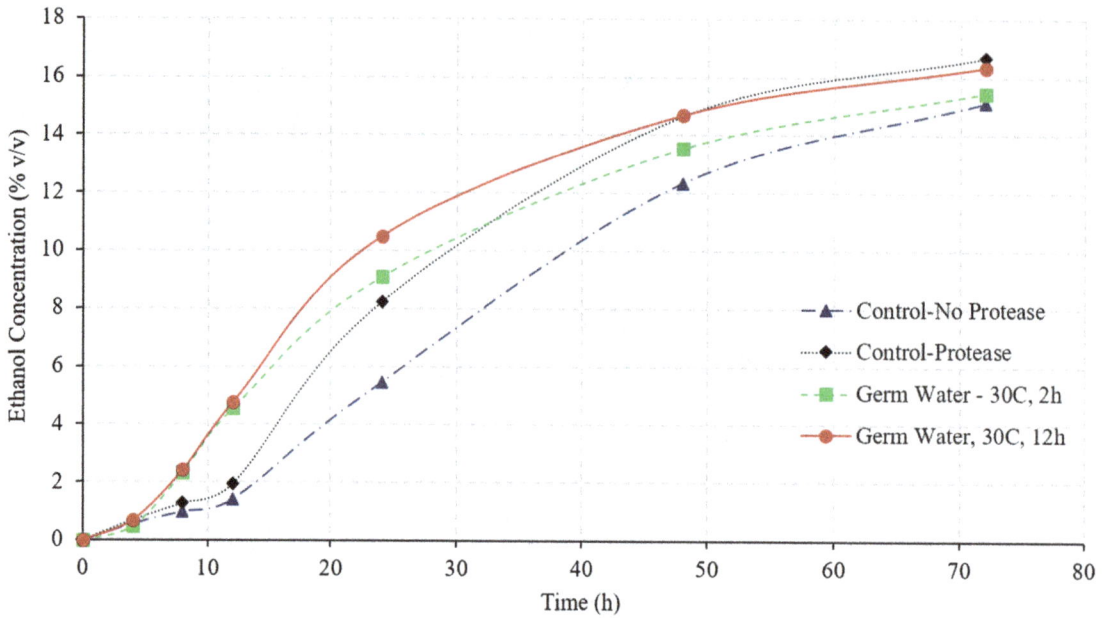

Fig. 6 Comparison of ethanol concentrations among control, treatment with protease addition, and treatment with germ water addition during fermentation of corn grits in GSH process

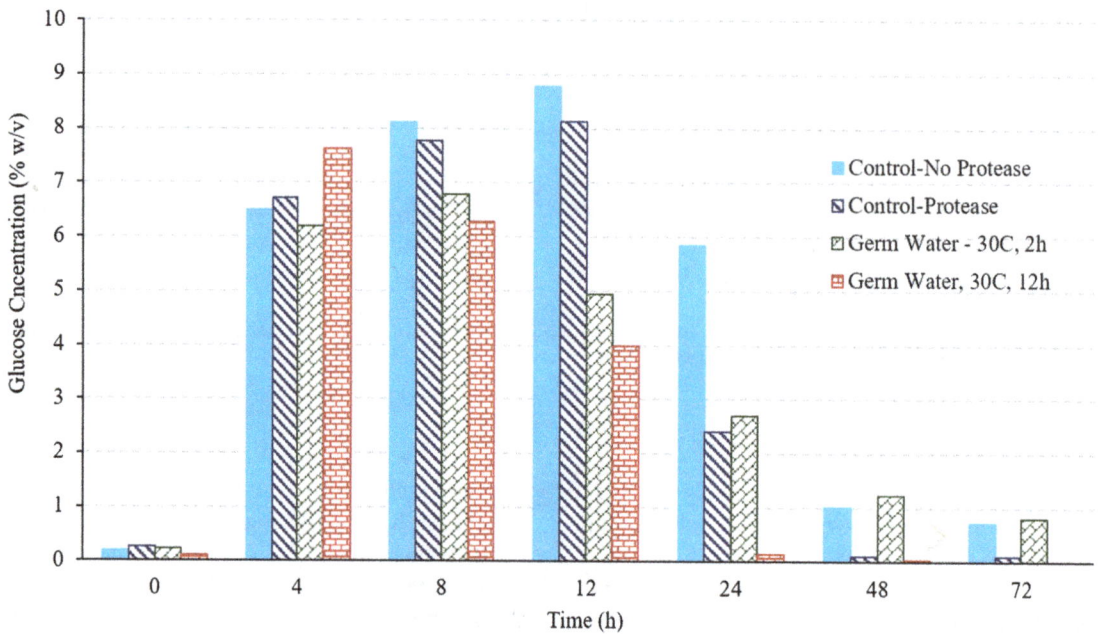

Fig. 7 Comparison of glucose concentrations among control, treatment with protease addition, and treatment with germ water addition during fermentation of corn grits in GSH process

enzymes loadings and synergistic action α-amylase and glucoamylase in the GSH process. Glycerol concentrations in control for GSH process were also about 38% lower than that in the case of conventional dry grind

process (0.93 vs. 1.51% w/v), which leads to higher ethanol production. Glycerol concentrations with germ water addition were 5% lower than that of control in GSH process (0.88 vs. 0.93% w/v). Similar to the observations in

Table 4 Ethanol yields and conversion efficiencies for all treatments in GSH process

Treatments	Final average ethanol concentration (%)	Average conversion efficiency (%)	Average fermentation rates (% v/v/h)
Control	15.10 c	76.75 c	0.245 e
Protease	16.63 a	84.58 a	0.378 d
Germ soak water 2 h 30 °C	15.47 b	76.60 c	0.431 c
Germ soak water 12 h 30 °C-PS	16.35 a	82.78 b	0.490 b
Germ soak water 12 h 30 °C-FS	16.63 a	84.09 ab	0.528 a
B-vitamins	16.43 a	83.91 ab	0.532 a
B-vitamins + germ soak water[a]	16.53 a	83.23 b	0.525 a

Means followed by the same letter in one column are statistically not different (at $P < 0.05$)

PS partial slurry, FS full slurry

[a] Partial slurry of 12 h 30 °C germ soak water

case of conventional process, the addition of protease enzymes resulted in complete fermentation; however, initial ethanol production rate was lower than that of with germ water addition (Fig. 6; Table 4). Ethanol yields of 0.52 and 0.53 L/kg dry grits were similar for germ water and protease addition but higher than that of control (0.48 L/kg dry grits).

Effect of water amount

Similar to the case of conventional process, the amount of germ soak water did not have a significant effect on the final ethanol concentration or conversion efficiency during GSHE process. However, the rate of ethanol production with full germ soak water was higher than the partial germ soak water (Fig. 8). This indicates that the micronutrients needed by the yeast were sufficient from the partial filtrate to obtain the maximum ethanol concentration at the end but higher nutrients in the full slurry lead the yeast to produce more ethanol at the beginning of the fermentation. There was complete fermentation at the end of 72 h as there was no residual glucose left.

Effect of B-vitamins

The effect of external nutrients (vitamins B12 and B-vitamin complex) is shown in Figs. 9 and 10. The final ethanol concentration at the end of 72 h was statistically similar for germ water alone, B-vitamins alone, and both germ water and B-vitamins (Table 4). The rate of fermentation, however, was higher for added B-vitamin treatment (B-vitamin alone and B-vitamin with germ soak water). After the first 24 h, glucose was not detected in the fermentation slurry, indicating the conversion of glucose to ethanol by yeast was at the same rate of its formation by GSH enzymes. In the treatment of B-vitamin with germ soak water, the buildup of glucose is seen to be minimum, which suggests high conversion efficiency of glucose to ethanol in yeast. As it has been mentioned before, B-vitamins are essential for the yeast metabolism and aid

in better functioning and increasing its stress tolerance (Branduardi et al. 2007; Laser 1941; Zhang et al. 2016). Final ethanol concentration of 16.5% was similar for all three treatments (germ water, B-vitamins, and combined germ water with B-vitamins). Starch-to-ethanol conversion efficiency was higher than 80% for all cases.

Composition of germ

Removal of soluble protein and micronutrients would potentially increase the oil content in germ and improves its economic value. Raw germ and germ obtained after soaking were analyzed for protein, oil, and fiber content. It can be observed from Fig. 11 that the oil concentrations in the germ increased by 29 and 36% for shorter and longer soaking conditions in comparison to that of raw germ. Since the market price of germ increases with its oil content (Johnston et al. 2005), the oil increase after soaking provides a huge advantage and makes this approach (germ soak water to improve fermentation) even more attractive. Compared to that of untreated germ, the protein concentrations of treated germ were about 1.5 and 3.7% lower for soaking conditions 30 °C, 2 h and 30 °C, 12 h, respectively. This indicates that during treatments, soluble micronutrients and proteins leached out in water, which when added to the fermentation broth, were available for the yeast to uptake and increased the fermentation efficiency.

Conclusions

This study investigated and optimized the use of nutrient-rich water from corn germ soaking to improve fermentation of corn grits in comparison to through the use of protease enzymes or B-vitamin additions. Optimum soaking time and amount of germ water required was determined corresponding to maximum ethanol yield in conventional dry grind and granular starch hydrolysis process. The addition of germ water from soaking conditions of 30 °C for 12 h resulted in complete fermentation

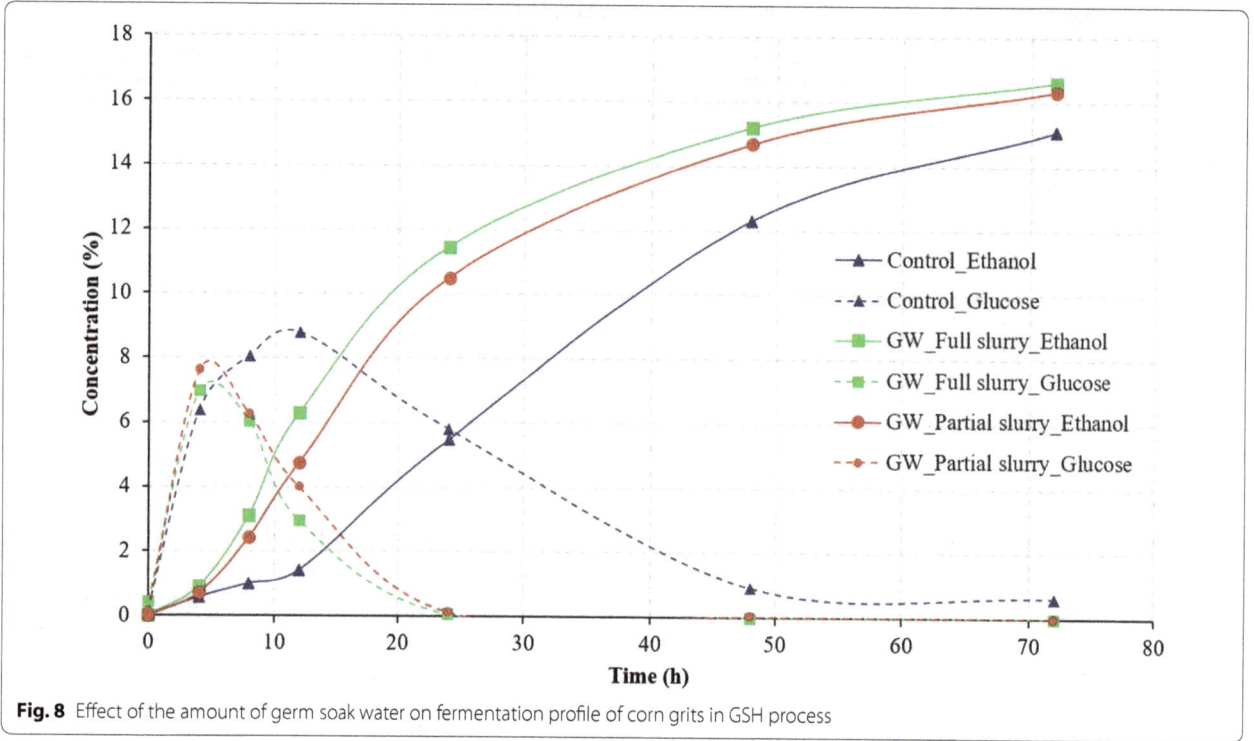

Fig. 8 Effect of the amount of germ soak water on fermentation profile of corn grits in GSH process

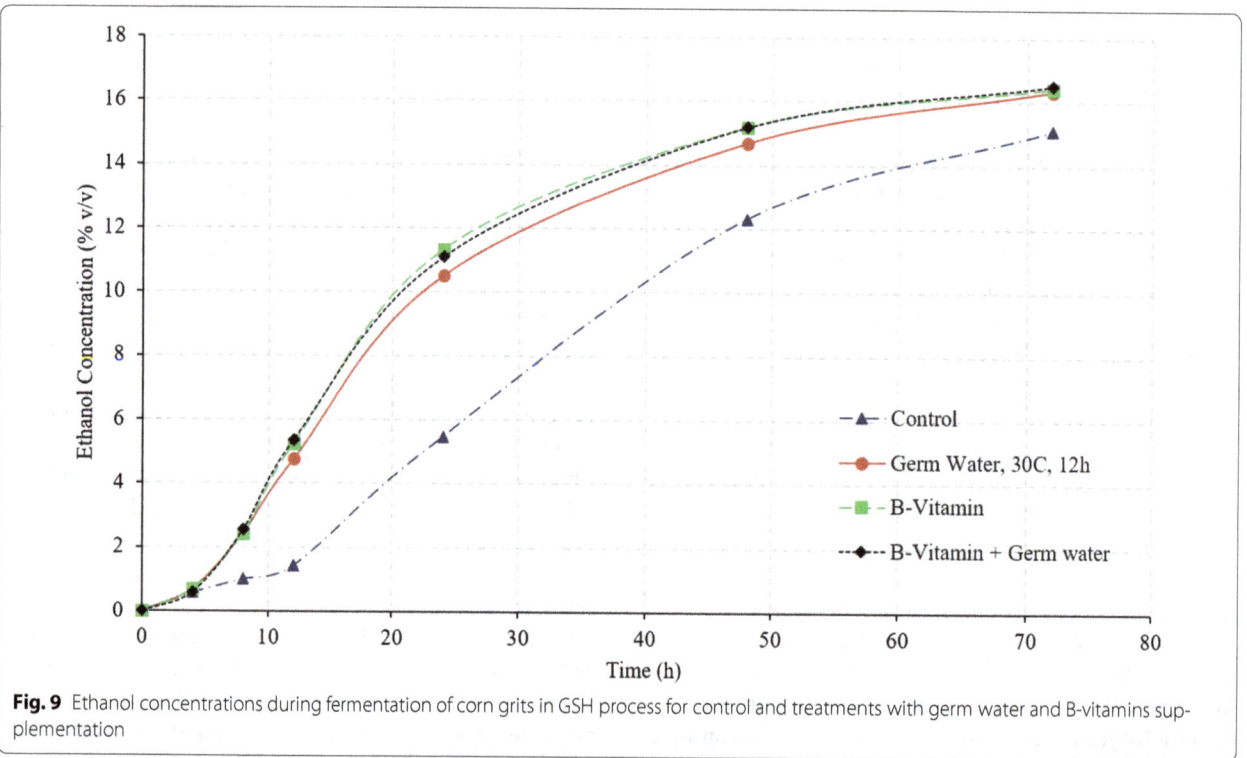

Fig. 9 Ethanol concentrations during fermentation of corn grits in GSH process for control and treatments with germ water and B-vitamins supplementation

for both conventional and GSH processes, compared to significant residual sugars for control. Final ethanol yields were 29 and 8% higher than that of control in case of conventional and GSH process, respectively. GHS enzymes have previously reported to work better than conventional dry grind enzymes. However, the

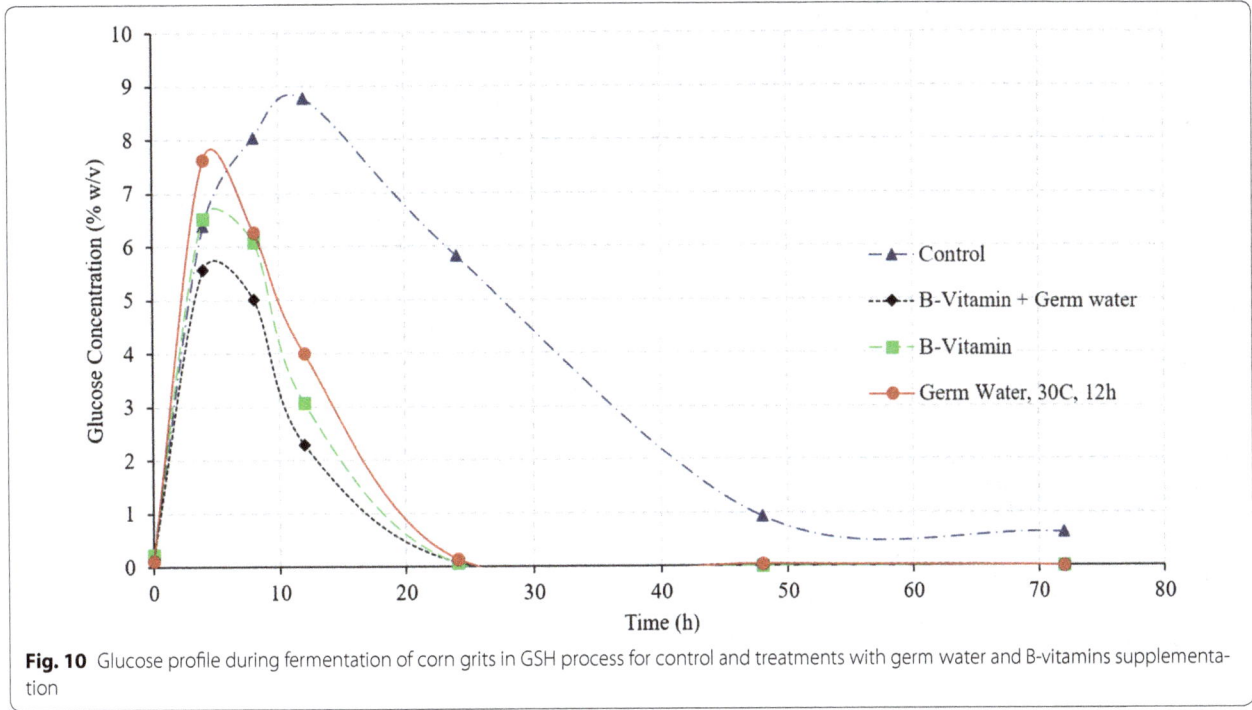

Fig. 10 Glucose profile during fermentation of corn grits in GSH process for control and treatments with germ water and B-vitamins supplementation

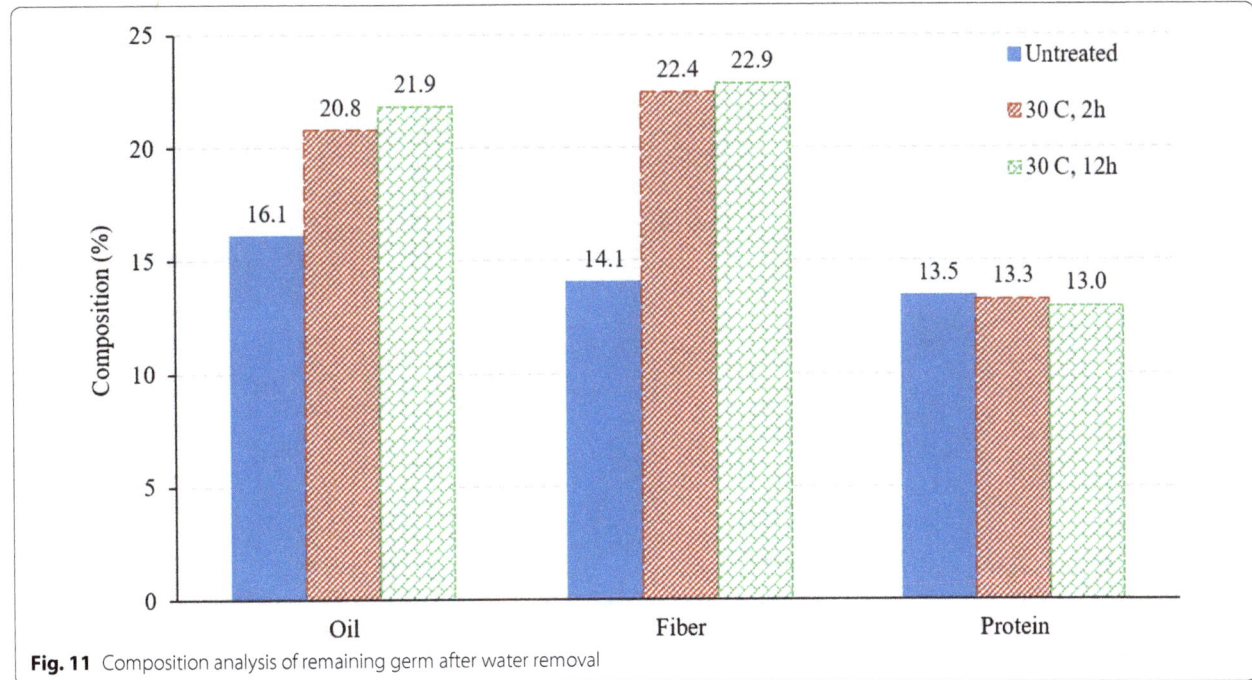

Fig. 11 Composition analysis of remaining germ after water removal

addition of germ water resulted in similar fermentation performance both GSHE and conventional enzymes. Initial ethanol production rates for samples supplemented with germ soak water were higher than that of samples supplemented with protease and similar to that from supplementation of B-vitamins for both processes. Due

to leaching of micronutrients and soluble proteins, soaking process improved the oil concentrations in the germ, which would enhance its economic value. Overall, the use of germ water from optimum soaking conditions can potentially eliminate the need for protease enzymes or expensive nutrients addition for efficient fermentation, and provide other advantages of higher oil concentrations in germ, and potential of acid use reduction in the process.

Authors' contributions
AJ, DK, and VS designed the study. AJ and DK conducted experiments, analyzed data, and prepared the manuscript. VS reviewed the results, helped in data analysis, and edited the manuscript. All authors read and approved the final manuscript.

Competing interests
The authors declare that they have no competing interests.

References

Branduardi P, Fossati T, Sauer M, Pagani R, Mattanovich D, Porro D (2007) Biosynthesis of vitamin C by yeast leads to increased stress resistance. PLoS ONE 2:e1092

Johnston DB, McAloon AJ (2014) Protease increases fermentation rate and ethanol yield in dry-grind ethanol production. Biores Technol 154:18–25

Johnston DB, McAloon AJ, Moreau RA, Hicks KB, Singh V (2005) Composition and economic comparison of germ fractions from modified corn processing technologies. J Am Oil Chem Soc 82:603–608

Kumar D, Singh V (2016) Dry-grind processing using amylase corn and superior yeast to reduce the exogenous enzyme requirements in bioethanol production. Biotechnol Biofuels 9:228. https://doi.org/10.1186/s13068-016-0648-1

Laser H (1941) The effect of thiamine (vitamin B1) on fermentation of yeast. Biochem J 35:488

Moreau RA, Powell MJ, Hicks KB (1996) Extraction and quantitative analysis of oil from commercial corn fiber. J Agric Food Chem 44:2149–2154

Murthy GS, Singh V, Johnston DB, Rausch KD, Tumbleson M (2006a) Evaluation and strategies to improve fermentation characteristics of modified dry-grind corn processes. Cereal Chem 83:455–459

Murthy GS, Singh V, Johnston DB, Rausch KD, Tumbleson M (2006b) Improvement in fermentation characteristics of degermed ground corn by lipid supplementation. J Ind Microbiol Biotechnol 33:655–660

Vidal BC Jr, Rausch KD, Tumbleson M, Singh V (2009) Protease treatment to improve ethanol fermentation in modified dry grind corn processes. Cereal Chem 86:323–328

Wang P (2008) Granular starch hydrolysis for fuel ethanol production. Dissertation, University of Illinois at Urbana-Champaign

White C (2012) Yeast nutrients make fermentations better. White Labs. http://www.jstrack.org/brewing/Yeast_nutrition_article.pdf. Accessed 15 June 2017

Zhang S, Qin X, Lu H, Wan M, Zhu Y (2016) The influence of vitamin E supplementation on yeast fermentation. J Inst Brew 122:289–292

8

Biological conversion of methanol by evolved *Escherichia coli* carrying a linear methanol assimilation pathway

Xiaolu Wang[1,2,3†], Yu Wang[1,2†], Jiao Liu[1,2†], Qinggang Li[1,2], Zhidan Zhang[1,2], Ping Zheng[1,2], Fuping Lu[3] and Jibin Sun[1,2*]

Abstract

Background: Methanol is regarded as a biorenewable platform feedstock because nearly all bioresources can be converted into methanol through syngas. Biological conversion of methanol using synthetic methylotrophs has thus gained worldwide attention.

Results: Herein, to endow *Escherichia coli* with the ability to utilize methanol, an artificial linear methanol assimilation pathway was assembled in vivo for the first time. Distinct from native cyclic methanol utilization pathways, such as ribulose monophosphate cycle, the linear pathway requires no formaldehyde acceptor and only consists of two enzymatic reactions: oxidation of methanol into formaldehyde by methanol dehydrogenase and carboligation of formaldehyde into dihydroxyacetone by formolase. After pathway engineering, genome replication engineering assisted continuous evolution was applied to improve methanol utilization. ^{13}C-methanol-labeling experiments showed that the engineered and evolved *E. coli* assimilated methanol into biomass.

Conclusions: This study demonstrates the usability of the linear methanol assimilation pathway in bioconversion of C1 resources such as methanol and methane.

Keywords: Methanol, C1 resource, Synthetic methylotroph, *Escherichia coli*

Background

Methanol is a C1 compound that can be synthesized either from petrochemical or renewable resources (Carvalho et al. 2017; Patel et al. 2016). Owning to its cost advantage and biocompatibility, methanol is regarded as an attractive feedstock for production of biochemicals and biofuels (Pfeifenschneider et al. 2017). Although native methylotrophs are capable of using C1 resources including methanol as carbon and energy sources, they are more challenging to engineer than genetically tractable hosts due to inefficient genetic-transfer systems and editing tools (Whitaker et al. 2015).

Recently, synthetic methylotrophs were constructed by introducing native methanol assimilation pathways into non-native methylotrophs such as *Escherichia coli* (Dai et al. 2017; Leßmeier et al. 2015; Müller et al. 2015; Rohlhill et al. 2017; Whitaker et al. 2017; Witthoff et al. 2015). To date, ribulose monophosphate (RuMP) cycle that utilizes ribulose-5-phosphate (Ru5P) as a formaldehyde acceptor is the only pathway used for synthetic methylotrophs. Despite the fact that Ru5P could be regenerated through pentose phosphate (PP) pathway, high coordination of heterologous RuMP cycle and native PP pathway is challenging (Whitaker et al. 2015). A computationally designed enzyme formolase (FLS) that can catalyze the carboligation of three formaldehyde molecules into one dihydroxyacetone (DHA) molecule was reported recently and used to construct an artificial carbon fixation pathway in vitro (Siegel et al. 2015). In the present study, an artificial linear methanol assimilation pathway based

*Correspondence: sun_jb@tib.cas.cn

†Xiaolu Wang, Yu Wang and Jiao Liu contributed equally to this work

[2] Tianjin Institute of Industrial Biotechnology, Chinese Academy of Sciences, Tianjin 300308, People's Republic of China

Full list of author information is available at the end of the article

on FLS was assembled in *E. coli*, and a combined strategy of metabolic engineering and adaptive evolution was applied to facilitate methanol utilization.

Methods

Bacterial strains, plasmids, and growth conditions

All bacterial strains and plasmids used in this study are listed in Additional file 1: Table S1. Gene expression and methanol utilization were performed in strain *E. coli* BW25113 Δ*frmA* (Ec-Δ*frmA*). The pTrc99A vector with a trc promoter was used for expression of the genes required for methanol utilization. Primers used for plasmid construction are listed in Additional file 1: Table S2. *E. coli* strains were cultured aerobically in lysogeny broth (LB) medium or M9 minimal medium supplemented with carbon sources at 30 or 37 °C. Detailed methods are described in Additional file 1: Additional methods.

Enzyme activity assay

Methanol dehydrogenase (MDH) and FLS activities were assayed using the methods described previously (Müller et al. 2015; Siegel et al. 2015). Detailed methods are described in Additional file 1: Additional methods.

Adaptive evolution

Detailed methods are described in Additional file 1: Additional methods.

Analysis of ^{13}C-labeling of proteinogenic amino acids

[13]C-labeling of proteinogenic amino acids was analyzed using a method described previously with some modifications (You et al. 2012). Detailed methods are described in Additional file 1: Additional methods.

Results and discussion

Assembly of linear methanol utilization pathway in *E. coli*

The linear methanol utilization pathway consists of two steps: oxidation of methanol into formaldehyde and carboligation of formaldehyde into DHA, which can be phosphorylated to dihydroxyacetone phosphate by dihydroxyacetone kinase and enter lower glycolysis (Fig. 1a). According to the calculation by eQuilibrator (Flamholz et al. 2012), the $\Delta_r G'^o$ values for the linear pathway and the RuMP pathway are 9.1 and -3.4 kJ/mol, respectively, suggesting that the RuMP pathway is more thermodynamically feasible. However, the product of the linear pathway can enter glycolysis and be metabolized quickly, providing a strong driven force for methanol utilization. Therefore, the linear pathway is also supposed to be feasible. To assemble the linear pathway in vivo, NAD$^+$-dependent MDH from *Bacillus methanolicus* and artificial FLS were overexpressed in *E. coli*. The multicopy plasmid pTrc99A with a strong trc promoter was used to achieve high-level expression of MDH and FLS since their specific activities are quite low (Krog et al. 2013; Siegel et al. 2015). Two recombinants Ec-Δ*frmA-mdh3*$_{MGA3}$-*fls* and Ec-Δ*frmA-mdh2*$_{PB1}$-*fls* carrying the *fls* gene and different *mdh* genes (Additional file 1: Table S3) were constructed. Enzyme activity assays demonstrated that both MDHs were functionally expressed. Strain Ec-Δ*frmA-mdh2*$_{PB1}$-*fls* showed higher methanol oxidation activity that was approximately twice as high as the

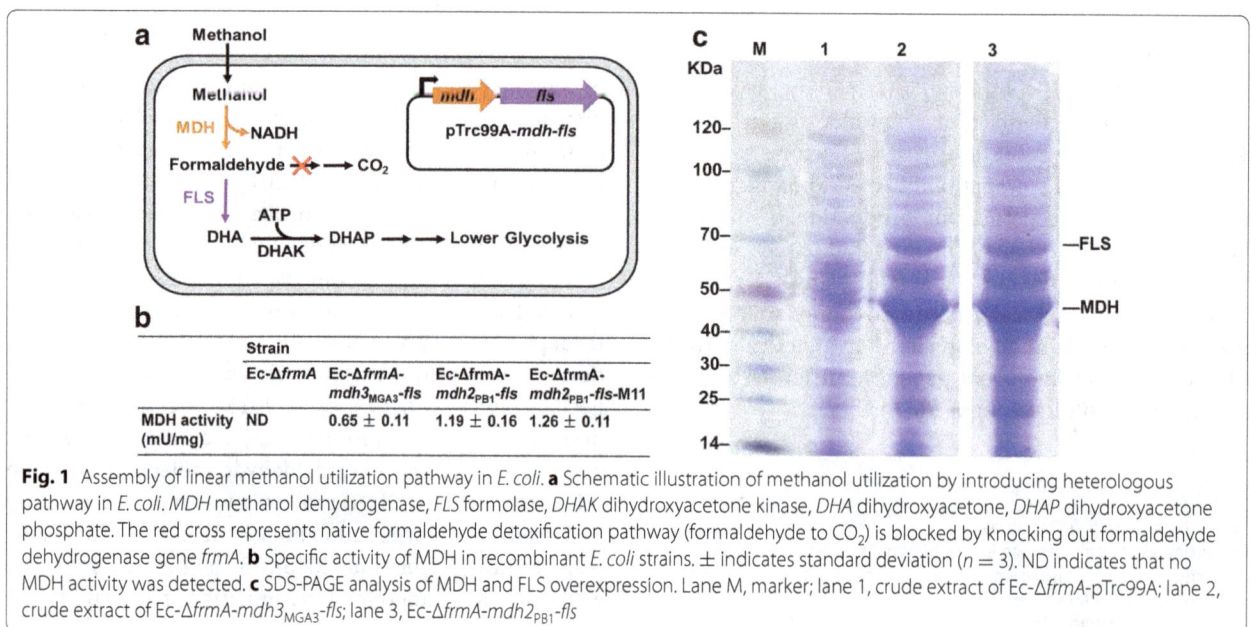

Fig. 1 Assembly of linear methanol utilization pathway in *E. coli*. **a** Schematic illustration of methanol utilization by introducing heterologous pathway in *E. coli*. *MDH* methanol dehydrogenase, *FLS* formolase, *DHAK* dihydroxyacetone kinase, *DHA* dihydroxyacetone, *DHAP* dihydroxyacetone phosphate. The red cross represents native formaldehyde detoxification pathway (formaldehyde to CO_2) is blocked by knocking out formaldehyde dehydrogenase gene *frmA*. **b** Specific activity of MDH in recombinant *E. coli* strains. ± indicates standard deviation ($n = 3$). ND indicates that no MDH activity was detected. **c** SDS-PAGE analysis of MDH and FLS overexpression. Lane M, marker; lane 1, crude extract of Ec-Δ*frmA*-pTrc99A; lane 2, crude extract of Ec-Δ*frmA-mdh3*$_{MGA3}$-*fls*; lane 3, Ec-Δ*frmA-mdh2*$_{PB1}$-*fls*

MDH activity of strain Ec-$\Delta frmA$-$mdh3_{MGA3}$-fls (Fig. 1b). FLS activity was determined by coupled reactions involving DHA formation from formaldehyde by FLS and DHA reduction by NAD^+-dependent glycerol dehydrogenase. However, the reverse activity of MDH that could reduce formaldehyde to methanol with NADH consumption interfered with the NADH-dependent DHA reduction. Therefore, FLS activity was not determined here, whereas SDS-PAGE analysis indicated that MDHs and FLS were successfully expressed (Fig. 1c). Strain Ec-$\Delta frmA$-$mdh2_{PB1}$-fls was used in the subsequent experiments for higher MDH activity.

Bioconversion of methanol into biomass of the engineered *E. coli*

Despite the equipment of linear methanol assimilation pathway, the engineered strain could not initiate growth in M9 minimal medium with methanol (approximately 8 g/L, 1% v/v) as the sole carbon source. Similar phenomena were observed in previous studies on RuMP-based synthetic methylotrophs and undefined supplements such as yeast extract and tryptone were added to initiate cell growth on methanol (Whitaker et al. 2017). Thus,

small amounts of yeast extract (1 g/L) were added in M9 minimal medium. Any improvements in cell growth in the presence of methanol might derive from the contribution of methanol assimilation. As controls, a $\Delta frmA$ strain containing the empty pTrc99A vector (strain Ec-$\Delta frmA$-pTrc99A) was cultivated using the aforementioned media. A methanol evaporation control without inoculation was also conducted.

As shown in Fig. 2a, approximately 1 g/L methanol evaporated away during the cultivation. When the control strain Ec-$\Delta frmA$-pTrc99A was cultivated using yeast extract and methanol as co-substrate, no additional methanol consumption but slightly decrease in cell growth was observed (Fig. 2a, b). We speculated that methanol might be oxidized to toxic intermediate formaldehyde by the non-specific activities of alcohol dehydrogenases of *E. coli*, which affected the cell growth negatively. Regarding to strain Ec-$\Delta frmA$-$mdh2_{PB1}$-fls, no significant increase in biomass was observed when methanol was added (Fig. 2c), whereas slightly more methanol (1.45 g/L) was consumed compared to the evaporation control, suggesting methanol utilization of the engineered strain. To further validate methanol utilization,

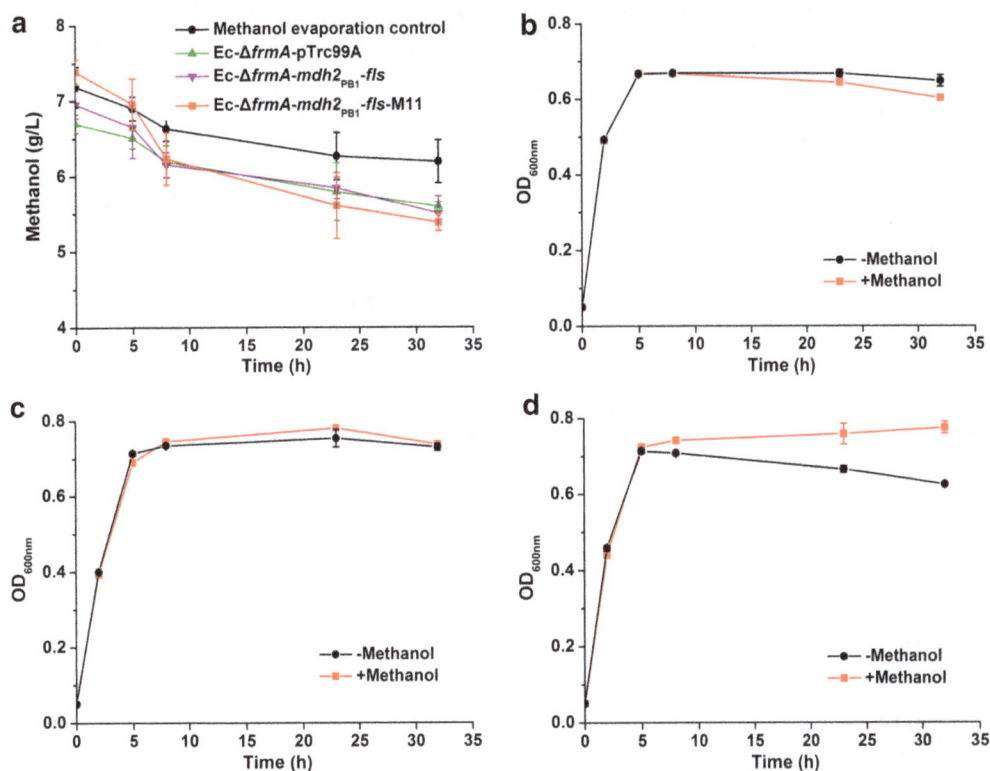

Fig. 2 Methanol consumption and growth characteristics of *E. coli* strains. **a** Methanol evaporation and consumption. **b** Cell growth of strain Ec-$\Delta frmA$-pTrc99A. **c** Cell growth of strain Ec-$\Delta frmA$-$mdh2_{PB1}$-fls. **d** Cell growth of strain Ec-$\Delta frmA$-$mdh2_{PB1}$-fls-M11. Cells were cultured in M9 minimal medium supplemented with 1 g/L yeast extract or M9 minimal medium supplemented with 1 g/L yeast extract and methanol. Error bars indicate standard deviation ($n = 3$)

^{13}C-methanol-labeling experiment was performed. When ^{13}C-methanol was used as a carbon source, ^{13}C-labeled amino acids in biomass including alanine, aspartic acid, glutamic acid, phenylalanine, proline, glycine, lysine, serine, threonine, tyrosine, and ^{13}C-labeled citric acid were detected (Fig. 3a; Additional file 1: Table S4). It has been reported that amino acids measurement could provide isotopic labeling information about eight crucial precursor metabolites in the central metabolism (You et al. 2012). The presented results showed that biosynthesis of key intermediates of glycolysis, PP pathway and TCA cycle, including 3-phosphoglycerate, phosphoenolpyruvate, pyruvate, acetyl-CoA, α-ketoglutarate, oxaloacetate, and erythrose 4-phosphate, withdrew carbon from ^{13}C-methanol.

Adaptive evolution of the engineered *E. coli* to improve methanol utilization

To further improve the microbial performance in methanol medium and screen methanol-utilizing mutants, adaptive evolution based on GREACE (genome replication engineering assisted continuous evolution) was conducted (Luan et al. 2013). A proofreading-defective element of the DNA polymerase of *E. coli* (ε subunit encoded by *dnaQ* gene) was expressed in strain Ec-Δ*frmA*-*mdh2*$_{PB1}$-*fls* to introduce random mutations into the genomic DNA during continuous passage cultivation in LB medium. For each passage, cells were transferred into M9 minimal medium supplemented with methanol to enrich potential mutants with improved cell growth on methanol. Mutants were then isolated from the culture and a mutant with the best cell growth on methanol was isolated and designated as Ec-Δ*frmA*-*mdh2*$_{PB1}$-*fls*-M11 (Additional file 1: Figure S1). When mutant Ec-Δ*frmA*-*mdh2*$_{PB1}$-*fls*-M11 was cultivated in

M9 minimal medium supplemented with 1 g/L yeast extract and methanol, 2 g/L methanol was consumed, which was more than that consumed by its parent strain Ec-Δ*frmA*-*mdh2*$_{PB1}$-*fls*. It was noticed that biomass of the mutant declined after 5 h and addition of methanol helps maintain the biomass (Fig. 2d). We speculated that such decline in cell growth was caused by the random mutations introduced by GREACE. ^{13}C-methanol-labeling experiment was then conducted and the results validated that mutant Ec-Δ*frmA*-*mdh2*$_{PB1}$-*fls*-M11 assimilated more ^{13}C-methanol into biomass (Fig. 3b; Additional file 1: Table S5). The results demonstrated that coupling of metabolic engineering and adaptive evolution was an enabling strategy to endow microorganisms with the ability to utilize methanol.

Synthetic methylotrophs have been constructed by heterogenous expressing MDH and RuMP genes (Pfeifenschneider et al. 2017). RuMP cycle depends on regenerating the formaldehyde acceptor Ru5P, which requires high coordination of many enzymes involved in formaldehyde assimilation and PP pathway (Whitaker et al. 2015). On the contrary, the linear formaldehyde assimilation pathway used in this study only requires one enzyme FLS and directly produces C3 intermediate DHA, which could be a great advantage for pathway engineering. Previous and the present studies revealed that constructing synthetic methylotrophs was far more complicated than complementing metabolic pathways where several crucial factors need to be considered, such as how to keep the intracellular formaldehyde concentration below the toxicity threshold (Witthoff et al. 2015) and how to balance the reducing equivalent generated by methanol oxidation (Price et al. 2016). In this case, combining metabolic engineering and adaptive evolution could be an easy strategy to prepare a desirable mutant that assimilates methanol efficiently.

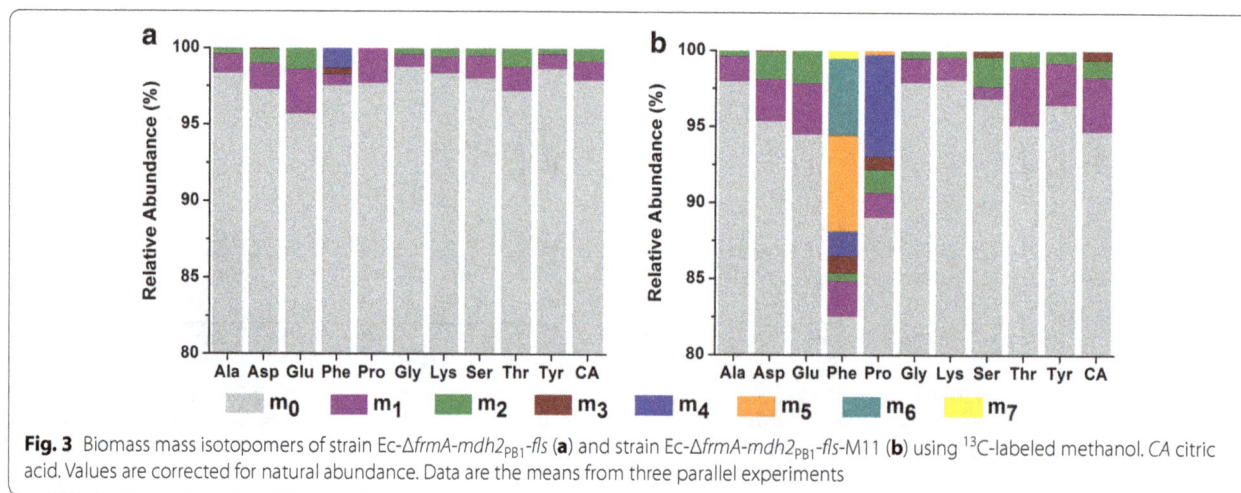

Fig. 3 Biomass mass isotopomers of strain Ec-Δ*frmA*-*mdh2*$_{PB1}$-*fls* (**a**) and strain Ec-Δ*frmA*-*mdh2*$_{PB1}$-*fls*-M11 (**b**) using ^{13}C-labeled methanol. *CA* citric acid. Values are corrected for natural abundance. Data are the means from three parallel experiments

By using such a combined strategy, improved methanol assimilation was obtained in the mutant Ec-Δ*frmA*-*mdh2*$_{PB1}$-*fls*-M11. Whole-genome resequencing revealed that no mutation was introduced into the plasmid, which was consistent with the unchanged MDH activity (Fig. 1b). Meanwhile, 66 missense, synonymous, and intergenic mutations that covered amino acid transport and metabolism, signal transduction, cell wall/membrane/envelope biogenesis, etc. were discovered (Additional file 2: Table S6). Further investigation of these mutations will likely elucidate key factors of methanol utilization in synthetic methylotrophs.

Conclusions

In this study, an artificial linear methanol assimilation pathway was functionally assembled in *E. coli*. Methanol utilization by the engineered strain was facilitated and further improved by adaptive evolution. ^{13}C-methanol-labeling experiment revealed the methanol incorporation into cellular biomass. This study is the first demonstration of applying the linear methanol assimilation pathway for biological conversion of methanol. The combined strategy of metabolic engineering and adaptive evolution is also a useful approach to endow platform strains with the ability to utilize other C1 resources such as the main component of natural gas, methane.

Additional files

Additional file 1: Additional methods. Figure S1. Adaptive evolution by genome replication engineering assisted continuous evolution (GREACE) and screening of methanol-utilizing mutants. **Table S1.** Bacterial strains and plasmids used in this study. **Table S2.** Sequences of primers used in this study. **Table S3.** Sequences of *mdh3*$_{MGA3}$, *mdh2*$_{PB1}$, and *fls* genes. **Table S4.** Biomass mass isotopomers of strain Ec-Δ*frmA*-*mdh2*$_{PB1}$-*fls* using ^{13}C-methanol as substrate. Values are corrected for natural abundance. **Table S5.** Biomass mass isotopomers of strain Ec-Δ*frmA*-*mdh2*$_{PB1}$-*fls*-M11 using ^{13}C-methanol as substrate. Values are corrected for natural abundance.

Additional file 2: Table S6. List of mutations identified using whole genome resequencing of strain Ec-Δ*frmA*-*mdh2*$_{PB1}$-*fls*-M11.

Abbreviations
RuMP: ribulose monophosphate; Ru5P: ribulose-5-phosphate; PP pathway: pentose phosphate pathway; MDH: methanol dehydrogenase; FLS: formolase; DHA: dihydroxyacetone.

Authors' contributions
YW, JL, and JS conceived and designed the experiments. XW, YW, QL, and ZZ performed the experiments. XW, YW, and JL analyzed the data. PZ, FL, and JS contributed reagents and analytic tools. YW, PZ, FL, and JS wrote the paper. All authors read and approved the final manuscript.

Author details
^{1} Key Laboratory of Systems Microbial Biotechnology, Chinese Academy of Sciences, Tianjin 300308, People's Republic of China. ^{2} Tianjin Institute of Industrial Biotechnology, Chinese Academy of Sciences, Tianjin 300308, People's Republic of China. ^{3} College of Biotechnology, Tianjin University of Science and Technology, Tianjin 300222, People's Republic of China.

Acknowledgements
Not applicable.

Competing interests
The authors declare that they have no competing interests.

Funding
This work was supported by the National Natural Science Foundation of China (31700044), the Key Research Program of the Chinese Academy of Sciences (ZDRW-ZS-2016-2), the first Special Support Plan for Talents Development and High-level Innovation and Entrepreneurship Team of the Tianjin Municipal City, and the Natural Science Foundation of Tianjin (14JCQNJC10000).

References
Carvalho L, Furusjo E, Kirtania K, Wetterlund E, Lundgren J, Anheden M, Wolf J (2017) Techno-economic assessment of catalytic gasification of biomass powders for methanol production. Bioresour Technol 237:167–177

Dai Z, Gu H, Zhang S, Xin F, Zhang W, Dong W, Ma J, Jia H, Jiang M (2017) Metabolic construction strategies for direct methanol utilization in *Saccharomyces cerevisiae*. Bioresour Technol. doi:10.1016/j.biortech.2017.05.100

Flamholz A, Noor E, Bar-Even A, Milo R (2012) eQuilibrator—the biochemical thermodynamics calculator. Nucleic Acids Res 40(Database issue):D770–D775

Krog A, Heggeset TM, Müller JE, Kupper CE, Schneider O, Vorholt JA, Ellingsen TE, Brautaset T (2013) Methylotrophic *Bacillus methanolicus* encodes two chromosomal and one plasmid born NAD$^+$ dependent methanol dehydrogenase paralogs with different catalytic and biochemical properties. PLoS ONE 8(3):e59188

Leßmeier L, Pfeifenschneider J, Carnicer M, Heux S, Portais JC, Wendisch VF (2015) Production of carbon-13-labeled cadaverine by engineered *Corynebacterium glutamicum* using carbon-13-labeled methanol as co-substrate. Appl Microbiol Biotechnol 99(23):10163–10176

Luan G, Cai Z, Li Y, Ma Y (2013) Genome replication engineering assisted continuous evolution (GREACE) to improve microbial tolerance for biofuels production. Biotechnol Biofuels 6(1):137

Müller JE, Meyer F, Litsanov B, Kiefer P, Potthoff E, Heux S, Quax WJ, Wendisch VF, Brautaset T, Portais JC, Vorholt JA (2015) Engineering *Escherichia coli* for methanol conversion. Metab Eng 28:190–201

Patel SK, Mardina P, Kim D, Kim SY, Kalia VC, Kim IW, Lee JK (2016) Improvement in methanol production by regulating the composition of synthetic gas mixture and raw biogas. Bioresour Technol 218:202–208

Pfeifenschneider J, Brautaset T, Wendisch VF (2017) Methanol as carbon substrate in the bio-economy: metabolic engineering of aerobic methylotrophic bacteria for production of value-added chemicals. Biofuels Bioprod Biorefin. doi:10.1002/bbb.1773

Price JV, Chen L, Whitaker WB, Papoutsakis E, Chen W (2016) Scaffoldless engineered enzyme assembly for enhanced methanol utilization. Proc Natl Acad Sci USA 113(45):12691–12696

Rohlhill J, Sandoval NR, Papoutsakis ET (2017) Sort-seq approach to engineering a formaldehyde-inducible promoter for dynamically regulated *Escherichia coli* growth on methanol. ACS Synth Biol 6(8):1584–1595

Siegel JB, Smith AL, Poust S, Wargacki AJ, Bar-Even A, Louw C, Shen BW, Eiben CB, Tran HM, Noor E, Gallaher JL, Bale J, Yoshikuni Y, Gelb MH, Keasling JD, Stoddard BL, Lidstrom ME, Baker D (2015) Computational protein design enables a novel one-carbon assimilation pathway. Proc Natl Acad Sci USA 112(12):3704–3709

Whitaker WB, Sandoval NR, Bennett RK, Fast AG, Papoutsakis ET (2015) Synthetic methylotrophy: engineering the production of biofuels and chemicals based on the biology of aerobic methanol utilization. Curr Opin Biotechnol 33:165–175

Whitaker WB, Jones JA, Bennett K, Gonzalez J, Vernacchio VR, Collins SM, Palmer MA, Schmidt S, Antoniewicz MR, Koffas MA, Papoutsakis ET (2017) Engineering the biological conversion of methanol to specialty chemicals in *Escherichia coli*. Metab Eng 39:49–59

Witthoff S, Schmitz K, Niedenfuhr S, Noh K, Noack S, Bott M, Marienhagen J (2015) Metabolic engineering of *Corynebacterium glutamicum* for methanol metabolism. Appl Environ Microbiol 81(6):2215–2225

9

Acetic acid mediated leaching of metals from lead-free solders

U. Jadhav[2], C. Su[1], M. Chakankar[1] and H. Hocheng[1*]

Abstract

Background: The replacement of lead (Pb)-bearing solders by several Pb-free solders is a subject of intense research in these days due to the toxic effects of Pb on the environment. However, the Pb-free solders contain metals such as silver (Ag), copper (Cu), and zinc (Zn). The increasing use of these Pb-free solders again increases the risk of release of Ag, Cu, and Zn metals into the environment. The Pb-free solders can, therefore, be used as a secondary source for the metals which will not only help in environmental protection but also for the resource recovery.

Results: This study reports a process to leach metals from hazardous soldering materials by acetic acid. Acetic acid was found more effective for metal recovery from the tin–copper (Sn–Cu) solder than tin–copper–silver (Sn–Cu–Ag) solder. Various process parameters were optimized for recovery of metals from Sn–Cu solder. It required 30 h for 100% recovery of Cu and Sn, respectively. The metal recovery increased gradually with an increase in acid concentration approaching complete recovery at an acid concentration of 80%. Effect of shaking speed and temperature on the recovery of metals from Sn–Cu solder was studied. The metal recovery decreased with an increase in solder weight.

Conclusion: The present study reveals an effective process to recycle the Pb-free solders. The low concentration of acetic acid was also found significant for metal leaching from solder. The research provides basic knowledge for recovery of metals from Pb-free solders.

Keywords: Industrial waste, Solders, Metal recovery, Acetic acid

Background

The tin–lead (Sn/Pb) solders are widely used in electronic equipment. However, lead is known to be a toxic material. It has health and environmental concern (Yoo et al. 2016). Therefore, there is an increasing demand in replacing Sn/Pb solders with Pb-free solders in the electronics industry (Cheng et al. 2011). Nowadays, more progress has been achieved in developing Pb-free solders. The tin–silver (Sn–Ag), tin–zinc (Sn–Zn), and tin–copper (Sn–Cu) solder alloys are replacing Pb-bearing solders (Abtew and Selvaduray 2000; Zeng and Tu 2002; Wu et al. 2004; Gao et al. 2012; Yang et al. 2016). The replacement of Pb-bearing solders by several Pb-free solders can avoid potential environmental risk from toxic Pb elements. However, excessive use of Pb-free solders releases metals, such as Ag, Cu, and Zn, into the environment. This would again pose a risk to ecosystems and human health (Cheng et al. 2011; Lim and Schoenung 2010). It is therefore essential to recover the metal values from the newly developed lead-free solders.

Several methods have been proposed to treat lead-free solders to recover their valuable metals. The existing processes for recycling spent solders include pyrometallurgy (Lee et al. 2007), hydrometallurgy (Yoo et al. 2012), and biohydrometallurgy (Hocheng et al. 2014). For pyrometallurgical processes, high energy consumption is unavoidable. Also, it emits toxic gases. The biohydrometallurgical processes have been gradually replacing the hydrometallurgical ones due to their higher efficiency, lower costs, and fewer industrial requirements (Li et al. 2010; Hocheng et al. 2014). But the treatment period for biohydrometallurgical process is long. Hydrometallurgy is a well-established process for the separation and recovery of metal from industrial wastes. Using hydrometallurgical

*Correspondence: hocheng@pme.nthu.edu.tw; hochengh@gmail.com
[1] Department of Power Mechanical Engineering, National Tsing Hua University, No. 101, Sec. 2, Kuang Fu Rd, 30013 Hsinchu, Taiwan ROC
Full list of author information is available at the end of the article

processes, a complete recovery of metals with high purity is possible. It requires low energy (Huang et al. 2009). The leaching of solders has been investigated using sodium hydroxide and sodium persulfate (Rhee et al. 1994), organic solvents (Takahashi et al. 2009), and nitric acid (HNO_3) (Yoo et al. 2012) as leaching agents. Notably, organic solvents are harmful to the environment (Yoo et al. 2012). The strong acid leachates release toxic gases like Cl_2, SO_3, and NO_x (Li et al. 2010, 2013). Also, the waste acid solution generated during the process is harmful to the environment. For the sustainable management of natural resources and to reduce environmental pollution, it is important to develop a simple recycling process to recover as much of the valuable metals as possible (Li et al. 2010, 2013; Shu et al. 2004). Organic acids could be an attractive extracting agent. With organic acids, the extraction can be performed at mildly acidic conditions (pH 3–5). A study reported the percolation leaching of the Cuban nickel tailings. The authors used tartaric acid and a mixture of tartaric and oxalic acids at different concentrations (Hernandez et al. 2007). Several studies reported the leaching of cobalt (Co) and lithium (Li) from spent lithium-ion batteries using various organic acids and hydrogen peroxide (Li et al. 2010, 2013). Merdoud et al. (2016) and Cameselle and Pena (2016) used organic acids as facilitating agents for electrokinetic decontamination of soils. Biswas and Mulaba-Bafubiandi (2016) used citric, oxalic, and gluconic acids for the leaching of copper (Cu) and cobalt (Co) from oxidized ore. They achieved maximum recovery of Cu and Co at 80 °C using 150 mM of citric acid. Their findings also suggest that the recovery of Cu and Co from the oxidized ore was highly dependent on the amenability of the ore mineral to organic acid attack. Suanon et al. (2016) used organic acid to improve the removal of metals from the soil because of its ready availability, relatively inexpensiveness, and environmentally benign nature. Also, the organic acids are biologically degradable (Veeken and Hamelers 1999). Considering these advantages of organic acids for metal leaching, the acetic acid was introduced as a leaching reagent in the present study. Acetic acid is commonly known as moderate and weak chelating agents (low molecular weight organic acid). It has been extensively used in the food industry, the medical field, and the manufacturing industry, among others. The present study was carried out to determine the effectiveness of acetic acid for recovery of metals from solders, and also, to determine various factors influencing metal dissolution from solder. Further, the results were compared with the biologically produced acetic acid using *Acetobacter* sp.

Methods

Materials

The solders Sn–Cu (63–37%) and Sn–Cu–Ag (60–37–3%) were supplied by Taiwan-Solid Enterprises Limited. For the experiments, the Sn–Cu and Sn–Cu–Ag solders were cut into 3 mm size. All the chemicals were purchased from Sigma Aldrich.

Microorganism and culture conditions

Acetobacter sp. TISTR 102 was obtained from BCRC strain collection, Taiwan, R.O.C. The strains were cultured in Yeast Glucose ethanol acetic acid (YGEA) medium for acetic acid production and the acid production rate was estimated by titration method as described by Beheshti Maal and Shafiei (2010). Briefly, 5 ml of culture was mixed with 20 ml of distilled water and mixed with 5 drops of phenolphthalein indicator. The solution was titrated against 0.5 N NaOH. The amount of acetic acid (g) produced in 100 ml medium was calculated as follows:

$$\text{Acetic acid (g/100 ml)} = \text{Volume of NaOH (ml)}$$
$$\text{used in titration} \times 0.03 \times 20.$$

Leaching of solder pieces by acetic acid

For leaching experiments, two types of lead-free solders were used viz., Sn–Cu and Sn–Cu–Ag. All samples were rinsed with distilled water, 75% ethanol, and dried before use. Acetic acid (100%) was used as a leaching agent. The leaching experiments were carried out according to Joksic et al. (2005) and Granata et al. (2011) with some modifications.

The effect of acetic acid volume on the solder leaching efficiency

In an initial experiment, the 100 mg of Sn–Cu and Sn–Cu–Ag solders were incubated with 10 ml acetic acid (100%) separately in 25 ml conical flask at 30 °C for 24 h. The flask was covered to avoid evaporation of acid. The aliquots of liquid samples were taken after 24 h, filtered, and analyzed for metal content by inductively coupled plasma resonance spectroscopy (ICP/OES).

The effect of incubation time on the solder leaching efficiency

The effect of incubation time on metal leaching from Sn–Cu solder was studied. The Sn–Cu solder pieces (100 mg) were covered with 10 ml of acetic acid (100%) in a 25 ml conical flask at 30 °C for 30 h. Over the course of the leaching process, the samples were taken at 10, 20, and 30 h and analyzed for metal content by ICP/OES.

The effect of various concentrations of acetic acid on the solder leaching efficiency

The effect of various concentrations of acetic acid on leaching of metals from Sn–Cu solder was studied. To study this parameter, two sets of experiments were carried out. In one set, the Sn–Cu solder pieces (100 mg) were covered with 10 ml of acetic acid (10–100%) in 25 ml conical flasks separately. In another set, the Sn–Cu solder pieces were covered with 10 ml of acetic acid (2–8%) in 25 ml conical flasks separately. These flasks were incubated at 150 rpm at 30 °C for 30 h.

The effect of acetic acid volume on the solder leaching efficiency

The Sn–Cu solder pieces (100 mg) were covered with variable volume (2–10 ml) of acetic acid (80%) in 25 ml conical flasks separately, to study the effect of volume of acetic acid on metal leaching. These flasks were incubated at 150 rpm at 30 °C for 30 h.

The effect of shaking speed on the solder leaching efficiency

Effect of variable shaking speeds on metal leaching from Sn–Cu solder was studied. The Sn–Cu solder pieces (100 mg) were covered with 4 ml of acetic acid (80%) in 25 ml conical flasks separately. These flasks were incubated at 0, 50, 100, 150, and 200 rpm at 30 °C for 30 h.

The effect of temperature on the solder leaching efficiency

Temperature optimum for metal recovery from Sn–Cu solder pieces (100 mg) was quantified. The solder pieces were covered with 4 ml of acetic acid (80%) in 25 ml conical flasks separately and incubated at various temperatures (30–50 °C) and 150 rpm for 30 h.

The effect of particle size on the solder leaching efficiency

Different sizes of Sn–Cu solder pieces (3, 5, and 7 mm) were covered with 4 ml acetic acid (80%) separately. These flasks were incubated at 150 rpm at 30 °C for 30 h.

The effect of solid–liquid ratio on the solder leaching efficiency

Variable weights (100–900 mg) of Sn–Cu solder pieces were covered with 4 ml acetic acid (80%) separately. These flasks were incubated at 150 rpm at 30 °C for 30 h.

The effect of biologically produced acetic acid on the solder leaching efficiency

The 3 mm Sn–Cu solder pieces (100 mg) were covered with 4 ml *Acetobacter* culture (grown at 30 °C at 150 rpm for 24 h) and incubated at 150 rpm at 30 °C for 30 h.

After exposure to the leaching solution, the samples were analyzed for metal content by ICP/OES for all the experiments. All the experiments were carried out in three sets.

Scanning electron microscope (SEM) analysis

The solder surface before and after treatment with acetic acid was analyzed by scanning electron microscope (SEM, JSM-5610 LV). The accelerating voltage used was 0.5–30 kV at a resolution of 10 μm (10 kV WD 11 mm).

Results and discussion

The basic experimental results for the metal leaching from Sn–Cu and Sn–Cu–Ag solders by acetic acid

In the present study, two types of solders Sn–Cu and Sn–Cu–Ag were used. Acetic acid was used as a leaching solution. The solders were incubated with acetic acid for 24 h. Table 1 shows an effect of acetic acid on metal leaching from Sn–Cu and Sn–Cu–Ag solders. It was observed that 38 ($\pm$ 0.22), 37 ($\pm$ 0.09), and 40 ($\pm$ 1.00) % metal leaching was obtained for Cu, Ag, and Sn, respectively, from Sn–Cu–Ag solder in 24 h at 100:10 (mg/ml) solid-to-liquid ratio. The reaction of Sn–Cu solder with acetic acid showed more metal leaching. A 54 ($\pm$ 0.15) and 61 ($\pm$ 0.95) % metal leaching was obtained for Cu and Sn, respectively, from Sn–Cu solder in 24 h at 100:10 (mg/ml) solid-to-liquid ratio. These results are in contrast to Hocheng et al. (2014). They observed more metal leaching for Sn–Cu–Ag solder as compared to Sn–Cu solder by *A. niger* culture supernatant. This discrepancy occurred due to the organic acid used for leaching of metals. The *A. niger* culture supernatant used by Hocheng et al. (2014) contained citric acid. The present study used acetic acid as a leaching agent for leaching of metals from solders. The ability of organic acids to coordinate with metal ions decides the amount of metals leached. Thus, the formation of more stable ligand makes the metal leaching much easier (Gao et al. 2002). A number of studies have verified that the organic acid with more carboxyl groups is beneficial to the heavy metal leaching since the formed ligand is more stable. Therefore, the leaching ability of citric acid with 3 carboxyl groups is supposed to be higher than that of other organic acids having 2 carboxyl groups. The acetic acid

Table 1 Metal removal from Sn–Cu–Ag (100 mg) and Sn–Cu (100 mg) solder by 10 ml acetic acid (100%) in 24 h at 150 rpm and 30 °C

Sr. no.	Solder type	Metal extraction (%)		
		Cu	Ag	Sn
1	Sn–Cu–Ag	38.10 ($\pm$ 0.22)	37.90 ($\pm$ 0.09)	40 ($\pm$ 1.00)
2	Sn–Cu	54.00 ($\pm$ 0.15)	–	61.20 ($\pm$ 0.95)

with only one carboxyl group possesses the lowest leaching ability (Gao et al. 2002; Qin et al. 2004; Schwab et al. 2008). However, the present study showed that the acetic acid possessed the remarkable leaching ability for Sn–Cu solder. It was presumed that the acid strength used for leaching experiments and the chemical bonding of Sn and Cu might have caused the entirely different results. Since more metal leaching was achieved for Sn–Cu solder, it was used for optimization of various process parameters in further experiments.

Effect of time on the efficiency of solder leaching by acetic acid

Figure 1 illustrates an effect of leaching time on the metal leaching from the Sn–Cu solder. The metal leaching increased with an increase in incubation time. In 20 h, 49 ($\pm$ 0.58) and 54 ($\pm$ 0.26) % metal leaching was observed for Cu and Sn, respectively. The 100% metal leaching was observed for Cu and Sn, respectively, with further 30 h incubation of Sn–Cu solder with acetic acid. These results are advantageous as compared to Hocheng et al. (2014) since they required 96 h for complete metal leaching from Sn–Cu solder by A. niger culture supernatant.

Effect of acetic acid concentration on solder leaching efficiency

Two separate set of experiments were conducted to study an effect of acid concentration on the metal leaching from Sn–Cu solder. In the first set, a gradual increase in metal leaching was observed with an increase in acid concentration approaching complete metal leaching at an acid concentration 80%. It is worthy to note that 68 ($\pm$ 0.58) and 70 ($\pm$ 0.72) % metal leaching was obtained for Cu and Sn, respectively, at 10% acid concentration

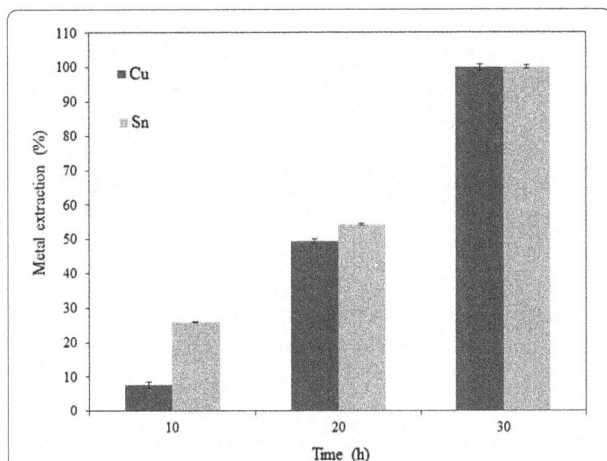

Fig. 1 Time period evaluation of metal recovery from Sn–Cu solder (100 mg) by 10 ml acetic acid (100%) at 150 rpm shaking speed and at 30 °C temperature

(Fig. 2a). These results suggest that the low concentrations of acetic acid are also effective for metal dissolution. A similar effect of an increase in the concentration of nitric acid on metal leaching from solder was reported by Yoo et al. (2012). The high concentration of leaching solution facilitates a more acidic environment which is essential for transmission and exchange of ions. This will help for the rapid extraction of metals (Kirpichtchikova et al. 2006; Mohanty and Mahindraker 2011). These results encourage the development of a two-step industrial process. The first step will produce the acetic acid using microorganisms. Further, in the second step, the acetic acid can be used for leaching of metals. In this way, a combination of hydro- and biohydrometallurgical processes can be applied for leaching of metals from solders. Several microorganisms produce acetic acid. *Acetobacter* and *Gluconacetobacter* are the main genera involved in the acetic acid production. Acetic acid bacteria (AAB) are strict aerobes that belong to Alphaproteobacteria. They have an ability to partially oxidize carbon sources into a corresponding organic compound, such as ethanol to acetic acid (Wang et al. 2013; Qi et al. 2014; Gullo et al. 2014). Acetic acid production usually requires lower capital investment and shorter start-up times. Furthermore, the raw materials (e.g., Corn, sugarcane, and sugar beet) are a renewable resource (Cheryan 2009). The acetic acid content differs depending on microbial culture; type of substrate used; and culture conditions. Several researchers reported production of 4 to 10% acetic acid (Sievers et al. 1992; Gullo et al. 2009; Mamlouk et al. 2011; Schleputz et al. 2013). Considering an amount of acetic acid produced by microorganisms, another set of experiment was carried out by varying the acetic acid concentration (2–10%). Again the metal leaching increased with an increase in acetic acid concentration. A 34 ($\pm$ 1.05) and 49 ($\pm$ 0.1)% metal leaching was achieved for Cu and Sn, respectively, using 4% acetic acid with 100/10 (mg/ml) solid-to-liquid ratio (Fig. 2b). These results suggest that both the hydro- and biohydrometallurgical processes can be established depending on the concentration of acetic acid used, for leaching of metals from Pb-free solders. Since in the present study, 80% acetic acid concentration was found optimum for leaching of 100% Cu and Sn, respectively, it was used in further studies to optimize process parameters.

The effect of acetic acid volume on the solder leaching efficiency

An experiment was carried out to study an effect of the volume of acetic acid (80%) on metal removal from Sn–Cu solder. It was found that 4 ml acetic acid was optimum for 100% metal leaching from Sn–Cu solder (100 mg) in 30 h (data not shown).

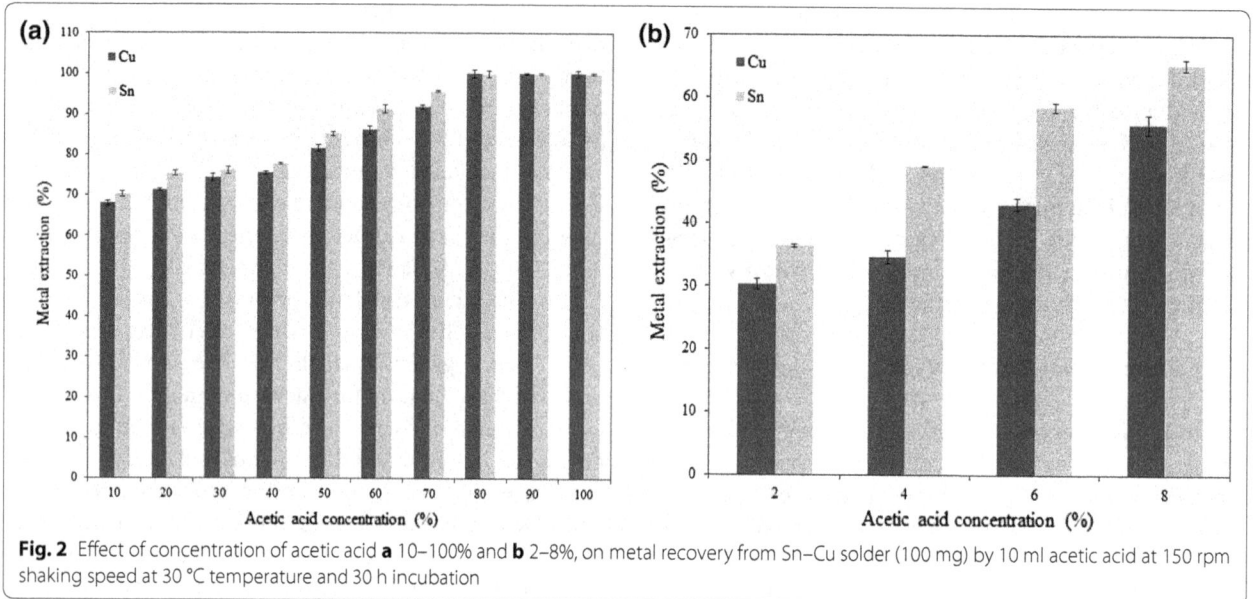

Fig. 2 Effect of concentration of acetic acid **a** 10–100% and **b** 2–8%, on metal recovery from Sn–Cu solder (100 mg) by 10 ml acetic acid at 150 rpm shaking speed at 30 °C temperature and 30 h incubation

The effects of shaking speed on solder leaching efficiency

Effect of shaking speed on metal removal from Sn–Cu solder was studied. It was observed that with an increase in shaking speed the metal leaching also increased. At static condition, only 2 ($\pm$ 0.16) and 5 ($\pm$ 0.6)% metal leaching was observed for Cu and Sn, respectively. With an increase in shaking speed to 100 rpm, the metal leaching also increased to 56 ($\pm$ 1.8) and 58 ($\pm$ 0.18)% for Cu and Sn, respectively. At shaking speeds 150 and 200 rpm, 100% metal leaching was observed for Cu and Sn, respectively (Fig. 3). Therefore, in further experiments, shaking speed of 150 rpm was used. The higher agitation speeds

keep mineral particles in a suspended state (Espiari et al. 2006). Therefore, more metal leaching was observed at 150 rpm. Hocheng et al. (2014) reported similar results, while Yoo et al. (2012) found that metal dissolution was independent of shaking speed.

The effect of temperature on solder leaching efficiency

Figure 4 shows an effect of leaching temperature on the metal leaching from Sn–Cu solders. The dissolution temperature was varied in the range of 30–50 °C, while all other parameters were kept constant. Complete metal leaching was observed at 30 °C. The metal leaching decreased with an increase in temperature after 30 °C. These results are contrary to Barakat (1999) and Yoo et al. (2012) who found that the higher temperatures yielded higher dissolution efficiencies. These authors used different acids. A loss of volume can occur for acetic acid at higher temperatures. This might have affected the metal leaching process.

The effect of particle size on the solder leaching efficiency

Different sizes of Sn–Cu solder pieces like 3, 5, and 7 mm were used to study the leaching efficiency. The obtained results suggest that the rate of metal leaching decreased with an increase in particle size due to decreased contact surface area. For the 7-mm size solder pieces, the 94 and 96% Cu and Sn leaching were observed, respectively (Fig. 5). Hasani et al. (2017) found that Pt dissolution from automotive catalytic converters increased significantly with decreasing particle size, due to smaller particles providing larger contact surface area between solid and the leaching reagent.

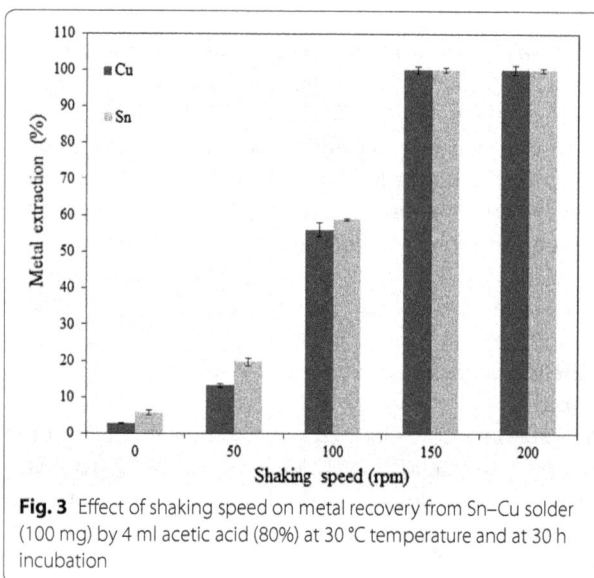

Fig. 3 Effect of shaking speed on metal recovery from Sn–Cu solder (100 mg) by 4 ml acetic acid (80%) at 30 °C temperature and at 30 h incubation

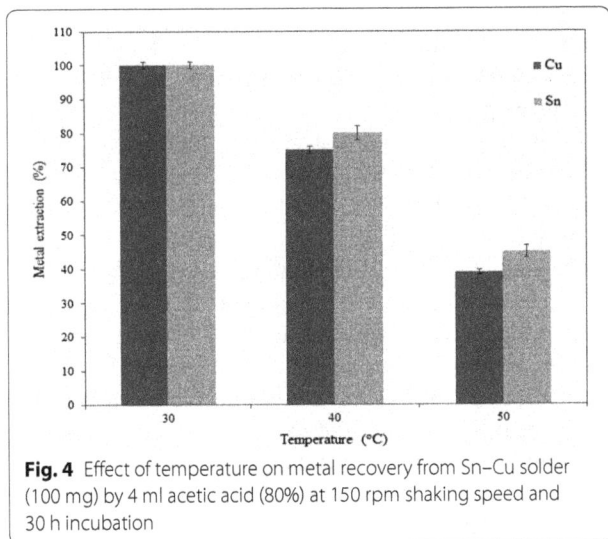

Fig. 4 Effect of temperature on metal recovery from Sn–Cu solder (100 mg) by 4 ml acetic acid (80%) at 150 rpm shaking speed and 30 h incubation

The effect of solder weight ratio on leaching efficiency

An effect of increasing solder weight, ranging from 100 to 900 mg, was investigated under the leaching conditions of 4 ml acetic acid, at 150 rpm, at 30 °C. The metal leaching decreased with an increase in solder weight. Still, it was possible to achieve 90 ($\pm$ 1.2) and 95 ($\pm$ 0.86)% metal leaching while using 500 mg solder (Fig. 6). These results are comparable with Yoo et al. (2012) and Hocheng et al. (2014). A lack of sufficient acid at an increased pulp density is responsible for the decrease in leaching efficiency (Aung and Ting 2005; Mehta et al. 2010; Hocheng et al. 2014).

The effect of microbially produced acetic acid on the solder leaching efficiency

The leaching efficiency of acetic acid was further compared with the acetic acid produced by *Acetobacter* sp. *Acetobacter* sp. was found to produce 5% acetic acid as estimated by the method described by Beheshti Maal and Shafiei (2010) (data not shown). It is in accordance with the literature which suggests the biological acetic acid production in the range of 4–10% (Sievers et al. 1992; Gullo et al. 2009; Mamlouk et al. 2011; Schleputz et al. 2013). *Acetobacter* sp. was grown for sufficient incubation period and the culture supernatant was collected. This culture supernatant was used for leaching of Cu and Sn from solder. The results suggest that 39 and 54% leaching of Cu and Sn were obtained, respectively, using *Acetobacter* sp. culture supernatant (Fig. 7). Although the obtained metal leaching efficiency using *Acetobacter* sp. culture supernatant is less as compared to 80% acetic acid, still a fairly good leaching efficiency is observed (Figs. 2a, 7). These results show that the *Acetobacter* sp. can play a significant role in bioleaching process for resource recovery not only from waste solder but also from other industrial wastes.

SEM analysis

Scanning electron microscopy was used to analyze the surface of solder during leaching process using 80% acetic acid (Fig. 8). Control samples show the smooth surface of solder particles. But the surface of solder after 10-h leaching treatment is very rough and deteriorated

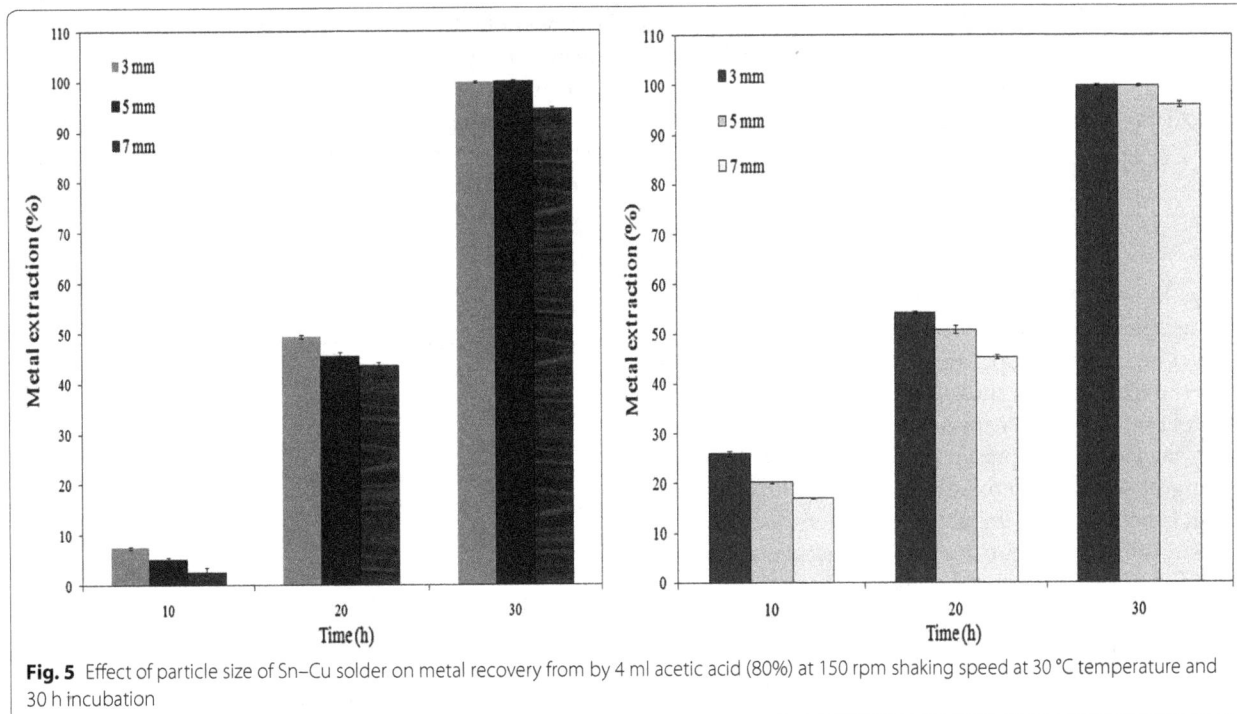

Fig. 5 Effect of particle size of Sn–Cu solder on metal recovery from by 4 ml acetic acid (80%) at 150 rpm shaking speed at 30 °C temperature and 30 h incubation

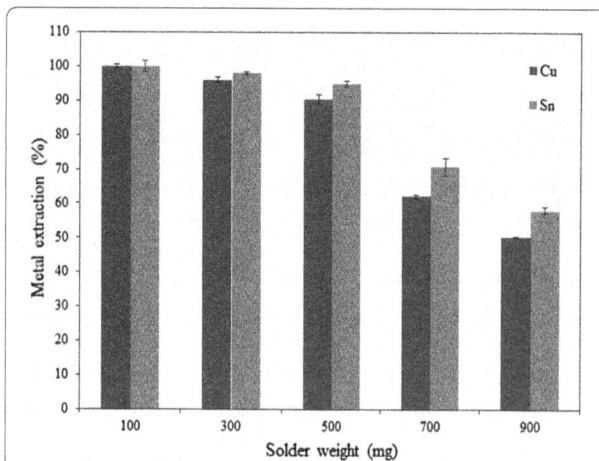

Fig. 6 Effect of increasing weight of Sn–Cu solder on metal recovery from by 4 ml acetic acid (80%) at 150 rpm shaking speed at 30 °C temperature and 30 h incubation

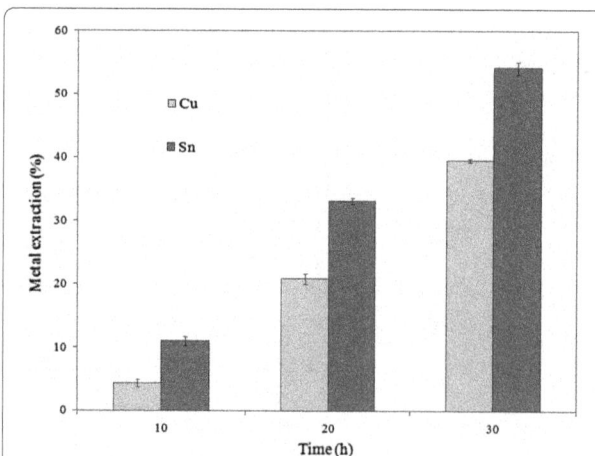

Fig. 7 Effect of microbially produced acetic acid on metal recovery from Sn–Cu solder (100 mg) at 150 rpm shaking speed and 30 h incubation

using 80% acetic acid at 150 rpm in 30 h at 100 mg/4 ml pulp density. The 10% acetic acid concentration was found useful as well for metal leaching (68 and 70% Cu and Sn, respectively).

Recent reports suggested that the alternative Pb-free solders are more harmful to the environment than Sn–Pb solder (Cheng et al. 2011). Unfortunately, most research until now is focused on leaching performance of heavy metal elements in lead-containing solder (Subramanian et al. 1995; Hin et al. 1997; Ramsay et al. 2002) and only a few studies have reported leaching behavior of alternatives solders (Yoo et al. 2012; Hocheng et al. 2014). A comparison of the present study with reported literature is shown in Table 2. Yoo et al. (2012) reported a hydrometallurgical process for leaching of metals from Sn–Cu–Ag solder. Hocheng et al. (2014) reported a bioleaching process for leaching of metals from Sn–Cu–Ag and Sn–Cu solders. A comparison of the present method with these reports clearly indicates that the efficiency of a process depends on the type of solder material, type, and concentration of lixiviant used. Yoo et al. (2012) required less time for metal leaching from Sn–Cu–Ag solder. But their process has certain limitations. A high temperature is required (75 °C) for their process. Also, they used nitric acid as a leaching agent (Table 2). This leads to a threat to the environment. NOx is released during the leaching process. This process also generates the wastewater containing high concentrations of inorganic acid (Li et al. 2010). The bioleaching process requires more time for metal leaching from Sn–Cu solder (Table 2). In this context, the present method is more advantageous since it is possible to carry out the process at 30 °C. Also, an organic acid (acetic acid) was introduced as leaching reagent. Acetic acid is a low molecular weight organic acid (weak chelating agent) (Gzar et al. 2014). Organic acids degrade easily under aerobic and anaerobic conditions as compared to HCl, HNO_3, and H_2SO_4. Due to this property of organic acids, the waste solutions remaining after the leaching process can be treated easily.

Conclusions

This study reveals an effective process to recycle the Pb-free solders. The metals in industrial waste are often present as oxides rather than sulfides. These metal oxides can only be leached by acids. Application of organic acid (acetic acid) for metal leaching provides a suitable and economic alternative. An increase in reaction time increased the leaching efficiency. The concentration of acetic acid greatly affects the leaching efficiency. The results suggest that the low concentrations of acetic acid are also effective for metal dissolution. The use of high temperature is not favorable for metal leaching process since the loss of acetic acid volume occurs at high

suggesting the removal of metals from the solder surface. Further, the 20 h sample shows the degraded and smooth surface suggesting nearly complete metal removal. Moreover, the EDS analysis shows only the presence of Sn in control samples and Cu appears to be present after treatment, suggesting the presence of Sn alone in the upper coating of solder and mixture of Sn and Cu in the core, which is released after the upper layer is eroded.

The results of present study showed that the acetic acid can be used for leaching of metals from lead-free solders. The metal leaching efficiency using acetic acid was better for Sn–Cu solder compared to Sn–Cu–Ag solder. The 100% metal leaching from Sn–Cu solder was achieved

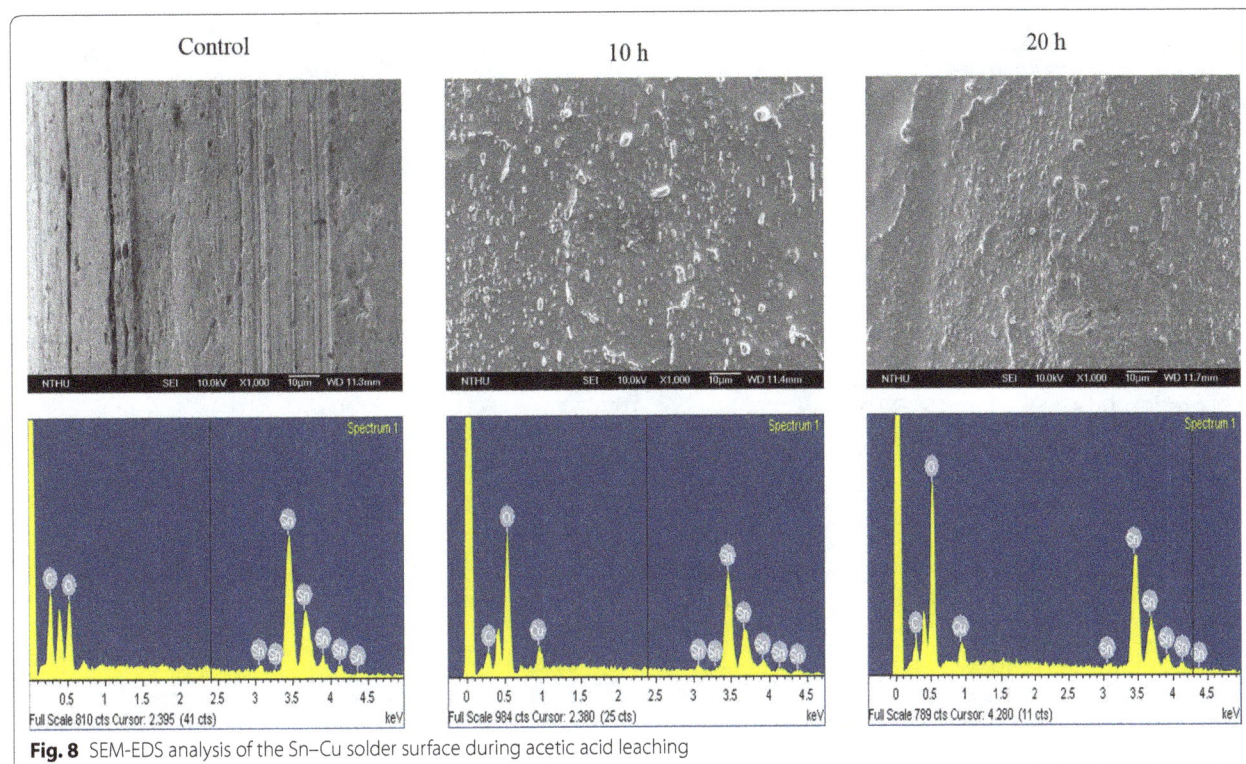

Fig. 8 SEM-EDS analysis of the Sn–Cu solder surface during acetic acid leaching

Table 2 Comparison of present study with existing literature

Sr. no.	Solder type	Lixiviant used	Concentration of lixiviant	Amount of solder used (g/L)	Temperature (°C)	Amount of metals recovered (%)	Time required (h)	References
1	Sn–Cu	Acetic acid	80%	25	30	100	30	Present study
2	Sn–Cu	*A. niger* culture supernatant containing citric acid	20 g/l	5	30	100	96	Hocheng et al. (2014)
3	Sn–Cu–Ag	*A. niger* culture supernatant containing citric acid	20 g/l	5	30	100	48	Hocheng et al. (2014)
4	Sn–Cu–Ag	Nitric acid	2 M	100	75	100	2	Yoo et al. (2012)

temperatures. But this factor is beneficial for establishing an industrial process due to less demand for energy. Also, a fair good amount of metal leaching was achieved at high pulp densities.

Authors' contributions

HH and UJ put forth the idea for the research carried out in the manuscript. SC helped in carrying out the experiments. MC was involved in editing and submission of the manuscript. All authors read and approved the final manuscript.

Author details

[1] Department of Power Mechanical Engineering, National Tsing Hua University, No. 101, Sec. 2, Kuang Fu Rd, 30013 Hsinchu, Taiwan ROC. [2] Department of Microbiology, Savitribai Phule Pune University, Pune 411007, India.

Acknowledgements and Funding

The current research is supported by National Science Council of Taiwan under Contract 100-2221-E-007-015-MY3.

Competing interests

The authors declare that they have no competing interests.

References

Abtew M, Selvaduray G (2000) Lead-free solders in microelectronics. Mater Sci Eng Res 27:95–141

Aung K, Ting Y (2005) Bioleaching of spent fluid catalytic cracking catalyst using *Aspergillus niger*. J Biotechnol 116:159–170

Barakat M (1999) Recovery of metal values from zinc solder dross. Waste Manag 19:503–507

Beheshti Maal K, Shafiei R (2010) Isolation and characterization of an *Acetobacter* strain from Iranian white–red cherry as a potential strain for cherry vinegar production in microbial biotechnology. Asian J Biotechnol 2:53–59

Biswas S, Mulaba-Bafubiandi AF (2016) Extraction of copper and cobalt from oxidized ore using organic acids hydrometallurgy conference: sustainable hydrometallurgical extraction of metals, Cape Town, 1–3 August 2016 Southern African Institute of Mining and Metallurgy

Cameselle C, Pena A (2016) Enhanced electromigration and electro-osmosis for the remediation of an agricultural soil contaminated with multiple heavy metals. Process Saf Environ Prot 104:209–217

Cheng C, Yang F, Zhao J, Wang L, Li X (2011) Leaching of heavy metal elements in solder alloys. Corros Sci 53:1738–1747

Cheryan M (2009) Acetic acid production. Appl Microbiol 1:13–17

Espiari S, Rashchi F, Sadrnezhaad S (2006) Hydrometallurgical treatment of tailings with high zinc content. Hydrometallurgy 82:54–62

Gao Y, He J, Ling W, Hu H, Liu F (2002) Effect of organic acids on Cu desorption in contaminated soils. China Environ Sci 22:244–248

Gao Y, Cheng C, Zhao J, Wang L, Li X (2012) Electrochemical corrosion of Sn-0.75Cu solder joints in NaCl solution. Trans Nonferr Metals Soc China 22:977–982

Granata G, Moscardini E, Furlani G, Pagnanelli F, Toro L (2011) Automobile shredded residue valorisation by hydrometallurgical metal recovery. J Hazard Mater 185:44–48

Gullo M, De-Vero L, Giudici P (2009) Succession of selected strains of *Acetobacter pasteurianus* and other acetic acid bacteria in traditional balsamic vinegar. Appl Environ Microbiol 75:2585–2589

Gullo M, Verzelloni E, Canonico M (2014) Aerobic submerged fermentation by acetic acid bacteria for vinegar production: process and biotechnological aspects. Process Biochem 49:1571–1579

Gzar H, Abdul-Hameed A, Yahya A (2014) Extraction of lead, cadmium and nickel from contaminated soil using acetic acid. Open J Soil Sci 4:207–214

Hasani M, Khodadadi A, Koleini SMJ, Saeedi AH, Perez Pacheco Y, Melendez AM (2017) Platinum leaching from automotive catalytic converters with aqua regia. IOP Conf Series J Phy Conf Series 786:012043

Hernandez CMF, Banz AN, Gock E (2007) Recovery of metals from Cuban nickel tailings by leaching with organic acids followed by precipitation and magnetic separation. J Hazard Mat B 139:25–30

Hin L, Torrents A, Davis A (1997) Lead corrosion control from lead, copper, lead solder, and brass coupons in drinking water employing free and combined chlorine. J Environ Sci Health 32:865–884

Hocheng H, Hong T, Jadhav U (2014) Microbial leaching of waste solder for recovery of metal. Appl Biochem Biotechnol 173:193–204

Huang K, Li J, Xu Z (2009) A novel process for recovering valuable metals from waste nickel–cadmium batteries. Environ Sci Technol 43:8974–8978

Joksic A, Katz S, Horvat M, Milacic R (2005) Comparison of single and sequential extraction procedures for assessing metal leaching from dredged coastal sediments. Water Air Soil Pollut 162:265–283

Kirpichtchikova T, Manceau A, Spadini L, Panfili F, Marcus M, Jacquet T (2006) Speciation and solubility of heavy metals in contaminated soil using X-ray microfluorescence, EXAFS spectroscopy, chemical extraction, and thermodynamic modeling. Geochim Cosmochim Acta 70:2163–2190

Lee J, Song H, Yoo J (2007) Present status of the recycling of waste electrical and electronic equipment in Korea resources. Resour Conser Recycl 50:380–397

Li L, Ge J, Chen R, Wu F, Chen S, Zhang X (2010) Environmental friendly leaching reagent for cobalt and lithium recovery from spent lithium-ion batteries. Waste Manag 30:2615–2621

Li L, Dunn J, Zhang X, Gaines L, Chen R, Wu F, Khalil A (2013) Recovery of metals from spent lithium-ion batteries with organic acids as leaching reagents and environmental assessment. J Power Sour 233:180–189

Lim S, Schoenung J (2010) Human health and ecological toxicity potentials due to heavy metal content in waste electronic devices with flat panel displays. J Hazard Mater 177:251–259

Mamlouk D, Hidalgo C, Torija M, Gullo M (2011) Evaluation and optimization of bacterial genomic DNA extraction for no-culture techniques applied to vinegars. Food Microbiol 28:1374–1379

Mehta K, Das C, Pandey B (2010) Leaching of copper, nickel and cobalt from Indian Ocean manganese nodules by *Aspergillus niger*. Hydrometallurgy 105:89–95

Merdoud O, Cameselle C, Boulakradeche MO, Akretche DE (2016) Removal of heavy metals from contaminated soil by electrodialytic remediation enhanced with organic acids. Environ Sci: Processes Impacts 18:1440–1448

Mohanty B, Mahindraker A (2011) Removal of heavy metals by screening followed by soil washing from contaminated soil. Int J Technol Eng System 2:290–293

Qi Z, Yang H, Xia X, Quan W, Wang W, Yu X (2014) Achieving high strength vinegar fermentation via regulating cellular growth status and aeration strategy. Process Biochem 49:1063–1070

Qin F, Shan X, Wei B (2004) Effects of low-molecular-weight organic acids and residence time on desorption of Cu, Cd, and Pb from soils. Chemosphere 57:253–263

Ramsay C, Lyons T, Hankin S (2002) Assessing exposure to lead in drinking water contaminated by corrosion of leaded solder. Epidemiology 13:624–630

Rhee K, Lee J, Lee C, Yang D, Chung K (1994) Proc. 2nd Int. Symp. Metall. Process. the Year 2000 and beyond and the 1994 TMS extraction and process metallurgy meeting 2, 515–526

Schleputz T, Gerhards J, Buchs J (2013) Ensuring constant oxygen supply during inoculation is essential to obtain reproducible results with obligatory aerobic acetic acid bacteria in vinegar production. Process Biochem 48:398–405

Schwab A, Zhu D, Banks M (2008) Influence of organic acids on the transport of heavy metals in soil. Chemosphere 72:986–994

Shu S, Lee J, Zhang Q, Saito F (2004) Co-grinding $LiCoO_2$ with PVC and water leaching of metal chlorides formed in ground product. Int J Mineral Process 74:373–378

Sievers M, Sellmer S, Teuber M (1992) *Acetobacter europaeus* sp. nov., a main component of industrial vinegar fermenters in central Europe. Syst Appl Microbiol 15:386–392

Suanon F, Sun Q, Dimon B, Mama D, Yu C-P (2016) Heavy metal removal from sludge with organic chelators: comparative study of *N,N*-bis(carboxymethyl) glutamic acid and citric acid. J Environ Manage 166:341–347

Subramanian K, Sastri V, Elboujdaini M (1995) Water contamination-impact of tin–lead solder. Water Res 29:1827–1836

Takahashi H, Tanaka T, Hamada M, Tohji K (2009) 10th Int. Symp. On east asian resource recycling technology, pp 761–764

Veeken A, Hamelers H (1999) Removal of heavy metals from sewage sludge by extraction with organic acids. Water Sci Technol 40:129–136

Wang Z, Yan M, Chen X, Li D, Qin L, Li Z, Yao J, Liang X (2013) Mixed culture of *Saccharomyces cerevisiae* and *Acetobacter pasteurianus* for acetic acid production. Biochem Eng J 79:41–45

Wu C, Yu D, Law C, Wang L (2004) Properties of lead-free solder alloys with rare earth element additions. Mater Sci Eng Res 44:1–44

Yang F, Zhang L, Liu ZQ, Zhong SJ, Ma J, Bao L (2016) Properties and microstructures of Sn–Bi–X lead-free solders. Adv Mat Sci Eng. doi:10.1155/2016/9265195

Yoo K, Lee J, Lee K, Kim B, Kim M, Kim S, Pandey B (2012) Recovery of Sn, Ag and Cu from waste Pb-free solder using nitric acid leaching. Mater Trans 53:2175–2180

Yoo K, Lee K, Jha MK, Lee JC, Cho K (2016) Preparation of nano-sized tin oxide powder from waste Pb-free solder by direct nitric acid leaching. J Nanosci Nanotechnol 16:11238–11241

Zeng K, Tu K (2002) Six cases of reliability study of Pb-free solder joints in electronic packaging technology. Mater Sci Eng Res 38:55–105

10

Ionic liquid-based enzyme-assisted extraction of chlorogenic acid from *Flos Lonicera Japonicae*

Yang Sun, Song Ding, He Huang and Yi Hu[*]

Abstract

Background: In recent years, ionic liquids and enzymes have been widely used in the separation and extraction processes of natural products. Chlorogenic acid (CGA) has important biological and pharmacological activities. It is significant to develop a green and efficient method to extract GCA from Flos Lonicera japonica (FLJ) by integrating the advantages of the ionic liquids and enzymes.

Results: The optimal type of enzyme and ionic liquid was screened. Pectinase in [C6mim] Br aqueous phase was demonstrated to be an ideal combination. The parameters including extraction time, extraction temperature, pH, enzyme amount, and IL concentration were optimized systematically. Scanning electronic microscopy of FLJ samples demonstrated that pectinase and ionic liquid disposal both obviously facilitated the extraction process by destroying the structure of cell wall. Circular dichroism spectroscopy showed that ionic liquid enhanced the activity of the pectinase by altering its secondary structure.

Conclusions: Compared with previous reported methods, ionic liquid-based enzyme-assisted extraction of GCA from FLJ was proved to be efficient and practical, offering a higher yield in a shorter time. A novel process was proposed for the extraction of active component from natural resources.

Keywords: Ionic liquid, Enzyme-assisted, Chlorogenic acid, *Flos Lonicera Japonicae*

Background

Flos Lonicera japonica (FLJ) is widely distributed in China and officially listed in the Chinese pharmacopoeia (Xiang and Ning 2008). It was reported to have various biological activities such as antiviral, antioxidant, and anti-inflammatory and widely used in treating exopathogenic wind-heat, epidemic febrile diseases and some infectious diseases such as SARS coronavirus and swine H1N1 flu virus (Shang et al. 2011; Wang and Weller 2006). Chlorogenic acid (CGA) is the major bioactive components of *FLJ*, and had been reported to have various biological and pharmacological activities such as antioxidant, anti-inflammatory, and anti-carcinogenic activities (Wu et al. 2012; Shin et al. 2015). A series of methods had been reported for the extraction bioactive CGA from plant materials. In conventional methods, CGA extraction was mainly achieved by water extraction (WE) (Zheng et al. 2006; Wu et al. 2006) or refluent ethanol extraction (REE) (Lan and Zhang 2005). Some disadvantage of these methods were fraction of oxidation, hydrolysis, and ionization due to excessive extraction time and relatively higher temperature (Duarte et al. 2010). Nowadays, microwave-assisted extraction (MAE) (Zhang et al. 2008) and ultrasound-assisted extraction (UAE) (Hu et al. 2009) had been developed as the feasible extraction methods. Although these two extraction methods greatly lessen the extraction time, it still can't meet the satisfaction of green chemistry for the consumption of organic solvents (Zhang et al. 2011).

In this work, we come up with a method to use ionic liquid aqueous phase to enhance the penetrability of the solvent over the extraction process and utilize the characteristics of enzymatic hydrolysis on cell wall to enhance

---
*Correspondence: huyi@njtech.edu.cn
School of Pharmaceutical Sciences, State Key Laboratory of Material-Oriented Chemical Engineering of Nanjing Tech University, Nanjing 211816, China

efficiency and extraction yield of CGA from *FLJ*. The optimized ILEAE method was investigated and the extraction mechanism by ILEAE was discussed by scanning electron microscopy and circular dichroism spectroscopy.

Experimental

Reagents and materials

Flos Lonicera japonica was obtained from Xinyi Honeysuckle Agricultural Development Co., Ltd. (Xin yi, China). *FLJ* was crushed and dried in the hot air oven before extraction. The standard sample of CGA was purchased from Aladdin (Shanghai, China) and its purity was ≥ 98%. All ionic liquids were bought from Shanghai Cheng Jie Chemical Co., Ltd. (Shanghai, China). Cellulase, pectinase, dextranase, and xylanase were purchased from Jiangsu Rui Yang Biotechnology Co., Ltd. (Jiangsu China). Other reagents were analytical grade and used without further treatment. Deionized water was used to prepare the sample solutions.

Analysis method

The chromatographic system (Dalian, China) consisted of ultimate 3000 autosampler, ultimate 3000 pump, and ultimate 3000 variable wavelength detector. Chromatographic separation was performed on a Amethyst C18-H (4.6 mm × 250 mm, 5 μm) reversed-phase column (Sepax Technologies, Inc.). For HPLC analysis, gradient elution was performed using methanol–water as the mobile phase. CGA was separated by UV detector at a wavelength of 327 nm. Injection volume was 10 ml, flow rate was 1.0 ml/min and column temperature was maintained at 25 °C. The CGA standard curve showed a good linear relationship within the scope of 0.01–1 mg/ml. The standard formula for extraction yield of CGA was $y = 0.999x + 1.296$ ($R = 0.999$, x: concentration, g/ml; y: the peak area).

Optimization of extraction systems

In the preliminary work, the components of extraction solution had been found having significant influence on extraction yield of GCA. Then, several experiments were carried out to investigate on solution components.

Screening the enzyme type

Calculated amounts of cellulase, pectinase dextranase, and xylanase were weighed 20 mg, respectively and an amount of 0.25 mol/l [C4mim]Br solvent was added to attain concentration of 0.5 mg/ml. The ionic liquid solvent pH 4.0 was prepared for further usage. 2.0 g dried samples were put into 40 ml of the enzymatic ionic liquid solvent and then put in a shaking bath at temperature of 40 °C for 2 h, individually. After the reaction, the extract was filtrated and diluted for subsequent HPLC analysis.

Screening the ionic liquid type

In the original step, five different anions including Br^-, Cl^-, $NO3^-$, BF_4^-, and PF_6^- and cations including $[C2mim]^+$, $[C4mim]^+$, $[C6mim]^+$, $[C8mim]^+$, and $[C10mim]^+$ with different length of carbon chains of glyoxaline were examined. Enzyme solution was added into 0.25 M ionic liquid solution to attain a concentration of 0.5 mg/ml, respectively. The ionic liquid solvents were prepared to adjust and attain solutions with pH of 4.0. 2.0 g dried samples were combined with 40 ml of the selected four enzymatic ionic liquid solution, individually and reacted under mixing 150 rpm (on a shaking bath) at temperatures of 40 °C for 2 h. After the reaction, the extract was filtrated and diluted for the further HPLC analysis.

Determination of pectinase activity

The pectinase activity was determined by hypoiodite sodium method according to Eq. (1). In this formula, B was $Na_2S_2O_3$ consumption of control sample; A was $Na_2S_2O_3$ consumption of control sample; N was equivalent concentration of $Na_2S_2O_3$; 0.51 was a constant meaning that 1 mg equivalent $Na_2S_2O_3$ equal to 0.51 mg equivalent free galacturonic acid; S was the total reaction liquid volume for keeping warm; E was used enzyme volume; t was the holding time; M was absorbing reaction quantity. The amount of enzyme produced by enzymatic reaction of 1 μg equivalent of galacturonic acid per minute was 1 enzyme activity unit (U).

$$\text{Pectinase activity} = \frac{(B - A) \times N \times 0.51 \times S}{E \times t \times M} \quad (1)$$

Ionic liquid-based enzyme-assisted extraction (ILEAE)

Pectinase was put into ionic liquids to attain IL-enzyme aqueous solution of concentrations (0.25–2.0 mg/ml). IL solutions were prepared to adjust pH values (3.0–5.0) for attaining the proper pH of extraction solution. 2.0 g dried sample was put into the IL-enzyme aqueous solvent and then it was put on a shaking bath for 0–4 h at 20–70 °C. After ILEAE, the extracts were filtrated and then diluted through a 0.45 μm filter for further HPLC analysis. All experimental steps were performed in duplicate.

Scanning electron micrographs (SEM)

TM3000 benchtop scanning electron microscopy (Hitachi Co., Ltd. Japan) was applied to examine the effect of IL-enzyme solutions on the structural changes of the plant cells after treatment. The samples were fixed on adhesive tape and examined under high vacuum condition at a voltage of 15.0 kV (10 μm, 1500 magnification).

Circular dichroism (CD) spectroscopy

Circular dichroism (CD) spectra were recorded on a JASCO-J810 Spectropolarimeter (Jasaco Co., Ltd. Japan) in a cell with 1 cm light path length at 25 °C. The scanning wavelength was 200–270 nm and the scanning rate was set at 50 nm/min. The CD spectra were expressed in terms of the average residue molar ellipticity in units of deg cm^2 dmol^{-1}. After obtaining CD spectra, the secondary structure of lipase was calculated using CDNN (version 2.1) which was distributed by Applied Photophysics.

Results and discussion

Sieving of enzyme and ionic liquid type

We considered the kinds of enzymes of literature commonly used in the enzymatic hydrolysis of the cell wall and finally selected four of the most representative enzymes including pectinase, cellulase, dextranase, and xylanase. The operation conditions were as follows: extraction temperature 40 °C, liquid–solid ratio 20 ml/g, enzyme concentration 0.5 mg/ml, pH 5.0, and extraction time 2 h in the 0.25 M [C4mim]Br aqueous solvent. From Fig. 1, the extraction yields were much higher than that without enzyme treatment, and the extraction yield of treatment with pectinase was higher than those with other enzyme types.

As a green solvent, ionic liquids had many potential applications and were applied in chemical industry on extraction of a variety of natural products (Jing et al. 2008; Yang et al. 2016), the component of the ionic liquids had a great effect on its physicochemical properties and may have a strong impact on the target analyte of the extracted yield. In order to sieve the better combination of anions and cations, both of them were investigated, respectively. We investigated several common anions on the basis of literature and finally screened out five anions including Br$^-$, Cl$^-$, NO$_3^-$, BF$_4^-$, and PF$_6^-$ (Zhang et al. 2011; Ma et al. 2011; Merlet et al. 2017). As can be seen from Fig. 2(a), four kinds of hydrophilic anions showed better extraction effect, while the extraction yield of hydrophobic PF6$^-$ was relatively poor. In the four hydrophilic anions, the halogen ions exhibit the best extraction yield especially Br$^-$, which may be due to the combination of the halogen ions with the H$^+$ in the cellulose to disintegrate the internal structure of cellulose. This result showed that the extraction rate of CGA depends on the type of anion which has been reported (Liu et al. 2016). Based the same cation of Br$^-$, different groups of cations including [C2mim]$^+$, [C4mim]$^+$, [C6mim]$^+$, [C8mim]$^+$, and [C10mim]$^+$ were examined and the results of the evaluation are shown in Fig. 2. From Fig. 2, the extraction yield of CGA from ethyl to hexyl is obviously increased,

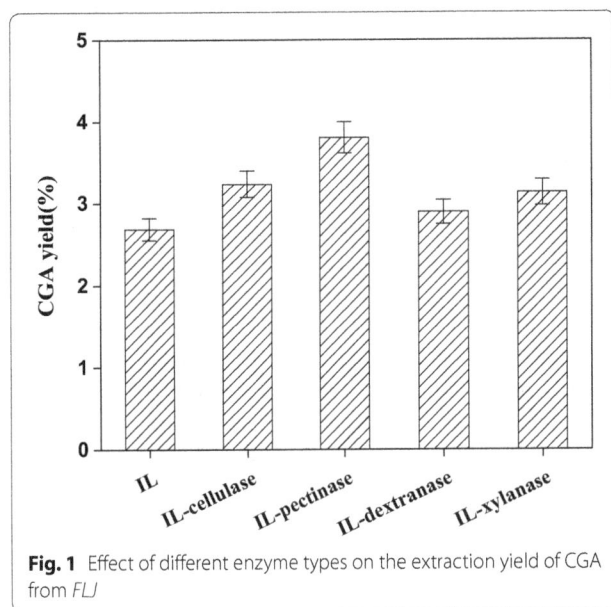

Fig. 1 Effect of different enzyme types on the extraction yield of CGA from *FLJ*

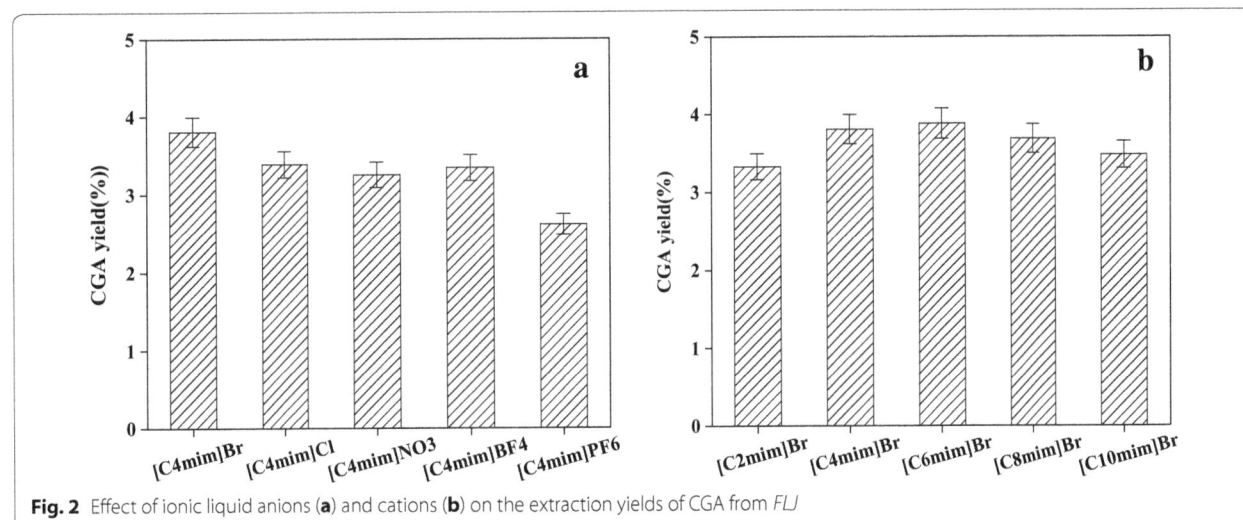

Fig. 2 Effect of ionic liquid anions (**a**) and cations (**b**) on the extraction yields of CGA from *FLJ*

it may be because CGA can be better dissolved in the lipophilic IL; the extraction yield of CGA decreased from hexyl to decylate may be due to the hydrogen bond and hydrophobic interaction of ionic liquids affecting the dissolution of CGA.

Optimization of the ILEAE process

In the initial process, we investigated the enzyme types and ionic liquid types. Subsequently, important factors affecting extraction yield including extraction time, extraction temperature, pH, enzyme concentration, and IL concentration were studied.

Effect of extraction time

The extraction time considerably affected the extraction rate of CGA. Under the conditions of 0.5 mg/ml pectinase and 0.25 M [C4mim]Br treatment, pH 3.0, and temperature 40 °C, the influence of extraction time on yields was studied. The results in Fig. 3a illustrated that when the extraction time increased from 20 to 40 min, the extraction yields increased obviously. And yet the time was changed from 40 min to 4 h, the results remain essentially unchanged. This may be due to the isomerization of the ester groups in the GCA structure and also explained that the long extraction time would make GCA hydrolyzed, which would lead to the decreased GCA extraction yield. It was reported that pectinase can catalyze hydrolysis of ester linkage and using *Eucommia ulmoides* as the raw material and IL-pectinase as the extraction medium, the extraction rate of CGA was close to 0 under 4 h in 0.25 M [C4mim]Br aqueous solution. We have carried on the experiment to verify, and discovered that long periods of extraction time can cause CGA to be broken down by the HPLC result.

Therefore, 40 min was chosen as appropriate time for subsequent experiments.

Effect of extraction temperature

To select a proper extraction temperature, we did a series of experiments to complete extraction process. The influence of extraction time on CGA extraction yield was studied under the conditions of 0.5 mg/ml pectinase and 0.25 M [C4mim]Br treatment, pH 3.0, and extraction time 40 min, As shown in Fig. 3b, the extraction yield reached the maximum at the optimum temperature of 40 °C and decreased with the increase of temperature. It may be the effect of temperature on the catalytic activity of the enzyme resulting in a decrease in the extraction rate of GCA.

Effect of pH value

Holding the conditions of 0.5 mg/ml pectinase and 0.25 M [C4mim]Br treatment, extraction time 40 min and extraction temperature 40 °C, the influence of pH value on CGA extraction yield was investigated. As shown in Fig. 3c, pectinase showed ability to collapse within a wide pH range. It can be noticed that the extraction yields varied unregularly in the range of pH 3–5. This result showed the activity of pectinase was susceptible to solvent pH. The extraction yield of CGA reach peak value around pH 4.0.

Effect of enzyme concentration

As shown in Fig. 3d, the influence of different enzyme concentrations on the extraction yields was investigated under the conditions of 0.25 M [C4mim]Br treatment, extraction time 40 min, pH 4.0, and temperature 40 °C. At the concentration of 1 ml/mg, the extraction yield reached the maximum, then the extraction yield of CGA was almost unchanged. Higher concentrations than 1 mg/ml couldn't increase extraction yields. Thus, 1 mg/ml was chosen as appropriate concentration for the extraction process.

Effect of IL concentration

As shown in Fig. 3e, under the conditions of 1 mg/ml pectinase treatment, pH 4.0, extraction time 40 min, and temperature 40 °C, the influence of different ionic liquid concentration on the yields of CGA was evaluated. Concentration of 0.75 mol/l gave a peak extraction yield about 6.06%, indicating that a concentration of 0.75 mol/l offered adequate amounts of pectinase for the disruption of cell wall. Higher concentrations than 0.75 mol/l did not increase yields. Thus, 0.75 mol/l was chosen as appropriate concentration for the extraction process.

Comparison with the reference methods

In order to compare our method with the previous reported methods, some representative literatures data were listed in Table 1. Obviously, our method exhibits a good extraction efficiency and higher yields were obtained.

Scanning electron micrographs (SEM) of different samples

In ILEAE method, the *FLJ* samples were detected by SEM for revealing the effect of ILEAE process and extraction mechanism on *FLJ* structure change. Figure 4 showed the micrographs of raw sample (a), the sample (b) treated only with pectinase, the sample (c) treated only with ionic liquids, and the sample (d) treated after ILEAE 2 h, respectively. From the sample (a), we found untreated samples showed a clear structure: thicker cell wall and intact cell structure. After pectinase treatment (b), the cell surface had undergone minor changes and some of the tissue damage occurred with protrusions in the middle of the leaves. From the sample (c), we found that after ionic liquid treatment, cell wall surface was significantly

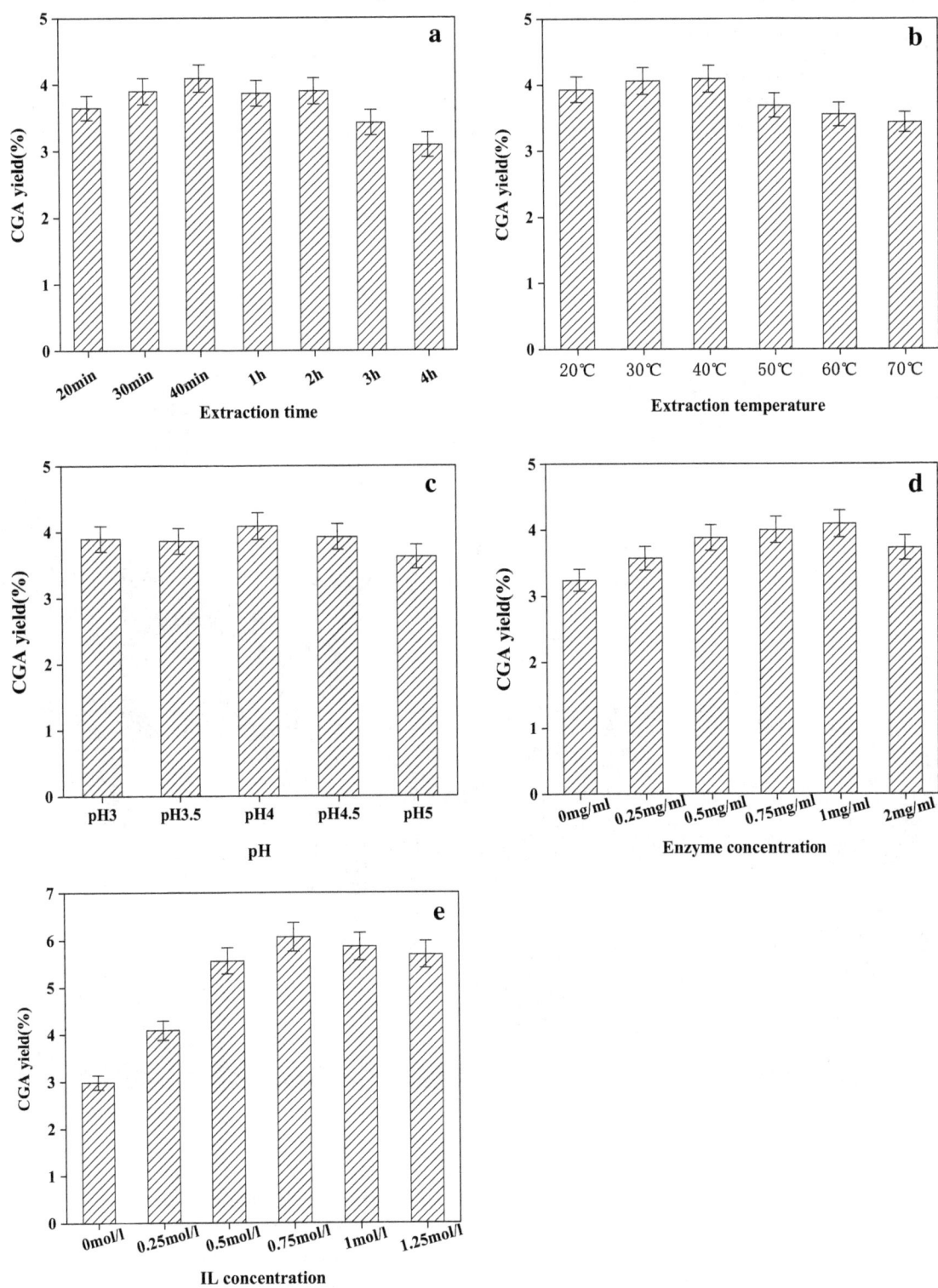

Fig. 3 Effects of extraction time (**a**), extraction temperature (**b**), pH (**c**), enzyme concentration (**d**), IL concentration (**e**) on the extraction yields of CGA from *FLJ*

Table 1 Comparison of ILEAE with the reference and conventional methods

Method	Optimized extraction parameters	CGA yield (%)
WE	Liquid ratio 10 ml/g, extraction temperature 75 °C, extraction time 1 h	2.29 (Yu et al. 2010)
	Liquid ratio 15 ml/g, extraction temperature 70 °C, extraction time 1 h	3.8 (Li et al. 2011)
REE	Ethanol concentration 80% (v/v),extraction temperature 70 °C, extraction time 80 min	3.63 (Li and Hou 2013)
	Ethanol concentration 75% (v/v), flocculation time 24 min, reflux time 1.5 h	3.81 (Xiao and Li 2011)
UAE	Soaking time 12 h, ultrasonic time 45 min, ethanol concentration 60%	5.62 (Lu et al. 2015)
MAE	Microwave power 300 W, microwave time 3 min, extraction temperature 60 °C, ethanol concentration 70% (v/v), extraction time 20 min	5.16 (Cao et al. 2008)
EAE	Dosage of pectinase 0.5%, operating temperature 45 °C, extraction time 120 min	4.55
ILEAE	0.75 M [C6mim]Br, extraction time 40 min, extraction temperature 70 °C, pH 4.0, 1 mg/ml pectinase	6.06

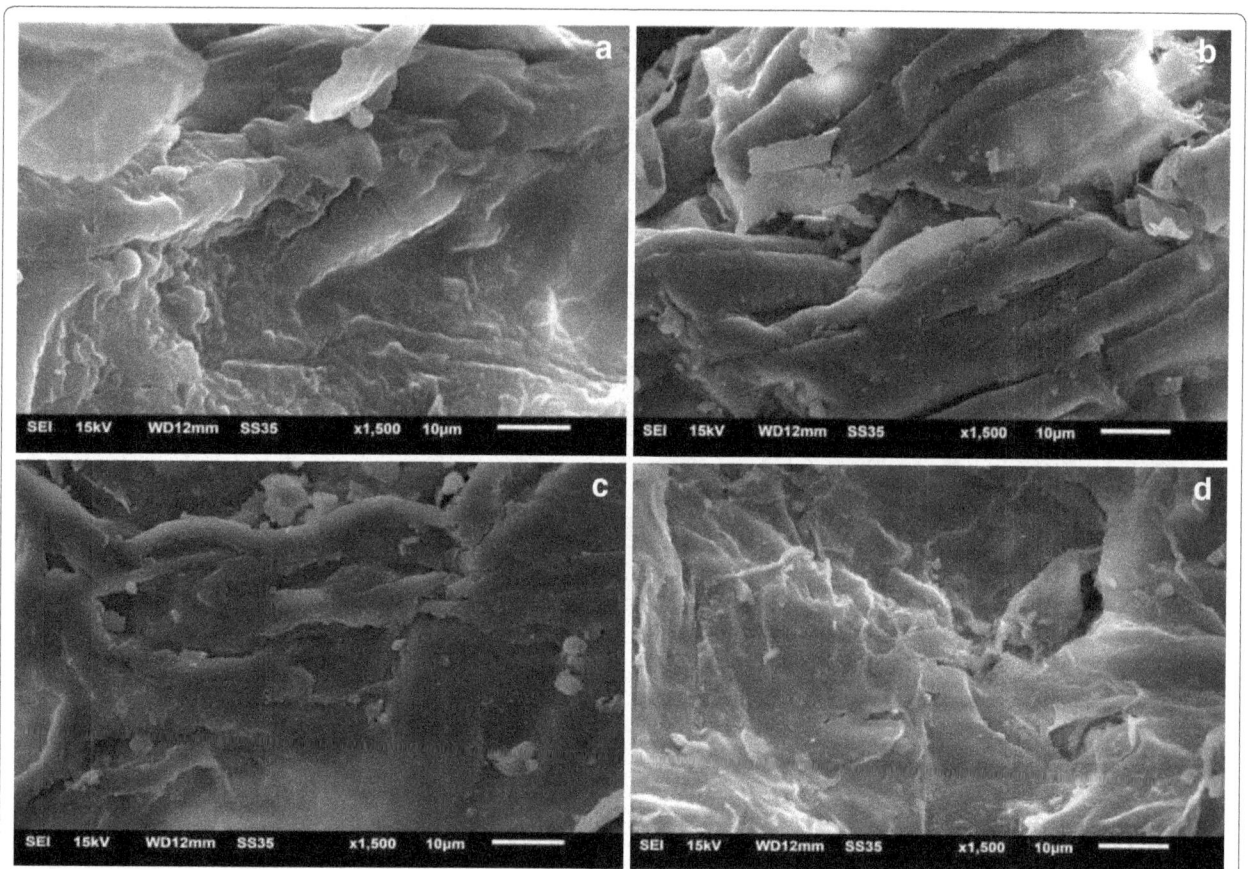

Fig. 4 The scanning electron micrographs of *Flos Lonicera Japonicae*. Untreated sample (**a**), treated only with pectinase (**b**), treated only with ionic liquids (**c**) and treated with ILEAE (**d**)

thinned, and the cell structure changed, which led to the target product exposed to the extraction solution. In the Fig. 4(d), after the mixed treatment of IL-enzyme, the cell wall surface tissue wrinkle serious and the cell wall thinner than a, b, c with the protrusions of the leaves disappeared. There is always a cell uplift in the leaves of the plant material, and the degree of protrusions in the treated sample can represent the dissolution and extraction capabilities of the liquid solution.

Circular dichroism (CD) spectroscopy analysis

Figure 5 showed the characteristic shape of the CD signal of the original pectinase, aqueous solution pectinase, and pectinase treated with ionic liquids which reflect the change of the secondary structure of the protein. Compared with the original pectinase, the sample treated with aqueous solution and the sample treated with ionic liquids had varying degrees of change on characteristic shape, which may cause changes of

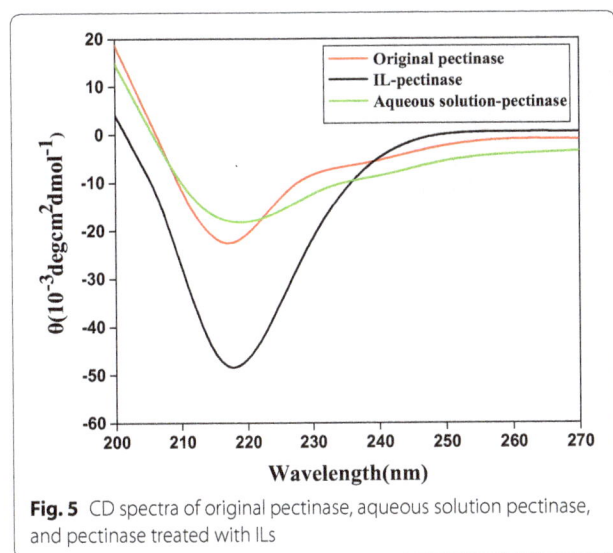

Fig. 5 CD spectra of original pectinase, aqueous solution pectinase, and pectinase treated with ILs

the secondary structure of the protein and the tertiary structure.

The results shown in Table 2 indicated the percentage change of α-helix, β-sheet, β-turn, and random crimp content in the secondary structure before and after pectinase treatment. Compared with the activity of the original pectinase, we found that the activity of aqueous solution pectinase decreased by 13.5% and the activity of IL- pectinase increased by 23.5%, respectively, and the CGA yield improved obviously. The results showed that the change of pectinase secondary structure resulted in enzyme activity increase which also make GCA extraction yield increased.

Conclusions

In this work of ILEAE, pectinase was successfully combined with ionic liquids for the extraction of CGA from *FLJ*. As far as we know, this is far-reaching significance investigation on developing a method for extraction of CGA from *FLJ* using ionic liquid-based enzyme-assisted solvent as extraction medium. From the final test results,

the CGA yield was enhanced obviously with the addition of pectinase and ionic liquids. Compared to reported extraction methods, the optimal ILEAE method reduced the extraction time significantly and offered higher extraction yield. By taking into account the practical application, the ILEAE showed great prospect and provided new ideas for the extraction and separation of other natural products.

Authors' contributions
YS wrote the article; YS and SD conducted the experiment; YH conceived the research; YH and HH provided experimental materials and guidance. All authors read and approved the final manuscript.

Acknowledgements
The authors thank anonymous reviewers for their insightful comments and careful corrections. We also acknowledge the support from the self-owned research project from key laboratory of material- oriented chemical engineering (Grant No. ZK201603), National Natural Science Foundation of China (21676143) and Qing Lan Project of Jiang Su Province and Program for Innovative Research Team in University of Jiangsu Province.

Competing interests
The authors declare that they have no competing interests.

Funding
The self-owned research project from key laboratory of material-oriented chemical engineering (Grant No. ZK201603), National Natural Science Foundation of China (21676143) and Qing Lan Project.

References
Cao Y, Li CJ, Xia ZN, Xu YQ, Zhang YM (2008) Optimization of ultrasound extraction technology of chlorogenic acid in Flos Lonicerae. Lishizhen Med Mater Med Res 19(12):2857–2858
Duarte GS, Pereira AA, Farah A (2010) Chlorogenic acids and other relevant compounds in Brazilian coffees processed by semi-dry and wet postharvesting methods. Food Chem 118(3):851–855
Hu F, Deng C, Liu Y, Zhang X (2009) Quantitative determination of chlorogenic acid in Honeysuckle using microwave-assisted extraction followed by nano-LC-ESI mass spectrometry. Talanta 77(4):1299
Jing F, Fan Y, Pei Y, Wu K, Wang J, Fan M (2008) Solvent extraction of selected endocrine-disrupting phenols using ionic liquids. Sep Purif Technol 61(3):324–331

Table 2 The percentage of secondary structure elements and enzyme activity

Sample	α-helix (%)	β-sheet (%)	β-turn (%)	Random coil (%)	Enzyme activity (u/mg)
Original pectinase	11	33	20	36	100[a]
Aqueous solution pectinase	9.6	29.2	21	40.2	86.1[b]
IL-pectinase	20	19	24	37	123.5[c]

[a] Untreated

[b] Treated under a condition of extraction time 40 min, extraction temperature 70 °C, pH 4.0, 1 mg/ml pectinase, and the solution was aqueous solution

[c] Treated under a condition of 0.75 M [C6mim]Br, extraction time 40 min, extraction temperature 70 °C, pH 4.0, 1 mg/ml pectinase, and the solution was IL aqueous solution

Lan WU, Zhang ZS (2005) Extraction and examination of chlorogenic acid from flos lonicerae. Food Sci 26(6):130–134

Li WL, Hou T (2013) Study on microwave assisted extraction of chlorogenic acid from Flos Lonicerae. Feed Anim Husb new feed 5:35–38

Li Z, Li N, Zhao TH (2011) Optimization of extraction process of chlorogenic acid from honeysuckle by orthogonal test. J Chengde Med College 28(1):12–14

Liu T, Sui X, Li L, Jie Z, Xin L, Li W (2016) Application of ionic liquids based enzyme-assisted extraction of chlorogenic acid from Eucommia ulmoides leaves. Anal Chim Acta 903:91–99

Lu DW, Zhang T, Li HY (2015) Study on extraction process of chlorogenic acid in Flos Lonicerae. Guangdong Chem Ind 42(19):40–41

Ma CH, Liu TT, Yang L, Zu YG, Chen X, Zhang L (2011) Ionic liquid-based microwave-assisted extraction of essential oil and biphenyl cyclooctene lignans from Schisandra chinensis Baill fruits. J Chromatogr A 1218(48):8573

Merlet G, Uribe F, Aravena C, Rodríguez M, Cabezas R, Quijada-Maldonado E (2017) Separation of fermentation products from ABE mixtures by perstraction using hydrophobic ionic liquids as extractants. J Membr Sci 537:333–343

Shang X, Pan H, Li M, Miao X, Ding H (2011) Lonicera japonica Thunb: ethnopharmacology, phytochemistry and pharmacology of an important traditional Chinese medicine. J Ethnopharmacol 138(1):1–21

Shin HS, Satsu H, Bae MJ, Zhao Z, Ogiwara H, Totsuka M, Shimizu M (2015) Anti-inflammatory effect of chlorogenic acid on the IL-8 production in Caco-2 cells and the dextran sulphate sodium-induced colitis symptoms in C57BL/6 mice. Food Chem 168:167–175

Wang L, Weller CL (2006) Recent advances in extraction of nutraceuticals from plants. Trends Food Sci Technol 17(6):300–312

Wu JW, Luo YJ, Wu X, Li XM (2006) Study on extraction process of honeysuckle aqueous extracts to optimize flocculation by uniform design method. Chin J Vet Drug 40(8):23–25

Wu SB, Meyer RS, Whitaker BD, Litt A, Kennelly EJ (2012) Antioxidant glucosylated caffeoylquinic acid derivatives in the invasive tropical soda apple, solanum viarum. J Nat Prod 75(12):2246

Xiang Z, Ning Z (2008) Scavenging and antioxidant properties of compound derived from chlorogenic acid in South-China honeysuckle. LWT Food Sci Technol 41(7):1189–1203

Xiao WP, Li J (2011) Extraction and determination of chlorogenic acid in Flos Lonicerae. J Anhui Agric Sci 39(35):21675–21676

Yang Z, Tan Z, Li F, Li X (2016) An effective method for the extraction and purification of chlorogenic acid from ramie (Boehmeria nivea L.) leaves using acidic ionic liquids. Ind Crops Prod 89:78–86

Yu J, Kuang LH, Lin FS (2010) Study on the water extraction technology of honeysuckle by orthogonal experiment. Drugs Clin 25(4):293–296

Zhang B, Yang R, Liu CZ (2008) Microwave-assisted extraction of chlorogenic acid from flower buds of Lonicera japonica, Thunb. Sep Purif Technol 62(2):480–483

Zhang L, Liu J, Zhang P, Yan S, He X, Chen F (2011) Ionic liquid-based ultrasound-assisted extraction of chlorogenic acid from Lonicera japonica, Thunb. Chromatographia 73(1–2):129–133

Zheng XQ, Jiang JF, Liu XL, Sun XY (2006) Extraction of chlorogenic acid from sunflower seeds by water solution and alcohol sedimentation. Food Sci 27(1):159–161

11

Plastic bag as horizontal photobioreactor on rocking platform driven by water power for culture of alkalihalophilic cyanobacterium

He Zhu, Chenba Zhu, Longyan Cheng and Zhanyou Chi[*]

Abstract

Background: Mixing in traditional algae culture system consumes intensive electricity. This should be replaced by nature force to reduce energy cost and, more importantly, to realize positive energy balance of algal biofuel production. This study aims to develop a horizontal photobioreactor, in which mixing can be provided with rocking movement driven by nature force.

Results: Simple boxes were used as small-scale horizontal photobioreactors on a rocking platform for culture of alkalihalophilic *Euhalothece* sp. ZM001. There was no CO_2 gas bubbling since 1.0 M $NaHCO_3$ supplied sufficient inorganic carbon in it. Effect of culture depth, rocking cycle, and light intensity to algal biomass production, pH change, and DO accumulation were investigated in this system. Biomass concentration of 2.73 g/L was achieved in culture with 2.5 cm depth, and maximum productivity of 17.06 g/m^2/day was obtained in culture with 10 cm depth. k_La in PBR with different culture depths and rocking cycles was measured, and it was from 0.57 to 33.49 h^{-1}, showing great variation. To test this system at large scale, a plastic bag with a surface area of 1 m^2 was placed on a rocking platform driven by water power, and it resulted in a biomass concentration of 1.88 g/L.

Conclusion: These results proved feasibility of a novel photobioreactor system driven by nature force, as well as low cost of manufacturing, and easy scaling-up.

Keywords: Microalgae, Bicarbonate, Mixing, Photobioreactor, Nature force

Background

Microalgae are promising for biofuel production to replace petroleum, but limited by high production cost. Photobioreactor (PBR) is the key component of algae culture system, and it is crucial for cost-effective algae culture process development. Traditional PBRs such as horizontal tubular and vertical flat panel PBRs have many drawbacks, including (1) high cost for manufacturing: the tubular PBR cost at small scale was 2400 €/m^2, and even if reduce this to 750 €/m^2 (or 5 €/L) at large scale, it would still account for 94% of total major equipment cost, and result in algal biomass production cost of 12.6 €/kg (Acien et al. 2012); (2) high energy consumption for mixing: the energy consumption for blower in flat panel PBR was 49 w/m^3, which accounted for 68% of total operating energy input (Tredici et al. 2015), and tubular PBR driven with centrifuge pump has even higher energy consumption rate (Jorquera et al. 2010); (3) high labor cost for installation and maintenance (Acien et al. 2012; Norsker et al. 2011); (4) difficulties to be scaled up (Acien Fernandez et al. 2013). These drawbacks directly or indirectly lead to microalgae's high production cost. This is corresponding to economical analysis on commercial scale microalgae culture, which revealed that major production cost come from photobioreactor (PBR) manufacturing, power consumption, CO_2 supply, and labor cost (Acien et al. 2012; Norsker et al. 2011; Tredici et al.

*Correspondence: chizhy@dlut.edu.cn
School of Life Science and Biotechnology, Dalian University of Technology, Dalian 116024, China

2015). To reduce these costs, innovative algal culture systems need to be developed, and it should have low cost for both PBR manufacturing and energy consumption for mixing. Besides, labor utilization for PBR installation and maintenance should be reduced to minimum, and inorganic carbon should be supplied with cost-effective way.

To explore a better approach for supplying carbon to microalgae culture, bicarbonate-based integrated carbon capture and algae production system (BICCAPS) has been developed in our previous study (Chi et al. 2011, 2013). This system uses high concentration of bicarbonate derived from carbon capture to provide sufficient inorganic carbon, and no continuous CO_2 sparging in algae culture process is necessary. This can avoid a series of problems, including high cost of CO_2 purification and transportation, difficulty of CO_2 storage, and low utilization rate due to outgassing. Another advantage of this system is that configuration of photobioreactor for BIC-CAPS can be very simple, since no sparging system is required. For example, a tissue culture flask (T-flask) was used as a simple PBR, and it well supported growth of alkalihalophilic cyanobacterium *Euhalothece* sp. in our previous study (Chisti 2013). If this T-flask is scaled up, it is actually a horizontal PBR, which is a hybrid design of PBR and open pond, and shares advantages from both, including low manufacturing cost, closed system, short light path, and easy scaling-up (Dogaris et al. 2015). Due to these merits, horizontal PBR has great potential application in future algae culture, and it is an excellent candidate for BICCAPS at large scale.

To develop an efficient algal culture system, sufficient mixing is obligatory in that it not only enhances mass transfer, but also improves frequency of algal cells' shifting between dark and light, which is crucial for improving photosynthesis efficiency (Abu-Ghosh et al. 2016). However, it consumes intensive energy to provide sufficient mixing. In traditional PBR systems, mixing is the largest portion of energy consumption, and it contributes significantly to total operating cost. For example, power consumption in horizontal tubular PBR and vertical flat panel PBR was estimated as 2500 and 53 w/m^3, respectively (Jorquera et al. 2010). Although raceway pond has a low estimated energy cost of 3.7 w/m^3 (Jorquera et al. 2010), this is actually not enough to provide sufficient mixing, which results in poor mass transfer, as well as limited cell movement in vertical direction (Mendoza et al. 2013a, b; de Godos et al. 2014). Intensive energy consumption for mixing also results in negative energy balance of algal biomass production, which is a great hurdle in commercialization of algal biofuel. Using electricity generated from renewable energy sources, such as solar or wind, may address this problem. However, it will

result in biofuel with even higher cost than using electricity from fossil fuel, since electricity generated from solar energy or wind is still more expensive than that of fossil fuel combustion by far (Davis and Martín 2014). Thus, it is obligatory to develop PBR systems driven by renewable energy with a cost-effective way.

In our previous study with T-flask as the PBR for BIC-CAPS, mixing was provided with an orbital shaker, and it well supported algae growth. This is obvious not practical for culture at large scale. However, it may be feasible to drive rocking movement of this simple PBR at large scale with nature forces such as falling water, wave, or wind (Kim et al. 2016). To investigate if this rocking movement can provide sufficient mixing, the effect of rocking cycle and culture depth was studied at different light intensities with small-scale horizontal PBR placed on a see-saw rocker. Biomass production, oxygen accumulation, and pH change were measured at these conditions. In addition, mass transfer coefficient (k_La) in this horizontal PBR was measured. At 1.0 m^2 scale, a plastic bag was used as low-cost PBR, which was placed on a rocking platform driven by water power. This system supported algae growth, indicating that sufficient mixing may be provided by horizontal PBR's rocking movement. This proved feasibility of a novel system with low cost of both PBR manufacturing and energy consumption.

Methods

Strain and medium

The alkalihalophilic cyanobacterium strain *Euhalothece* sp. ZM001 was purchased from Culture Collection of Autotrophic Organisms (CCALA), Academy of Sciences of the Czech Republic. The seed was cultured in M-medium, which contains Na_2CO_3, 53 g/L; $NaHCO_3$, 42 g/L; NaCl, 50 g/L; KCl, 2 g/L; Na_2SO_4, 1.4 g/L; KNO_3, 2.5 g/L; $K_2HPO_4 \cdot 3H_2O$, 0.5 g/L; $FeCl_3$, 0.0003 g/L; EDTA, 0.0005 g/L; and 1 mL of the A5 trace element solution (Mikhodyuk et al. 2008). The pH of M-medium was 10.5. It was inoculated into concentrated medium (C-medium) for cultivation, which contains Na_2CO_3, 53 g/L; $NaHCO_3$, 42 g/L; KCl, 10 g/L; Na_2SO_4, 7 g/L; KNO_3, 12.5 g/L; $K_2HPO_4 \cdot 3H_2O$, 2.5 g/L; $FeCl_3$, 0.0015 g/L; EDTA, 0.0025 g/L; and 5 mL of the A5 trace element solution. The pH of C-medium was 9.5.

Small-scale horizontal PBR for algae culture

Simple square boxes made from Acryline at the size of 12 cm × 12 cm were used as horizontal PBRs in this experiment, which were placed on a see-saw rocker to provide mixing. To investigate different mixing intensities, the see-saw rocker was controlled to finish one up/down rocking cycle every 1 s or every 2 s. Culture depth of 2.5, 5.0, 7.5, and 10 cm were investigated with this system. To prevent evaporation, the reactor was sealed with transparent membrane on the top, with two holes left to release oxygen. LED panel with 70% red and 30% blue light was placed above the PBRs. Light intensities investigated were 70, 135, and 340 µmol/m^2/s, which were measured at the surface of culture medium. The whole culture system was placed in an incubator, which was controlled at 35 °C (Chi et al. 2013).

Dry cell weight measurement

To measure biomass concentration, 40 mL of sampled cell suspension was centrifuged at 10,000 rpm for 15 min. The cell pellets were then washed twice with 1% (w/v) salt water, to avoid osmotic pressure change caused by fresh water. After that, this salt water was added to the cell pellets to make a final volume of 5 mL, and placed in a petri dish for drying at 105 °C for 4 h. Then, dry cell weight was calculated by subtracting salt weight from the total dry weight (Chi et al. 2014).

Dissolved oxygen (DO) measurement

A non-invasive Fibox 4 fiber optic oxygen transmitter (PreSens Precision Sensing GmbH, Germany) was used to measure DO in PBRs. Since DO value in some experiments were read as over saturation (higher than 34 mg/L), the accuracy of this instrument was verified by following procedure: the culture medium used in this experiment was bubbled with pure oxygen for half an hour, and the reading DO value was 32.7 mg/L. Thus, this measure error was 4.0%, and acceptable.

To measure DO at different depths in the PBR, oxygen sensor spots were placed on the side wall at different distances from PBR bottom. For the culture depth of 2.5 cm, one sensor spot was placed 1.25 cm above the bottom. For culture depth of 5.0, 7.5, and 10 cm, two sensor spots were placed: one was 1.0 cm below the medium surface, and the other was 1.0 cm above PBR bottom.

$k_L a$ measurement

$k_L a$ was determined with a dynamic method. PBR was filled with culture medium, and oxygen was bubbled through a diffuser at the bottom of PBR. It kept bubbling until DO reached 27.3 mg/L. Change of DO over time was measured with a probe at the liquid surface until DO reached 13.6 mg/L. Change of DO over time was assumed to be a function of $k_L a$ and driving force $\left([O_2^*] - [O_2]\right)$ in Eq. 1.

$$\frac{d[O_2]}{dt} = k_L a\left([O_2^*] - [O_2]\right). \tag{1}$$

By integrating this equation between time zero and time t, the mass transfer coefficient $k_L a$ can be obtained by Eq. 2.

$$\ln\left(\frac{[O_2^*] - [O_2]_{t=0}}{[O_2^*] - [O_2]}\right) = k_L a \cdot t. \tag{2}$$

Test of oxygen tolerance capability of *Euhalothece* sp.

Euhalothece sp. ZM001 was cultured in bubble column reactors with diameter of 5 cm, and the total volume was 500 mL. The room temperature was maintained at 25 °C with air conditioner. The cultures were placed in front of a LED panel with white light, and the light intensity was 135 µmol/m^2/s. Three reactors bubbled with pure oxygen, and the other three bubbled with air. Biomass production, DO, and pH were measured during the culture process.

Culture with plastic bag as horizontal PBR at 1 m^2 scale

A plastic bag with a size of 1 m × 1 m was used as the horizontal PBR. It was placed on a rocking platform driven with water power (Fig. 1). This platform has a size of 1 m × 1 m × 0.3 m (*L: W: H*), with steel as frame and thick PVC membrane as bottom and side walls. A water tank was attached to one side of its frame, and water keeps flowing into the tank. When this water tank is half-full, this side of platform was pushed down by the weight of water. Once water tank is full, the water was discharged until empty, which is controlled by a pressure-sensing valve. Then, this side of rocking platform was lifted up. In this way, this platform generates a continuous rocking movement. The rocking cycle of this platform was measured as 60 s. Culture depth of 5.0 cm was used in the outdoor culture with plastic bag, and this culture was conducted from September 6 to September 26, 2016 at Dalian City, China.

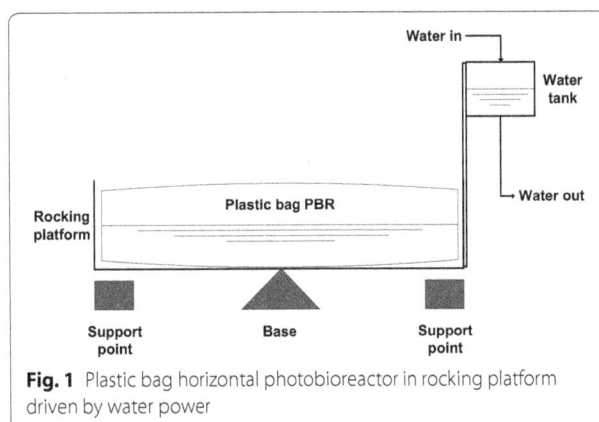

Fig. 1 Plastic bag horizontal photobioreactor in rocking platform driven by water power

Results

Mixing driven by rocking movement of horizontal PBR is quite different from traditional way such as paddle wheel or circulation pump. Rocking cycle should be a decisive factor to mixing. Thus, the effect of rocking cycle to biomass production and DO accumulation was investigated with small-scale horizontal PBRs at different light intensities. Culture depth of 2.5, 5.0, 7.5, and 10.0 cm were used in these experiments since it provides light path at the same level as traditional PBRs.

Culture at high light intensity and short rocking cycle

High light intensity of 340 $\mu mol/m^2/s$ and short rocking cycle of 1 s was tested at first. With 7-day culture, the final biomass concentration reached 2.73, 1.76, 1.06, and 0.98 g/L for 2.5, 5.0, 7.5, and 10 cm culture depths, respectively (Fig. 2a). There were pH drifting during the culture, but pH was no more than 11.0 in all experiments. As an extremely alkaliphilic cyanobacterium, *Euhalothece* sp. ZM001 grow well in this pH range (Acien Fernandez et al. 2013).

DO in culture with 2.5 cm depth was kept below 12.1 mg/L during the whole culture process, indicating that it has excellent oxygen mass transfer. Compared with this, 5.0 cm culture accumulated higher DO level than this, which was from 15.0 to 29.8 mg/L. DO in 7.5 cm culture accumulated to an even higher level, from 18.2 to 36.4 mg/L. It is interesting that DO in 10 cm culture accumulated to a level only a little higher than that of 2.5 cm, but much lower than that of 7.5 and 5.0 cm. The maximum DO observed during the culture process was 12.1, 29.8, 36.4, and 20.8 mg/L for 2.5, 5.0, 7.5, and 10 cm of culture depth, respectively (Fig. 2b).

There was no much difference between DO at surface and bottom for 5.0 cm culture. However, the DO at surface was obviously higher than bottom for cultures with 7.5 and 10 cm depth. For 10 cm, this difference was significant when cell growth was in exponential phase ($p = 0.057$). This should attribute to the fact that cells closer to surface receive more light and produce more oxygen. This is actually a non-negligible character of horizontal PBR, since oxygen has to be released from water surface. In this viewpoint, intensive mixing making homogeneous condition may not be necessary, since it brings oxygen from surface to bottom, which is actually more difficult to be released.

Culture at high light intensity and long rocking cycle

High light intensity of 340 $\mu mol/m^2/s$ and longer rocking cycle of 2 s was tested in this experiment. The results show that maximum DO observed during the whole culture process was 29.3, 39.6, 47.3, and 49.6 mg/L for 2.5, 5.0, 7.5, and 10 cm of culture depth, respectively (Fig. 3b).

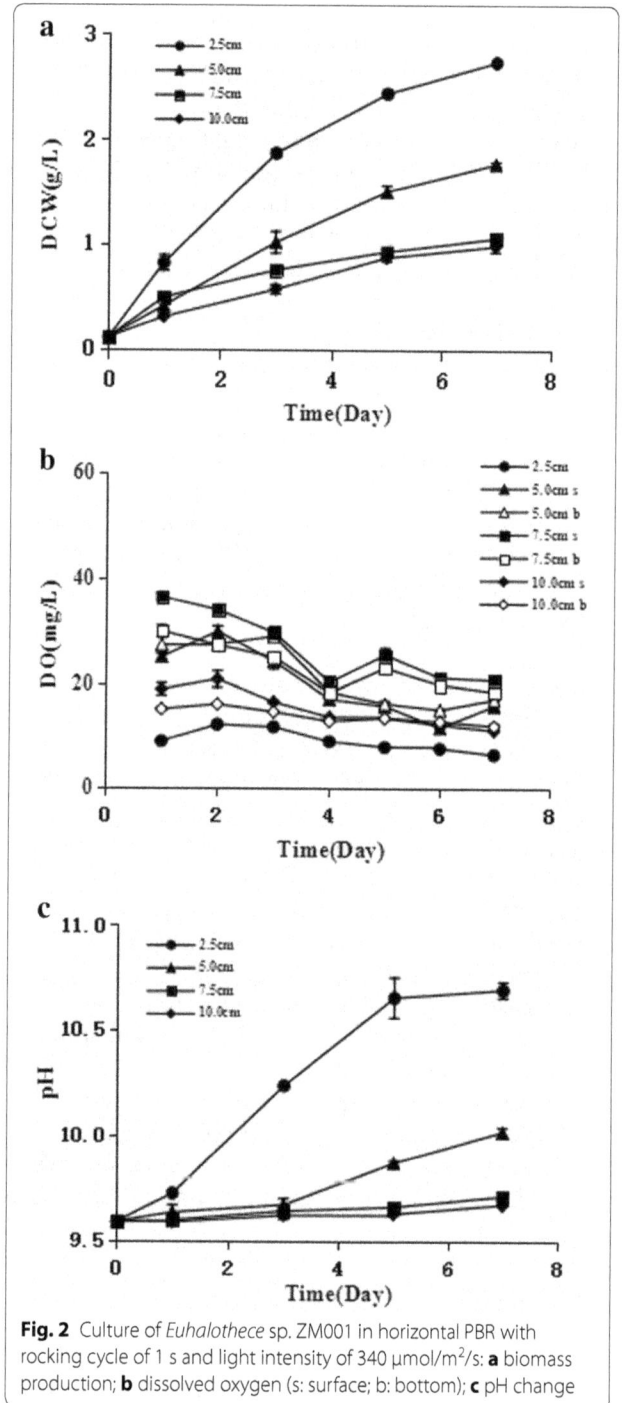

Fig. 2 Culture of *Euhalothece* sp. ZM001 in horizontal PBR with rocking cycle of 1 s and light intensity of 340 μmol/m²/s: **a** biomass production; **b** dissolved oxygen (s: surface; b: bottom); **c** pH change

Compared with experiment above, DO was accumulated to a much higher level and it was often over-saturated, indicating that 2-s rocking cycle was not enough to provide sufficient mixing.

Final biomass concentration in cultures with 2.5 and 5.0 cm depth was 2.18 and 1.42 g/L, respectively, which was lower than its counterpart with 1 s rocking cycle.

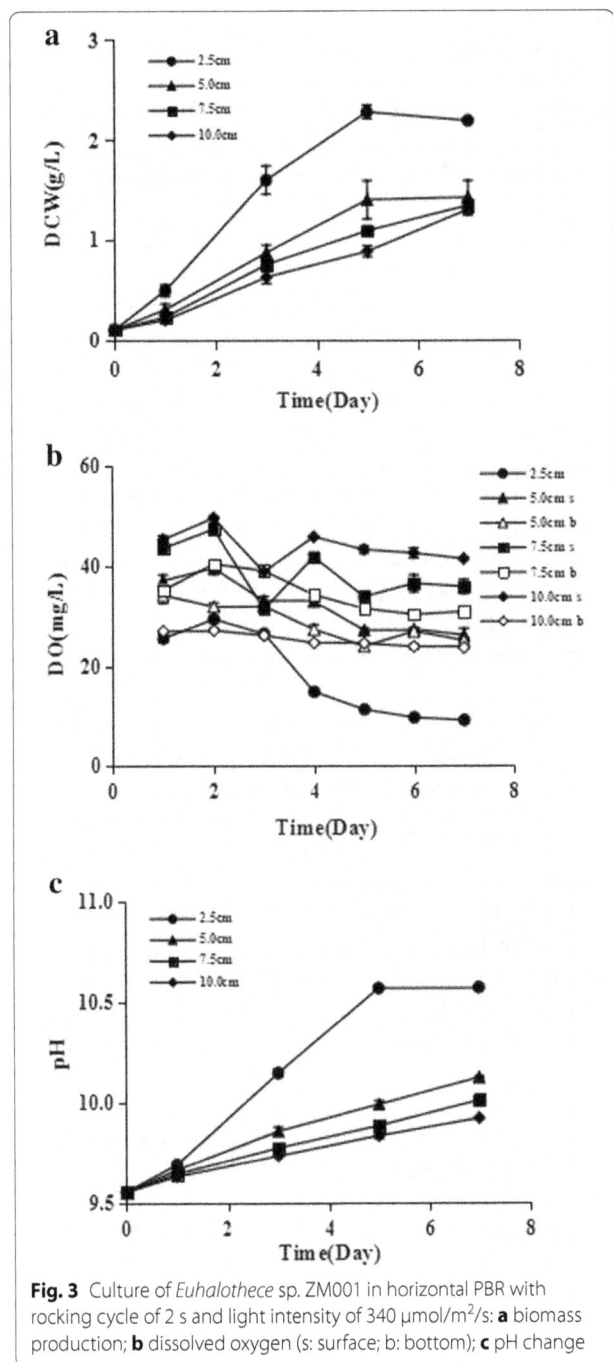

Fig. 3 Culture of *Euhalothece* sp. ZM001 in horizontal PBR with rocking cycle of 2 s and light intensity of 340 μmol/m²/s: **a** biomass production; **b** dissolved oxygen (s: surface; b: bottom); **c** pH change

why they resulted in lower biomass production than that of longer rocking cycle in this experiment.

Similar to above experiment with short rocking cycle, there was significant difference between DO at surface and bottom ($p < 0.05$) for the all culture days in culture with 10 cm depth. The maximum difference was 22.6 mg/L in day 2, and this difference is even greater than that in above experiment with rocking cycle of 1 s (Fig. 3b). Compared with 10 cm culture, the maximum difference of DO between surface and bottom in 5.0 and 7.5 cm cultures was much less (Fig. 3b).

Culture at medium light intensity and long rocking cycle
Medium light intensity of 135 μmol/m²/s and 2-s rocking cycle was tested in this experiment. At this condition, final biomass concentration reached 2.12, 1.15, 0.75, and 0.70 g/L for 2.5, 5.0, 7.5, and 10 cm culture depth, respectively (Fig. 4a). Final biomass concentration was obviously lower than that with high light intensity for 5.0, 7.5, and 10 cm cultures, but not 2.5 cm.

The maximum DO observed in this experiment was 23.0, 30.0, 37.9, and 52.2 mg/L for 2.5, 5.0, 7.5, and 10 cm cultures, respectively. The DO difference between surface and bottom in 5.0 and 7.5 cm cultures was small, which was no more than 5.2 and 5.1 mg/L, respectively. However, this difference in 10 cm culture was still high (Fig. 4b).

Culture at low light intensity and long rocking cycle
In this experiment, culture was tested with low light intensity of 70 μmol/m²/s and rocking cycle of 2 s. The microalgae final biomass concentration was 1.47, 0.99, 0.62, and 0.62 g/L for 2.5, 5.0, 7.5, and 10 cm culture depth, respectively (Fig. 5a). These final biomass concentrations were much lower than their counterpart with high light intensity, since low light intensity input less energy to the culture. The maximum DO was 23.9, 26.5, 32.1, and 23.4 mg/L for 2.5, 5.0, 7.5, and 10 cm cultures (Fig. 5b), indicating less oxygen is accumulated. Also, there was no significant difference between DO of surface and bottom for all the culture depths. DO was at the level of 20–30 mg/L during the whole culture process for all culture depths.

Oxygen tolerance of *Euhalothece* sp. ZM001
It is interesting that *Euhalothece* sp. ZM001 resulted in growth even when DO was over-saturated in the above experiments. To verify its tolerance to oxygen, it was cultured in bubble column with pure oxygen sparging. As the result show in Fig. 6, the final biomass concentration was 1.61 and 3.16 g/L for cultures bubbled with pure oxygen and air, respectively. When continuous bubbling

However, in culture depth of 7.5 and 10 cm, the final biomass concentration was 1.34 and 1.29 g/L, respectively, which was higher than its counterpart with 1-s rocking cycle. It was observed that there was no good circulation flow in 7.5 and 10.0 cm culture when rocking cycle of 1 s was applied. This may result in limited cell movement between light and dark zone, and probably be the reason

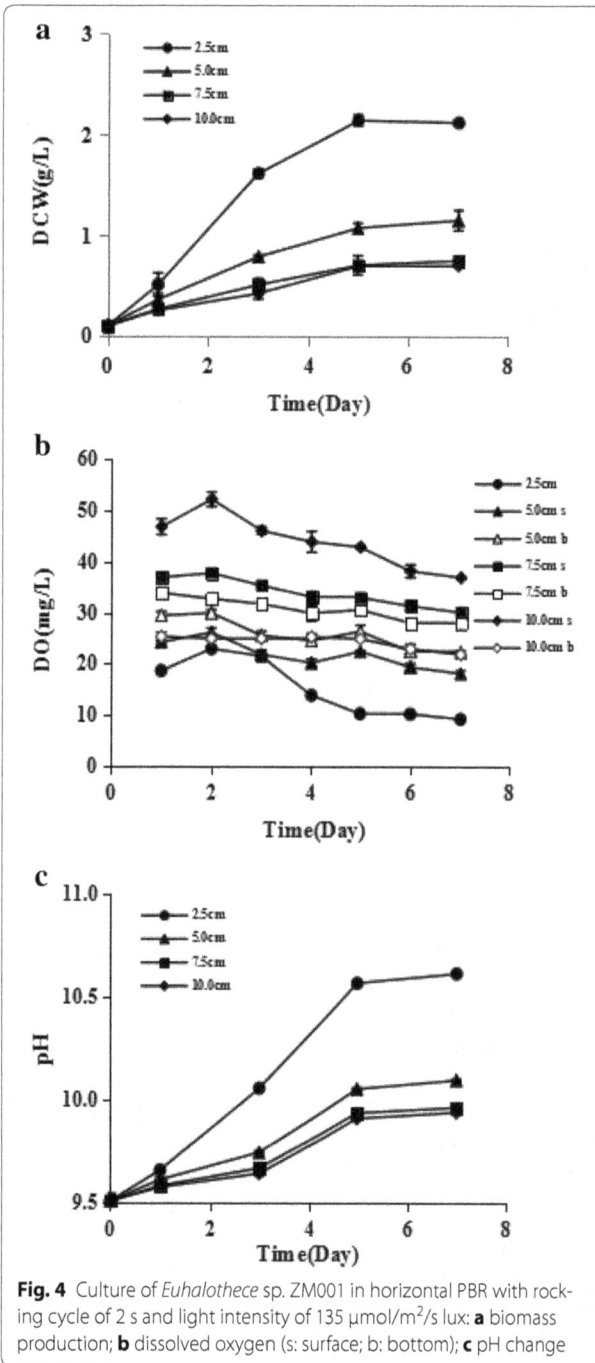

Fig. 4 Culture of *Euhalothece* sp. ZM001 in horizontal PBR with rocking cycle of 2 s and light intensity of 135 μmol/m²/s lux: **a** biomass production; **b** dissolved oxygen (s: surface; b: bottom); **c** pH change

k_La measurement at different culture conditions

To investigate mass transfer rate at different culture depths and rocking cycles, the k_La in each condition is measured and shown in Table 1. With 1-s rocking cycle, the 2.5 cm culture depth had the highest k_La of 33.49 h^{-1}. This result is corresponding to DO accumulation, which was always below 12.1 mg/L in culture at light intensity of 340 μmol/m²/s. Compared with this, k_La of 5.0 and 10 cm cultures was 5.76 and 3.86 h^{-1}, and the maximum DO was 29.8, and 20.8 mg/L, respectively (Fig. 1b). 7.5 cm culture had lowest k_La of 1.04 h^{-1}, and this is the direct reason why it resulted in high DO level, and the maximum observed DO was 36.4 mg/L (Fig. 1b).

For the rocking cycle of 2 s, 2.5 cm depth had a low k_La of 1.50 h^{-1}, and the maximum DO observed at 340 μmol/m²/s light intensity was 29.3 mg/L. Compared with this, 5.0, 7.5, and 10 cm cultures with lower k_La all resulted in high level of DO accumulation, and the maximum observed DO was 39.6, 47.3, and 49.6 mg/L, respectively. The high DO directly affect algae growth, as indicated in oxygen tolerance experiment, and led to lower biomass concentration.

It can be learned from Table 1 that k_La increases with the decrease of culture depth. This should be attributed to two reasons. On one hand, lower culture depth has a larger surface/volume ratio, which increases the value of "a" in k_La. On the other hand, thinner water in horizontal PBR has better circulation flow, and it increase the value of "k_L" in k_La. However, 7.5 cm depth has a much lower k_La than 10 cm at both rocking cycles, and this should be attributing to its poor circulation flow, which was observed during this k_La measurement test.

Outdoor culture at 1 m² scale

At the scale of 1 m², rocking cycle of the platform driven by water power was measured as 60 s, which is quite different from that of small-scale PBR. This cycle is dependent to time spent for water flow into the tank and discharge. *Euhalothece* sp. ZM001 was inoculated into the plastic bag PBR at 0.1 g/L. After experiencing a long lag phase, this culture resulted in a final biomass concentration of 1.88 g/L (Fig. 7). The average daily biomass productivity can be calculated as 0.08 g/L/day or 4.2 g/m²/day. During the culture process, the high temperature was from 24 to 36 °C, and the low temperature during the night was from 14 to 22 °C. *Euhalothece* sp. ZM001 has optimal growth between 35 and 40 °C. This temperature range experienced in outdoor culture was actually not ideal and resulted in low average productivity. However, it is notable that the maximum daily biomass productivity was 0.4 g/L/day or 20 g/m²/day, which is obtained from day 13 to 15 (Fig. 7).

with oxygen, DO in the whole culture process was kept from 32.6 to 39.7 mg/L. The pH of these two cultures was always kept below 10.5, indicating that the difference between the two cultures' biomass production was not caused by pH. These results showed that high DO does affect growth of *Euhalothece* sp. ZM001, but it still grew to a density of 1.69 g/L (13 days), indicating it as an excellent species tolerant to high oxygen level.

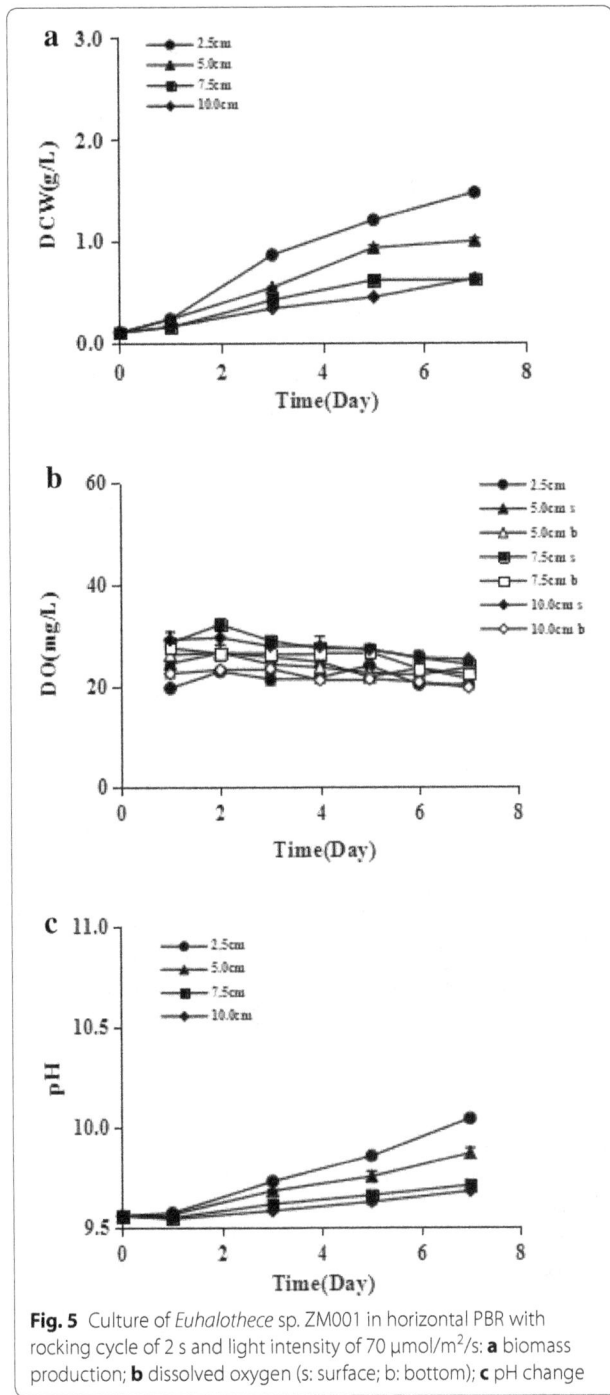

Fig. 5 Culture of *Euhalothece* sp. ZM001 in horizontal PBR with rocking cycle of 2 s and light intensity of 70 µmol/m²/s: **a** biomass production; **b** dissolved oxygen (s: surface; b: bottom); **c** pH change

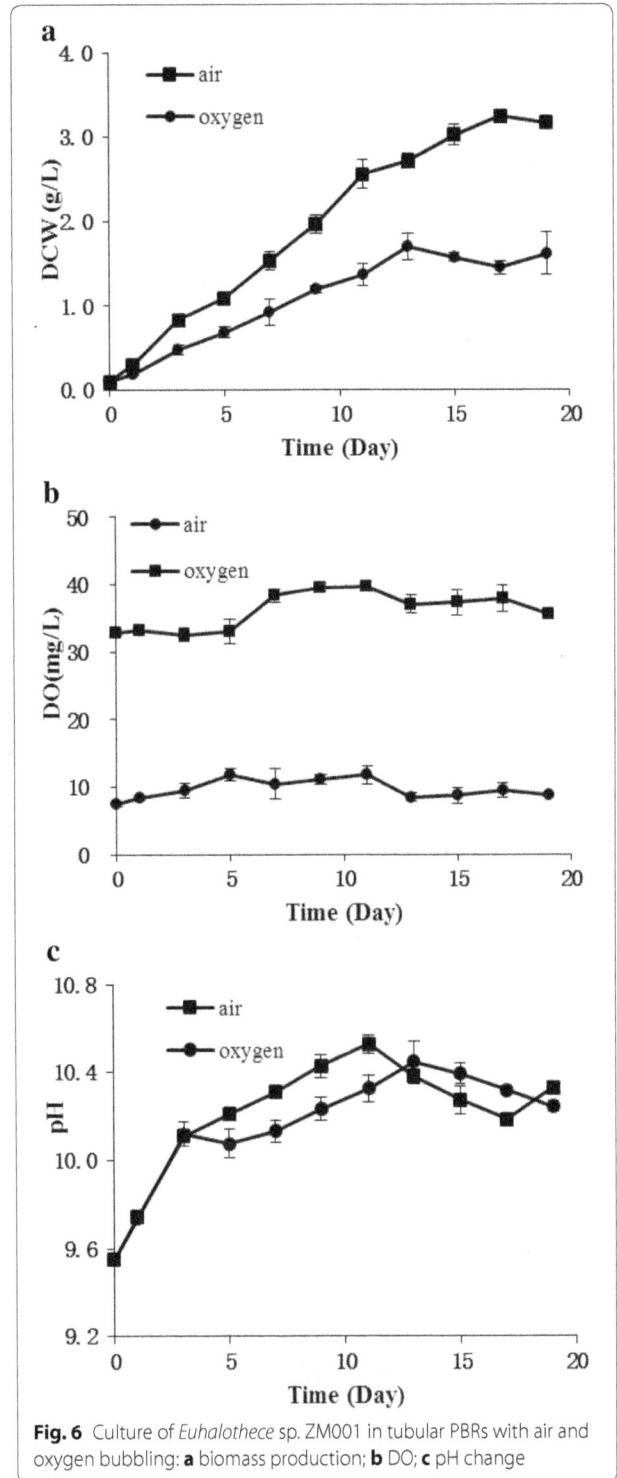

Fig. 6 Culture of *Euhalothece* sp. ZM001 in tubular PBRs with air and oxygen bubbling: **a** biomass production; **b** DO; **c** pH change

pH change during this culture was from 9.56 to 9.89. With rocking cycle of 60 s, the $k_L a$ was measured as 0.89 h^{-1}. Due to this low mass transfer rate, the DO accumulated to a high level during the day time, and the recorded high was 35.6 mg/L. As an oxygen-tolerant strain, it still resulted in growth to 1.88 g/L final biomass concentration at this high DO level. This indicates that horizontal PBR mixed with rocking movement driven by water power is feasible to support growth of alkalihalophilic cyanobacterium of *Euhalothece* sp. ZM001, but the oxygen transfer rate should be improved to obtain higher productivity.

Table 1 $k_L a$ **at different culture depths and rocking cycles in small-scale horizontal PBR**

Culture depth (cm)	$k_L a$ (h^{-1})	
	1-s rocking cycle	2-s rocking cycle
2.5	33.49	1.50
5.0	5.76	1.15
7.5	1.04	0.57
10	3.86	0.84

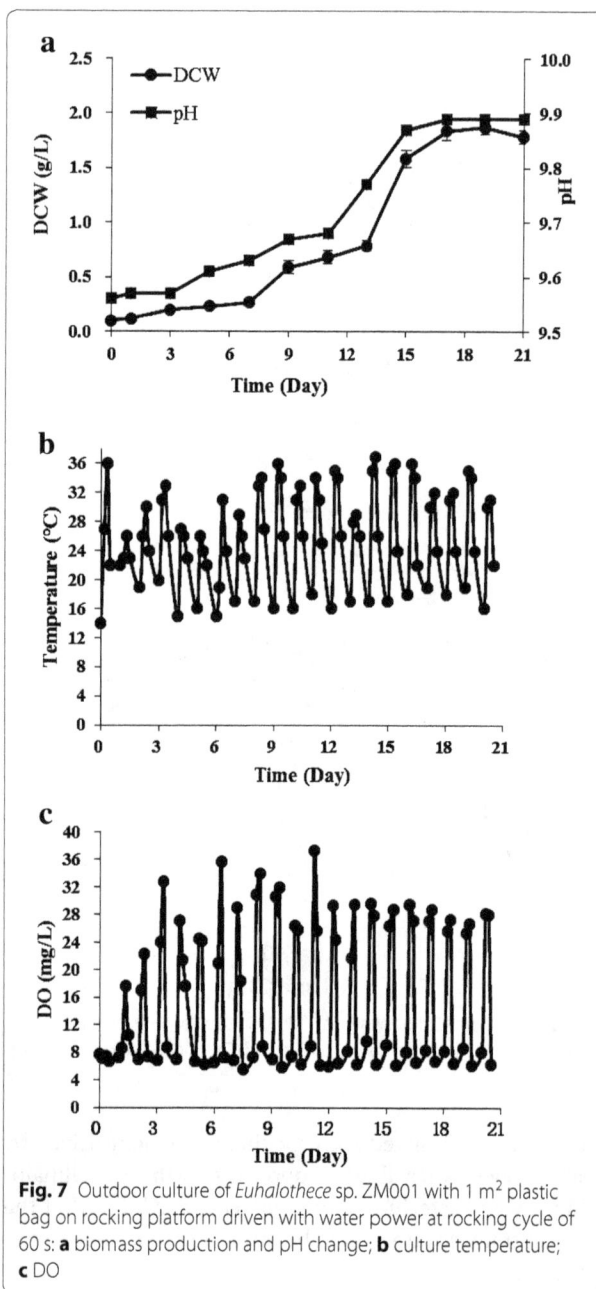

Fig. 7 Outdoor culture of *Euhalothece* sp. ZM001 with 1 m^2 plastic bag on rocking platform driven with water power at rocking cycle of 60 s: **a** biomass production and pH change; **b** culture temperature; **c** DO

Discussion

This study aims to develop a horizontal PBR system mixed with rocking movement driven by nature force, and these preliminary results showed the feasibility of it. Plastic bag can be used as horizontal PBR for this system, and culture depth no more than 10 cm will not face high water pressure problems as experienced in vertical PBRs. Plastic bag PBR has the advantage of low cost, and it can be simply disposed when facing fouling problem. The shape of plastic bag horizontal PBR culture system developed in this study is similar to closed pond. However, it is difficult to construct a pond with culture depth of only 10 cm or less, and it is difficult to use paddle wheel to drive the mixing in such shallow water. Instead, mixing in horizontal PBR can be driven by rocking movement, and much less culture depth can be used. This short light path resulted in much higher biomass concentration than closed pond, as shown in this study.

Since high concentration of bicarbonate supplied sufficient carbon "once for all" at beginning of the culture, it is not necessary to continuously bubbling CO_2 during algae culture process. This not only allows a simpler PBR design, but also reduces the cost of gas pipeline construction in massive culture, as well as the related cost for maintenance and labor utilization. It should be noted that using bicarbonate as the carbon source would not have a higher cost than bubbling CO_2, since bicarbonate only work as a carrier of CO_2 and there is no net consumption of it. When algal culture process finished, part of bicarbonate is converted into carbonate, and it can be recycled to absorb more CO_2 and supply into the next batch of algal culture (Chi et al. 2011; Chisti 2013). Thus, this culture system would has a low algal biomass production cost on the whole, since it systematically reduced the PBR manufacturing cost, mixing energy cost, carbon supplying cost, and labor utilization cost.

As a preliminary study, the algal biomass productivity of this system is still low. The average areal productivities for the cultures at small scale are summarized and shown in Table 2. In the indoor culture, the highest areal productivity was 17.06 g/m^2/day, which is obtained in the culture with 10 cm depth, 340 μmol/m^2/s light intensity, and 2-s rocking cycle. In the outdoor culture at the scale of 1 m^2, the average productivity was 4.2 g/m^2/day. However, the maximum daily biomass productivity was 0.4 g/L/day or 20 g/m^2/day (Fig. 7), indicating that the productivity could be great improved with further study.

Since high concentration of bicarbonate supplied plenty of inorganic carbon, the mass transfer of carbon is no longer a problem for this system. The problem for oxygen, however, is still to be studied, since it may cause photo-inhibition when accumulated to high level (Peng et al. 2013). In this study, there was great variation on

Table 2 Average productivity of different culture conditions (g/m²/day)

	2.5 cm	5.0 cm	7.5 cm	10 cm
340 µmol/m²/s, 1 s	9.39	11.89	10.24	12.58
340 µmol/m²/s, 2 s	7.44	9.40	13.24	17.06
135 µmol/m²/s, 2 s	7.20	7.48	6.96	8.62
70 µmol/m²/s, 2 s	4.89	6.39	5.54	7.50

k_La values in small-scale horizontal PBRs. The maximum k_La measured was 33.49 h^{-1} in culture with 2.5 cm depth and rocking cycle of 1 s. This value is at the same level as bubble columns, which was from 16.2 to 50.4 h^{-1} (Peng et al. 2013). With the same rocking cycle, the k_La in 5 and 10 cm cultures was 5.76 and 3.86 h^{-1}. These values are comparable with horizontal tubular PBR, which was reported as from 3.6 to 10.8 h^{-1} (Peng et al. 2013). However, with still the same rocking cycle of 1 s, k_La in 7.5 cm culture was only 1.04 h^{-1}. This is at the same level as raceway pond, which is reported as 0.87 and 0.94 h^{-1} in the straight and curved channel part (Mendoza et al. 2013a, b). When longer rocking cycle of 2 s was used, the ranges of k_La were from 0.57 to 1.50 h^{-1}, and the DO often accumulated to very high level, indicating that it actually did not provide sufficient mixing. This is also true for the culture at 1.0 m² scale, which had a k_La of only 0.89 h^{-1} at the rocking cycle of 60 s.

Although high biomass concentration was obtained in culture of oxygen-tolerant *Euhalothece* sp., it is actually not practical to culture oxygen-sensitive algal strains if mixing is not sufficient. Thus, the capability of oxygen mass transfer in horizontal PBR should be improved in future study. This may be realized by optimizing the angle and cycle of horizontal PBR's rocking movement. Also, the length of PBR would be a significant factor to mixing condition, since it can significantly change the flow field. In addition, baffles in horizontal PBR or open pond can change vertical flow, which was reported to improve mass transfer and frequency of light/dark cycle, and resulted in significant higher biomass concentration than its control (Zhang et al. 2015). Thus, mixing and mass transfer in horizontal PBR driven by rocking movement deserve intensive study in future. An interesting phenomenon observed in this study is that there was great difference between DO at the culture surface and bottom when culture depth was 7.5 cm or greater. This may also occur in open pond system, and special attention should be paid to this issue in study on mixing in horizontal PBR, as well as its design and operation.

Different algae species have different capabilities of tolerance to oxygen. For instance, when DO was elevated to a level higher than 13 mg/L, the final biomass concentration was substantially reduced in culture of *Chlorella sorokiniana* (Ugwu et al. 2007), while cell death occurred only when DO reached 36 mg/L for *Spirulina platensis* Vonshak et al. (1996). It is interesting that *Euhalothece* sp. ZM001 has extraordinary tolerance capability to high DO. It resulted in growth to 1.69 g/L even with pure oxygen bubbling. Due to this character, it may be an excellent algae strain to be applied in life support system for oxygen production in spacecraft or submarine. Besides, this algae strain also has excellent capability of tolerant to high temperature, high pH, and high salt concentration (Chi et al. 2011; Chisti 2013). The tolerance mechanism is a great research interest, which is worthy of studying in depth.

This study tested horizontal PBR at the scale of 1 m², and its scaling-up requires rocking platform with much larger size. A possible solution is to make rocking platform with the size of raceway pond, which is usually several meters in width and several tens of meters in length. This would result in rocking platform at the size of several hundred square meters. Driving rocking movement of this large-scale platform would not be a big problem, since falling water can provide intensive mechanical power and was used to drive heavy machines before electricity is popularized. Using hundreds of these rocking platforms would realize massive cultures at the scale of hectors. Construction of rocking platforms at this scale may be costly, but it would not be greater than horizontal tubular or vertical flat plate PBR systems at the same scale, in authors' opinion. If hydraulic power is used to drive this system, the cultivation field should be built at the site with plenty of water resources. Also, this horizontal PBR can be placed on water surface of ocean, river, or lake, driving rocking movement with wave, which is actually an ongoing research conducted by the authors. In addition, its rocking movement may be driven with many other nature forces, such as tide or wind, as long as they can provide mechanical power to drive a back/forth movement. There were actually many mature technologies to realize this before electricity power is popularized. Using nature force to drive the mixing would significantly improve energy balance for algal biomass production, which is a great hurdle in the commercialization of algal biofuel.

Conclusion

The algal culture system developed in this study proved feasibility of using nature force to drive movement of rocking platform to realize mixing in horizontal PBR. Low-cost plastic bag can be used in this system to reduce the manufacturing cost of horizontal PBR. This PBR well supported growth of alkalihalophilic microalgae *Euhalothece* sp. ZM001 without CO_2 gas bubbling.

Abbreviations

BICCAPS: bicarbonate-based integrated carbon capture and algae production system; DO: dissolved oxygen; k_La: mass transfer coefficient; PBR: photobioreactor.

Authors' contributions

HZ conducted most microalgae culture work; CZ conducted analysis work on mass transfer coefficient measuring; LC participated in making photobioreactor and conducted part of microalgae culture work. ZC is the corresponding author. All authors read and approved the final manuscript.

Acknowledgements

Not applicable.

Competing interests

The authors declare that they have no competing interests.

Funding

This research was supported by the "Fundamental Research Funds for the Central Universities" [DUT14RC(3)065].

References

Abu-Ghosh S, Fixler D, Dubinsky Z, Iluz D (2016) Flashing light in microalgae biotechnology. Bioresour Technol 203:357–363

Acien Fernandez FG, Fernandez Sevilla JM, Molina Grima E (2013) Photobioreactors for the production of microalgae. Rev Environ Sci Bio-Technol 12(2):131–151

Acien FG, Fernandez JM, Magan JJ, Molina E (2012) Production cost of a real microalgae production plant and strategies to reduce it. Biotechnol Adv 30(6):1344–1353

Chi Z, O'Fallon JV, Chen S (2011) Bicarbonate produced from carbon capture for algae culture. Trends Biotechnol 29(11):537–541

Chi Z, Xie Y, Elloy F, Zheng Y, Hu Y, Chen S (2013) Bicarbonate-based Integrated Carbon Capture and Algae Production System with alkalihalophilic cyanobacterium. Bioresour Technol 133:513–521

Chi Z, Elloy F, Xie Y, Hu Y, Chen S (2014) Selection of microalgae and cyanobacteria strains for bicarbonate-based integrated carbon capture and algae production system. Appl Biochem Biotechnol 172(1):447–457

Chisti Y (2013) The problems with algal fuels. Biotechnol Bioeng 110(9):2318–2319

Davis W, Martín M (2014) Optimal year-round operation for methane production from CO_2 and water using wind and/or solar energy. J Clean Prod 80:252–261

de Godos I, Mendoza JL, Acien FG, Molina E, Banks CJ, Heaven S, Rogalla F (2014) Evaluation of carbon dioxide mass transfer in raceway reactors for microalgae culture using flue gases. Bioresour Technol 153:307–314

Dogaris I, Welch M, Meiser A, Walmsley L, Philippidis G (2015) A novel horizontal photobioreactor for high-density cultivation of microalgae. Bioresour Technol 198:316–324

Jorquera O, Kiperstok A, Sales EA, Embirucu M, Ghirardi ML (2010) Comparative energy life-cycle analyses of microalgal biomass production in open ponds and photobioreactors. Bioresour Technol 101(4):1406–1413

Kim ZH, Park H, Hong SJ, Lim SM, Lee CG (2016) Development of a floating photobioreactor with internal partitions for efficient utilization of ocean wave into improved mass transfer and algal culture mixing. Bioprocess Biosyst Eng 39(5):713–723

Mendoza JL, Granados MR, de Godos I, Acien FG, Molina E, Banks C, Heaven S (2013a) Fluid-dynamic characterization of real-scale raceway reactors for microalgae production. Biomass Bioenergy 54:267–275

Mendoza JL, Granados MR, de Godos I, Acien FG, Molina E, Heaven S, Banks CJ (2013b) Oxygen transfer and evolution in microalgal culture in open raceways. Bioresour Technol 137:188–195

Mikhodyuk OS, Gerasimenko LM, Akimov VN, Ivanovsky RN, Zavarzin GA (2008) Ecophysiology and polymorphism of the unicellular extremely natronophilic cyanobacterium *Euhalothece* sp. Z-M001 from Lake Magadi. Microbiology 77(6):717–725

Norsker N-H, Barbosa MJ, Vermue MH, Wijffels RH (2011) Microalgal production—a close look at the economics. Biotechnol Adv 29(1):24–27

Peng L, Lan CQ, Zhang Z (2013) Evolution, detrimental effects, and removal of oxygen in microalga cultures: a review. Environ Prog Sustain Energy 32(4):982–988

Tredici MR, Bassi N, Prussi M, Biondi N, Rodolfi L, Chini Zittelli G, Sampietro G (2015) Energy balance of algal biomass production in a 1-ha "Green Wall Panel" plant: how to produce algal biomass in a closed reactor achieving a high Net Energy Ratio. Appl Energy 154:1103–1111

Ugwu CU, Aoyagi H, Uchiyama H (2007) Influence of irradiance, dissolved oxygen concentration, and temperature on the growth of *Chlorella sorokiniana*. Photosynthetica 45(2):309–311

Vonshak A, Torzillo G, Accolla P, Tomaselli L (1996) Light and oxygen stress in *Spirulina platensis* (cyanobacteria) grown outdoors in tubular reactors. Physiol Plant 97:175–179

Zhang Q, Xue S, Yan C, Wu X, Wen S, Cong W (2015) Installation of flow deflectors and wing baffles to reduce dead zone and enhance flashing light effect in an open raceway pond. Bioresour Technol 198:150–156

12

Impact of moisture content on instant catapult steam explosion pretreatment of sweet potato vine

Li-Yang Liu[2,3], Jin-Cheng Qin[1,3], Kai Li[1], Muhammad Aamer Mehmood[1,4*] and Chen-Guang Liu[1*]

Abstract

Background: Lignocellulose originating from renewable and sustainable biomass is a promising alternative resource to produce biofuel. However, the complex component, especially high moisture content, leads to a higher cost of transportation and processing. The instant catapult steam explosion (ICSE) pretreatment can exploit the intracellular water of lignocellulosic materials and convert into vapors leading towards the breakdown of the feedstock during the explosion process. However, it is necessary to study the impact of moisture content on the pretreatment.

Results: The sugar yield of wet feedstock after ICSE pretreatment reached 88.05%, which was higher when compared to dried and untreated biomass. The utilization of wet feedstock decreased the production of inhibitor and improved the carbohydrate content in ICSE-treated biomass. There occurred a shrinkage of feedstock after drying process and the mechanical breakage upon ICSE pretreatment. Moreover, not all water was converted into vapor to cause breakage in the lignocellulose.

Conclusion: ICSE has shown to be preferably suitable to pretreat wet sweet potato vine with high moisture content, either fresh or soaked biomass that has been dried before. By using these materials, it would have a higher sugar yield and lower inhibitor production after pretreatment. Based on these advantaged aspects of ICSE platform, two potential strategies are proposed to improve the economic and environmental impacts of pretreatment.

Keywords: Pretreatment, Moisture content, Instant catapult steam explosion, Sweet potato vine, Enzymatic hydrolysis

Background

Nowadays, lignocellulosic biomass offers a promising alternative to produce biofuel owing to its abundance, renewability, and sustainability. However, the recalcitrant nature of biomass requires additional pretreatment steps to make it susceptible to cellulolytic enzymes (Mosier et al. 2005). Generally, pretreatment efficiency depends on low moisture content and small particle size of the biomass which also would reduce the cost of transportation (Vidal et al. 2011). However, traditional drying process is an energy intensive process and the solar irradiance is limited in countries like China as compared to South Asia or Africa; thus, the air-drying process will occupy vast agriculture area preventing in-time crop rotation. So, to keep the crop rotation framework intact, most of lignocellulosic biomass is burnt on the field during harvesting seasons in northern China, causing atmosphere pollution and public health hazards (Tan et al. 2010; Qu et al. 2012). Alternatively, this low-cost biomass may be subjected to efficient pretreatment which do not involve drying step.

Fortunately, steam explosion pretreatment in principle requires higher moisture content in lignocellulose biomass, which may improve the efficiency of pretreatment. The biomass is usually treated under high pressure for several seconds to minutes to prompt the hydrolysis of

*Correspondence: draamer@gcuf.edu.pk; cg.liu@sjtu.edu.cn
[1] State Key Laboratory of Microbial Metabolism, School of Life Sciences and Biotechnology, Shanghai Jiao Tong University, Shanghai 200240, China [4] Department of Bioinformatics and Biotechnology, Government College University Faisalabad, Faisalabad 38000, Pakistan
Full list of author information is available at the end of the article

hemicellulose content and then release the pressure with a short time to break the microstructure of lignocellulose (Galbe and Zacchi 2012). Previous researchers adopted steam explosion to pretreat wet corn stalk and found that higher initial moisture content would improve enzymatic hydrolysis of lignocellulose and reduce 20% of steam consumption, because water in feedstock presented a buffering effect on reaction during steam explosion process (Sui and Chen 2015).

However, the traditional steam explosion instrument will produce abundant of inhibitors such as acetic acid, furfural, coumaric acid, and 5-hydroxymethylfurfural (5-HMF). Moreover, the longer pretreatment time will consume more energy (Cullis et al. 2004). To offset these drawbacks, instant Catapult Steam Explosion (ICSE) was invented. The short de-pressure time (0.0825 s) and pretreatment duration (1–5 min) provide ICSE huge mechanical force to destruct the biomass structure and less energy consumption with simple operation than traditional steam explosion (Gong et al. 2012). Previously, it has been demonstrated that ICSE can improve the enzymatic degradability of lignocellulose with few inhibitors and less energy consumption (Liu et al. 2014a). ICSE also has positive impacts on the efficiency of subsequent chemical pretreatments involved with organic solvent, acid, and alkali (Liu et al. 2014b).

Sweet potato (*Ipomoea batatas*) belongs to the family *Convolvulaceae*, containing abundant starch in the root. So far, China being the largest producer of sweet potato in the world has planted over 6.2 million hectares, and produces over 71 million tons of sweet potatoes, which accounts for the 67% of the global production (Fao 2016). The high sugar content and easily cultivated in saline and alkali land of sweet potato allow it to be used as an excellent resource to produce biofuels (Xia et al. 2013). Although significant advancements have been made on the conversion of sweet potato yet the progress made is not up to the desired levels. Many sweet potato vines including leaves, stems, and petiole are disposed in the farming field or used as low-value product like animal feed (Tian et al. 2009). Comparing with other lignocellulose resources such as corn stover and rice stalk, sweet potato vine contains higher moisture content and extractives (Jibril et al. 1999). These characters may be helpful to study the effects of lignocellulosic moisture content on ICSE pretreatment.

In present study, sweet potato vines were used to study the effects of moisture content on ICSE pretreatment by investigating the composition, inhibitors production, sugar yield, and thermal stability, for different feedstock including fresh feedstock, naturally dried feedstock, manually dried feedstock, and soaked dried feedstock. It was studied as a simple and practical method to obtain wet lignocellulose for both industry and lab.

Methods

Biomass collection and management

Fresh sweet potato vines were collected from a field near the city of Dalian in China during the summer, which contained 16.8% cellulose, 9.6% hemicellulose, 42.89% lignin and ash, 17.6% protein, and 1.5% fat after dried. Collected materials were cut into 1- to 2-cm fragments by scissors, and stored at − 20 °C for further use.

Fresh feedstocks (FF) were treated in different ways to obtain materials with different moisture contents. FF was dried by natural environment under the sun for 1 week (NDF), which was close to the natural condition before collecting the samples from the field. FF was also dried at 65 °C to the constant weight. This manually dried feedstock (MDF) is widely accepted for the pretreatment.

Samples with coordinating water were also prepared by soaking 20 g MDF into 200 mL of tap water at room temperature at a short time for 2 h (SF-2), medium time (30 h, SF-30), and a long time (60 h, SF-60). To stop the soaking processing and stabilize the moisture content, the Buchner funnels with filter paper were used to separate the liquid and wet feedstock. Water on the surface of wet feedstock was cleaned by filter paper, following to put in the desiccator to stabilize their moisture content for 24 h. All feedstocks were sealed and stored at − 20 °C.

Chemicals and enzymes

Analytical grade glucose, xylose, furfural, 5-hydroxymethyl furfural (5-HMF), acetic acid, and coumaric acid were purchased from Sangon Biotech Crop. (Shanghai, China) and Solarbio Life Science Corp. (Beijing, China). Cellulase (GENENCOR accelerase 1500, cellulase enzyme activity: 105 FPU/mL) was kindly donated by Dupoint Genicor Science Corp. (Shanghai, China).

ICSE pretreatment

20 g of biomass including FF, NDF, MDF, SF-2, SF-30, and SF-60 were loaded directly into 400-mL chamber of the ICSE equipment (QBS-80B Steam Explosion Test Bed, Henan Hebei, Zhengdao Corp.). Steam with pressure (3.25 MPa) and temperature (240 °C) was prepared in advance and sent into the chamber to make the pretreated pressure be stabilized at 2.8 MPa for 90 s by controlling the quantity of flow. After this pretreatment, the ICSE-treated sweet potato vines were released from the chamber by depressurization in 0.0825 s, causing treated material to explode into a stainless-steel cyclone. The slurry (treated material) was collected by scoop into the plastic bag, and then sealed and stored at − 20 °C for further compositional and enzymatic hydrolysis analysis (Liu et al. 2014a).

Moisture content analysis

The moisture content (C_{m_1}) of materials before and after ICSE pretreatment was calculated via the equation

$$C_{m_1} = (X_1 - X_2)/X_1,$$

where X_1 is the mass before oven drying and X_2 is the mass after drying in the 65 °C oven to get a constant weigh.

Inhibitor and sugar analysis after ICSE pretreatment

5 grams of each of the ICSE-treated materials were centrifuged in 1957×g for 10 min. The supernatants were used for inhibitor and sugar concentration (C_i) analysis by high-performance liquid chromatography analysis. The solid components were dried at 65 °C to get a constant weight (X_3). After the analysis of liquor solution, the inhibitor and sugar yield (mg/g) after ICSE pretreatment were calculated by the equation: inhibitor yield = $C_i \times (5 - X_3)/X_3$. All experiments were performed in triplicate and the results are presented as mean and standard derivation.

Compositional analysis

Untreated MDF and solid components of ICSE-pretreated samples were totally dried and finely ground to less than 40-mesh. 100 mg of each sample was mixed with 1 mL of 72% (w/v) sulfuric acid, and placed in the 50-mL colorimetric tube. The tubes were incubated at 30 °C water bath for 1 h with stirring every 10 min to ensure the intensive mixing. After 1 h, final concentration of acid was brought to 4% by adding 28 mL of deionized water, and was put in the autoclave at 121 °C for 1 h. After that, all tubes were left to cool down to room temperature, and the treated samples were filtered by the Buchner funnel with Whatman No.1 filter paper. The supernatants of each tube were used to analyze the glucose, cellobiose, xylose, and arabinose, representing cellulose and hemicellulose, respectively. The solid components were thoroughly washed and dried until it reached a constant weight, representing lignin and ash content. All experiments were performed in triplicate and the results are presented as mean and standard deviations (Sluiter et al. 2012).

Enzymatic hydrolysis

Untreated and the solid fractions of ICSE-pretreated samples were accurately weighted (300 mg) and placed in the 50-mL colorimetric tubes, suspended in 30 mL buffer solution (acetic acid, pH 4.8) containing the enzyme at the ratio of 30 FPU/g. The mixture was incubated at 50 °C for 48 h followed by a centrifugation. Supernatant was subjected to sugar content analysis, as described

previously (Liu et al. 2014a, b). The sugar yield was calculated by Eq. (1) below:

$$\text{Sugar yield(\%)} = \frac{\left[(\text{glucose} \times 0.9) + (\text{xylose} \times 0.8)\right]}{\text{carbohydrate in biomass}} \times 100. \tag{1}$$

Glucose and xylose were analyzed by the HPLC. All experiments were performed in triplicate, and results are presented as mean and standard deviations.

High-performance liquid chromatography analysis

The supernatant from ICSE-pretreated samples and enzymatic hydrolysis were filtered through 0.45-µm filter and 20 µL of each sample was loaded to the ion exclusion column (300 mm × 7.8 mm, Bio-Rad, Hercules, Aminex HPX-87H) at 50 °C in HPLC system equipped with a refractive index detector and UV detector (Waters, MA, USA), for the quantitation of inhibitors and saccharides. Sulfuric acid 0.01 mol/L was used as mobile phase at flow rate 0.5 mL/min (Liu et al. 2014a).

Scanning electron microscopy (SEM)

The microscopic morphology changes in the untreated and ICSE-pretreated samples after dried by lyophilization were observed using SEM. Each of the samples was placed on the aluminum sample platform and scanned by environmental scanning electron microscopy (Quanta 450, FEI, USA20 kV), using 20 kV as described previously (Liu et al. 2014a).

Thermal gravimetric analysis (TGA)

About 10 mg of freeze-dried MDF, ICSE-treated FF, and ICSE-treated MDF were placed in the platinum crucibles for thermal degradation analyses using TGA Q500 (TA Corporation, USA). Samples were heated from 25 to 500 °C at a constant heating rate of 10 °C/min under nitrogen atmosphere at the flow rate of 50 mL/min (Carrier et al. 2011).

Results and discussion

The change of moisture content in different feedstock

Moisture content of any plant biomass is an important parameter for its subsequent usage as a feedstock for bioenergy. To evaluate the effect of moisture content on pretreatment and enzymatic hydrolysis, five forms of samples were prepared (Table 1). FF normally contains 80.50% to sustain the growth of plants. NDF shown to contain 63.57% moisture content which was obtained through exposure of harvested FF under the sun for 7 days. MDF was subjected to soaking to regain the moisture content. Interestingly, the moisture content of SF immediately came up to 62.93% through submerging

Table 1 The moisture content and the compositional analyses of liquid phase after ICSE pretreatment

Feedstocks	FF	NDF	MDF	SF-2	SF-30	SF-60
Moisture content (%)						
Untreated	80.50	63.57	0.97	62.93	83.38	84.24
Pretreated	96.95	96.04	93.30	95.62	97.90	97.29
Liquid contents						
Pretreated (mg/g)						
Glucose	0.7865 (0.0649)	0.3798 (0.0123)	0.2638 (0.0279)	0.0284 (0.0091)	0.0008 (0.0012)	0.0000 (0.0000)
Xylose	0.8858 (0.0577)	0.5393 (0.0292)	0.4733 (0.0320)	0.0682 (0.0137)	0.0031 (0.0020)	0.0006 (0.0000)
Acetic acid	0.0768 (0.0049)	0.0926 (0.0007)	0.1293 (0.0019)	0.0496 (0.0015)	0.0674 (0.0517)	0.0000 (0.0000)
Coumaric acid	0.2405 (0.0827)	0.4769 (0.0303)	0.7838 (0.0149)	0.2797 (0.0136)	0.0659 (0.1110)	0.0241 (0.0000)
5-HMF	0.0702 (0.0247)	0.0492 (0.0325)	0.0418 (0.0021)	0.0231 (0.0080)	0.0059 (0.0040)	0.0041 (0.0008)
Furfural	24.17 (2.18)	17.41 (0.74)	34.81 (0.96)	18.76 (1.04)	5.37 (1.40)	5.42 (0.24)

Data were shown as "mean (standard deviation)"

MDF for 2 h and finally stabilized at around 84.24% at 60 h. The moisture content of SF-60 was close to that of FF, which may be attributed to water absorption by favorable hydrophilic component and the sponge-like structure of lignocellulose (Dhakal et al. 2007). In addition, it is worth mentioning that, to achieve the same available biomass, the weight of MDF was one-fifth of FF, so the drying process could significantly reduce the total weight of lignocellulose, which is very important for biomass transportation (Axelsson et al. 2012). Moreover, soaking is a convenient way to regain the water, if required, after the easy transportation of dried biomass to the biofuel industry.

ICSE-pretreated sample contained higher moisture content ranging from 93.30 to 97.90%, which was caused by the abundant liquefied water condensed from the steam and stored in the pores of feedstock after the explosion (Table 1). Since the initial moisture content of feedstock also affects the final moisture content, it is obvious to get similar trends of moisture content after pretreatment. However, the ICSE pretreatment demonstrated its capability to smooth the difference of moisture content among all samples, which nullify the impact of drying on ICSE.

Composition and inhibitors in pretreated feedstocks

The pretreatment with high temperature that can contribute to enhance the available sugars upon hydrolysis is preferred in the bioconversion of biomass to biofuels. However, under high-temperature pretreatment, the lignocellulose will be partially degraded to inhibitors such as organic acid, furfural, 5-HMF, which drastically hinder the efficiency of enzymatic hydrolysis and fermentation process (Jönsson et al. 2013). The 5-HMF, furfural, and coumaric acid, respectively, represent the cellulose, hemicelluloses, and lignin, and reflect the compositional

Table 2 The compositional analysis of untreated MDF and the solid contents after ICSE pretreated

	Cellulose/%	Hemicellulose/%	Lignin and ash/%
Untreated			
MDF	16.83 (1.86)	9.60 (1.20)	42.88 (0.67)
Pretreated			
FF	40.25 (1.62)	10.27 (0.13)	22.39 (0.52)
NDF	41.10 (2.18)	10.92 (0.90)	19.66 (2.91)
MDF	41.96 (4.66)	8.23 (1.40)	24.76 (1.86)
SF-2	42.28 (1.75)	10.95 (1.96)	23.88 (4.41)
SF-30	48.69 (1.35)	15.04 (0.92)	20.40 (1.85)
SF-60	48.38 (0.31)	15.95 (0.86)	21.28 (2.77)

Data were shown as "mean (standard deviation)"

change of untreated and pretreated biomass (Vander et al. 2014).

Table 2 shows that ICSE improved the cellulose content from 16.83% (untreated MDF) to 40.25–48.69% (ICSE treated), accompanied with the reduction of "lignin" from 42.88% (untreated MDF) to 19.66–24.76% (ICSE treated) and the change of hemicelluloses content from 9.60% (untreated MDF) to 8.23–15.95% (ICSE treated). The "lignin" content in the solid phase was higher than that of references' data, since the protein, fats, and extractives may mix with lignin to impact the acid insoluble lignin and the acid soluble lignin (Jibril et al. 1999). In spite of considerable degradation of hemicelluloses under high temperature, the removal of abundant organic extractives and proteins helped hemicelluloses maintain the content percentage and even better (Zhan et al. 2013; Rocha et al. 2012). This may also explain the significant reduction of "lignin" content in feedstocks after ICSE pretreatment, since extractives, fats, or proteins in detected "lignin" might easily to be hydrolyzed within higher temperature

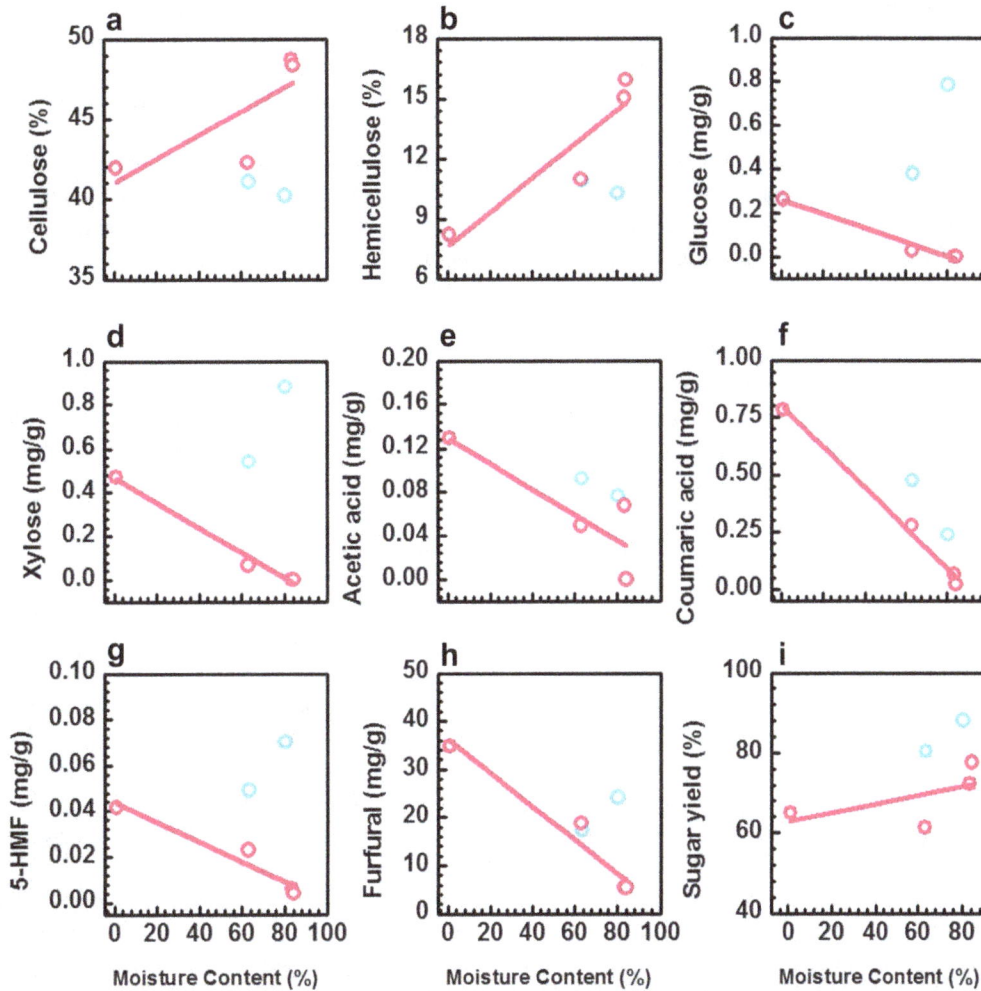

Fig. 1 Effects of moisture content on components, inhibitors, and sugar yield. Carbohydrates (**a** cellulose, **b** hemicellulose) in the solid fraction, sugar (**c** glucose, **d** xylose), and inhibitors (**e** acetic acid, **f** coumaric acid, G 5-HMF, **h** furfural) in the liquid fraction, and sugar yield (**i**) of ICSE-pretreated feedstocks after enzymatic hydrolysis. Cyan plots: FF and NDF; pink plots: MDF, SF-2, SF-30, SF-60

during ICSE pretreatment. Figure 1a, b shows the positive correlation for linear fitting between cellulose (hemicelluloses) and the moisture content in MDF, SF-2, SF-30, and SF-60 after ICSE pretreatment. Those four feedstocks marked in pink color share the oven-drying process in common. The oven-drying process with rapid dehydration of feedstock under higher temperature resulted in significant structure change from FF (Fig. 4). Therefore, it is worth noticing that two cyan scatters representing FF and NDF were non-compliance with the line correlation among those four feedstocks, since the plant's intracellular components would impact the yields of sugar and inhibitor at non-soaking conditions.

Figure 1c, d illustrates a negative correlation between glucose (xylose) in the liquid phase and the moisture content. The R^2-value was more than 0.98 which strongly validates the reliability of linear relationships. Soaking

will dissolve low molecule sugar from feedstock, so the feedstock after long-time soaking lost a plenty of soluble sugar. Though the portion of glucose and xylose was derived from the thermal degradation of cellulose and hemicelluloses, ICSE mainly acts as physical pretreatment and produces little monosaccharides. Therefore, soaking is the prior factor to reduce the glucose and xylose in liquid phase. Interestingly, FF contained mostly glucose and xylose, which were nutrients for sweet potato cell in the vine after harvesting. After 7 days drying, the glucose and xylose decreased due to continuous respiration by living cells in NDF.

The inevitable high-temperature process during ICSE pretreatment converts some lignocellulose to inhibitory by-products such as 5-HMF, furfural, acetic acid, and coumaric acid from hexose, pentose, and lignin, respectively (Table 1). The correlation of inhibitor and moisture content was consistent with the relation between sugar and moisture content (Fig. 1). Among these inhibitors, the coumaric acid and furfural produced from lignin and hemicelluloses, respectively, were higher than others, which reflected that ICSE process could more easily hydrolyze lignin and hemicelluloses than cellulose (Liu et al. 2014a). Since soaking is analogous to washing and can rinse monosaccharides and oligosaccharides out of the material surface (Frederick et al. 2014), longer soaking time diminished the surface residual sugars and subsequently decreased the accumulation of inhibitors during ICSE pretreatment (Table 1).

The pretreatment efficiency is not only determined by the moisture content of feedstock, but also by the method of feedstock handling. Therefore, the structure of feedstock should be analyzed for further understanding of this process.

Thermal degradation analysis of untreated and ICSE-pretreated feedstocks

Thermal stability of the biomass is analyzed by using TGA, a widely adopted technique to determine the thermal degradation of plant biomass (Carrier et al. 2011; Sanchez-Silva et al. 2012). In general, the differential-TG (DTG) curves of lignocellulose biomass often show three main peaks: one reflects the evaporation of extractives or water at 100–200 °C, the others stand for degradation of cellulose and lignin at 300–350 and 300–500 °C, respectively. The DTG curve usually exhibits shoulders in the temperature range of 200–300 °C, corresponding to the hemicelluloses degradation. The relative intensities of each peak could be used to calculate the quantities of hemicelluloses, cellulose, and lignin present in the lignocellulose (Carrier et al. 2011).

Interestingly, the respective hemicelluloses shoulder of MDF's DTG curve almost disappeared after ICSE pretreatment and the height of its maximum peak increased 12.5%

and moved to a higher temperature due to the degradation of hemicelluloses and the related lifting of cellulose (Fig. 2). Noticeably, it is highly consistent for the temperature and area of DTG curves' shoulder with a range from 200 to 350 °C between ICSE-FF and ICSE-MDF due to their minor variance of carbohydrate content, less than 0.66%. This finding was further confirmed by the compositional analysis (Table 2). In summary, ICSE pretreatment degraded the hemicelluloses content and nullified the impact of drying process on the change of lignocellulosic composition when compared to wet feedstock.

Impact of pretreatment on enzymatic hydrolysis

The enzymatic hydrolysis is the key step to obtain the fermentable sugars from lignocellulosic biomass. Before pretreatment, handling operation enhanced the structural and compositional change of feedstock (Table 2 and Fig. 3), which can be confirmed by sugar yield after enzymatic hydrolysis. NDF and MDF owned better sugar yield than FF, and soaking duration further improved the sugar yield. The soaking process could wash away some chemical composition that might inhibit enzymolysis or lead to invalid enzymatic adsorption (Frederick et al. 2014). In addition, the hornification effects due to the drying process for MDF might lead the plant to have a lower accessibility to enzyme or chemical reagents, which might also inhibit the enzymatic hydrolysis (Fernandes et al. 2004).

A considerable pretreatment method should ensure that pretreated feedstocks are suitable for robust saccharification (Agbor et al. 2011). Figure 1 illustrates that ICSE pretreatment significantly enhanced the enzyme hydrolysis when compared to the untreated feedstocks. Especially, the sugar yield of ICSE-FF was shown a fourfold increment and reached to the maximum value (88.05%) of all samples. At this condition, sugar concentration after enzymatic hydrolysis is about 5.18 g/L. Similar improvements were observed in NDF and soaked biomass. These

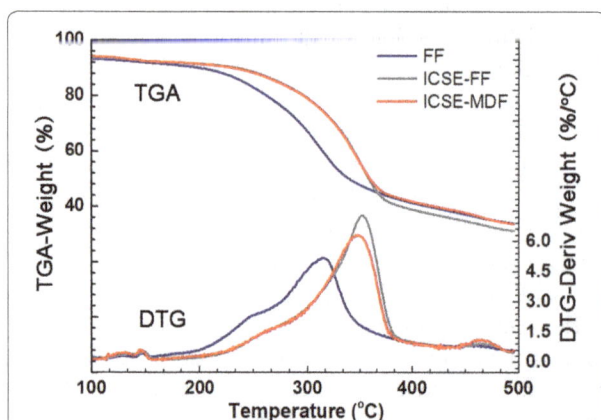

Fig. 2 TGA and DTG analysis of FF, ICSE-FF, and ICSE-MDF

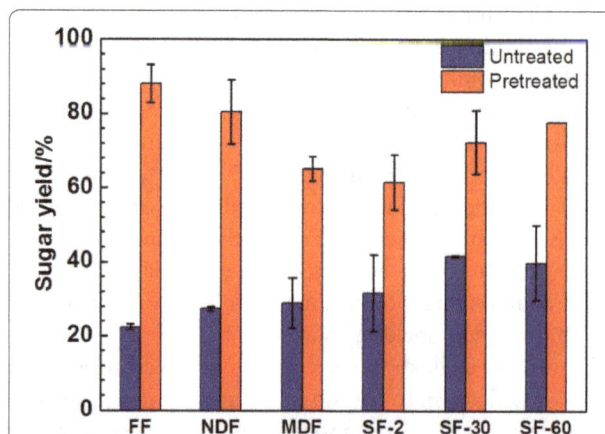

Fig. 3 Sugar yields of untreated and pretreated feedstock

results indicate that higher moisture content is helpful to improve further enzymatic hydrolysis after ICSE pretreatment (Figs. 1i and 3); Sui and Chen drew the same conclusion by using normal steam explosion pretreatment on corn stover (Sui and Chen 2015). However, this impact was less obvious on the samples which were soaked for short durations, i.e., 2 h. In addition, whether it is treated or not, the sugar yield of SF-30 was close to SF-60. So, 30 h is enough to improve the degradability of lignocellulose. Though the moisture content of NDF and SF-2 was similar (about 63%), their sugar yields were quite different due to the hornification effect. Soaking process enabled lignocellulose to absorb moisture, but the shrinkage of internal fibers could not be fully recovered. It also demonstrated that the moisture content was not the only factor to predominate the enzymatic hydrolysis, but structure of feedstocks might be more pivotal on accessibility of cellulase than expected (Fernandes et al. 2004).

The morphological and micro-structural analysis on feedstock

So far, compared with MDF, ICSE pretreatment had more efficient impacts on FF and SF. The macroscopic appearance of feedstocks under different conditions is shown in Fig. 4; both FF and SF-60 were full of water, which seem to be more resilient than dried feedstock. Similar with wood industry, the morphological structure of dried wood could get back in shape after absorbing some water (Wang et al. 2006). Feedstock was destructed by steam explosion which turned into more uniform and reduced the particle size. As an order of destruction, FF occupied first place followed by dried feedstocks. It was interesting to note that soaked feedstock produced non-uniform particles when subjected to ICSE and some of the biomass components were unable to be reached by hot steam. The similar findings were observed previously, because the hornification effects would lead to different biomass structures between FF and SF and potentially impact ICSE pretreatment (Fernandes et al. 2004). Soaking could not infiltrate to all parts due to mass transfer limitation at short time, which formed the un-soaked tissue mainly located in inner part of biomass (Borrega and Kärenlampi 2010).

The sample used for SEM micrographs was dried by freeze-drying that preserved cell structure (Fig. 4), whereas the NDF and MDF undergone a drying process at ambient and moderately high temperature, and their cell wall becomes very vulnerable due to dehydration. As shown in Fig. 4, the pores in FF were biggest among three feedstocks, which facilitate the steam or hot water penetrating biomass and enhancing explosion force to

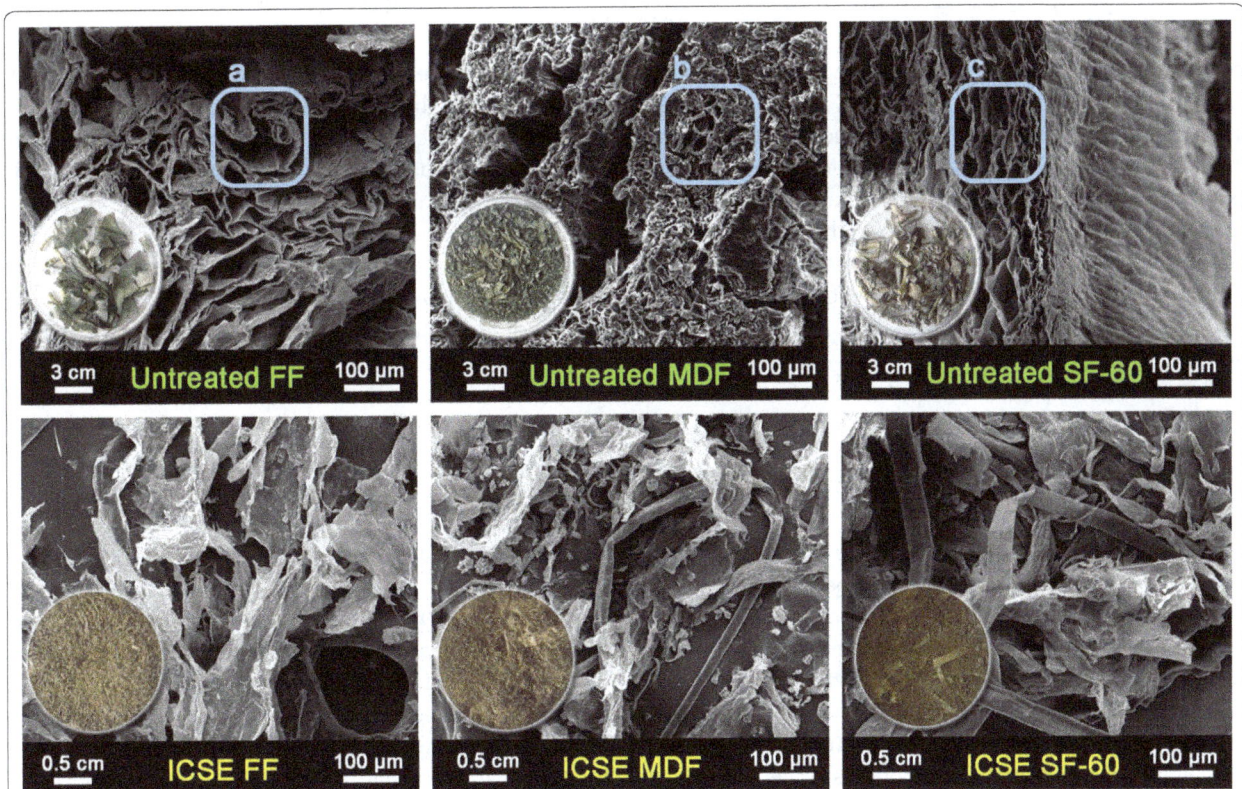

Fig. 4 Photographs (round in the left bottom) and SEM microscopic appearance of biomass. The feedstocks included untreated and ICSE-pretreated sweet potato vines for FF, MDF and SF-60 h. **a–c** mark the pores of lignocellulose

degrade components. When the feedstock was totally dried, all pores in MDF shrank to about one twenty of area of pores in FF due the hornification effects (Fernandes et al. 2004). It blocked the entrance of steam and hot water into pores and lead to a lower accessibility of vapors. If the MDF was soaked in the water, the rehydrating feedstocks tend to return previous status with big pores. Nevertheless, the irreversible ruins on cell wall by drying process have occurred, which leads to the pores of SF much bigger than MDF's, but still smaller than FF's.

After ICSE pretreatment, all feedstocks were similar because of their similar moisture content (Table 1), but there was still obvious different observations by naked eyes. The particles of ICSE-pretreated FF were the smallest and distributed evenly. Inversely, the particles of ICSE-pretreated MDF were not homogeneous and the particle size was bigger than ICSE-pretreated FF particles. Some big debris could be found in pretreated MDF and SF-60.

The role of water and steam in ICSE pretreatment
Given the basic principle of steam explosion, the mechanical force is generated by the expanding gas (water vapors). So, changes of gas volume during the working temperature range may be helpful to enhance the destruction impact of the steam explosion. For

$$m_S = \frac{1 \text{ kg} \times (230 - 25)\text{K} \times (1.7 \text{ kJ/kg} \cdot \text{K} \times (1 - C_m) + 4.37 \text{ kJ/kg} \cdot \text{K} \times C_m)}{1812.6 \text{ kJ/kg}}. \tag{4}$$

convenient evaluation, if pressure changes are ignored, 1 L air at 25 °C would be 1.7 L at 230 °C (Eq. 2). However, 1 L water at 25 °C would be transformed to 2100 L steam at 230 °C (Eq. 3). Therefore, higher moisture content potentially produces more inner steam for mechanical work which subsequently can destroy the compact structure of biomass. But the steam which is out of feedstock only can convert their thermal energy to kinetic energy.

$$\text{Air volume} = \frac{T_2 P_1}{T_1 P_2} V1 - \frac{(230 + 273)\text{K}}{(25 + 273)\text{K}} \times 1\text{L} - 1.7\text{L} \tag{2}$$

$$\text{Steam volume} = \frac{1000\text{g}}{\frac{18\text{g}}{\text{mol}}} \times 37.8 \frac{\text{L}}{\text{mol}} = 2100\text{L} \tag{3}$$

In the small pores of biomass, the steam explosion process followed two steps. The first part is heating: the gas and liquid were heated from room temperature to setting temperature about 230 °C. The water in the biomass remained as liquid phase because the working pressure was always higher than saturated vapor pressure. So, the liquid hot water dissolved partial lignocelluloses and other water-soluble components. The next step was flash depressurizing. When the pressure in working vessel is instantly released to atmosphere pressure,

the vaporization from hot water generated huge volume steam subsequently causing biomass degradation.

Though the presence of higher moisture content seems attractive for steam explosion, however, the high moisture content of biomass would require more energy due to higher specific heat capacity of water. On the other hand, it would not possible for all liquid water to transform into steam during the explosion if the moisture content is high, even with heavier inputs of energy. Roughly calculation for the energy input and steam generation during steam explosion can be undertaken as Eq. 4, which showed the climbing amount of steam (m_s) with the increase of C_m, where, 1 kg biomass with C_m moisture content was heated from 25 to 230 °C (1.8 MPa). Since the change of specific heat capacity is little within this temperature range, the average values of specific heat capacity for water and feedstock are 4.37 and 1.7 kJ/(kg K), respectively. The vaporization heat of water at 230 °C is 1812.6 kJ/kg. When the C_m is lower than 27.50%, theoretically, the biomass should become completely dry subjected to ICSE. But what happened, the condensed water from heating steam increased the moisture content of pretreated biomass. Therefore, the moisture content of wet feedstock helped to remove the hemicelluloses and other water-soluble components, and generated more inner steam for improved explosion impact (Sui and Chen 2015).

Effects of moisture content on lignocellulose pretreatment
Due to the high moisture content of lignocellulose, researchers had studied its effects on various pretreatment methods such as ionic liquid pretreatment, grinding, and steam explosion (Table 3). Grinding of wet lignocellulose would consume more energy and finally have a bigger particle size than dried feedstock (Barakat et al. 2015). Mixing water with ionic liquid would lead to a lower pretreatment efficiency, though it would reduce reagents and the whole pretreatment cost (Shi et al. 2014a).

Fortunately, the moisture content of lignocellulose would improve pretreated efficiency of steam explosion or supercritical CO_2 pretreatment. Compared with dried feedstock, wet lignocellulose would prohibit the pyrolysis of hemicelluloses and facilitates the following enzymatic hydrolysis during steam explosion pretreatment (Sui and Chen 2015). In addition, previous works about some special steam explosion, such as the pre-soaking lignocellulosic biomass with SO_2 before the pretreatment (Cullis et al. 2004) and the addition of ammonia in the steam water (Moniruzzaman et al. 1997), showed higher sugar yield than ones obtained from dried feedstock and in agreement with this study. Considering the high cost of lignocellulosic pretreatment, the steam explosion is a better option. It has an appealing trait, which does not require too much refining operation before pretreatment and allows simplifying the handling on raw feedstock.

Though researchers studied moisture effects on pretreatment efficiency, few of them studied impacts of intracellular moisture. Here, FF showed 88.05% sugar yield after ICSE pretreatment which was competitive when compared to other resources and steam explosion pretreatment. Conclusively, the sweet potato vines and ICSE would be an excellent resource and pretreatment process for the preparation of high-value substrates for microbial fermentation.

The strategies for feedstock collection and transportation

The collection, storage, transportation, and pretreatment of biomass are energy consuming processes. The moisture content has significant impacts on these processes (Kudakasseril et al. 2013). In US, cellulosic ethanol plant uses common feedstock like agriculture residue, which are naturally dried for 1–2 months in the fields and bunched by rotary baler. The dry feedstock could save the cost of transportation (Axelsson et al. 2012). However, this path is not suitable for cellulosic plants in China, because the agricultural residues cannot be allowed to stay too long in the field due to the following farming, and the wet feedstock with high moisture content is heavy enough to cost high transportation charges (Shi et al. 2014b).

The research has demonstrated that the fresh and soaked biomass (pre-dried) performed better after ICSE

pretreatment than dried feedstock, considering the dramatical improvement of carbohydrate content and sugar yield, and the lower production of inhibitors. Therefore, two strategies based on ICSE pretreatment were proposed (Fig. 5) for agricultural countries which are forced to use high-density cultivation such as China and Pakistan to fulfill the requirements of their huge populations.

The first strategy follows the ordinary handling process before pretreatment, but the dry biomass needs to be soaked to regain the moisture content and to wash out the monosaccharides. The high moisture content will benefit the ICSE performance with the improvement of sugar yield. Attractively the soaked water can also be recycled for next soaking, which concentrates the soluble sugar form raw biomass and avoid the accumulation of inhibitors during thermal pretreatment.

In the second proposed strategy, fresh biomass may be directly subjected to ICSE at an adjacent working station near the farm. The liquid phase of pretreated biomass can be either returned to field as fertilizer or utilized for biogas production, since it contains plentiful organic compounds. The second strategy is more suitable to the populous countries, including China, where crop rotation is required either due to seasonal issues or due to heavy requirements of food and feed. For example, in eastern China, the crop residue should be moved quickly within 1 month for the following cultivation; thus, there is no time available for field drying (Rasmussen et al. 1980). Moreover, ICSE can efficiently decrease particle size by destroying the compact structure and elevate the sugar content, when fresh biomass is used. Moreover,

Table 3 Summary for the effects of moisture content on lignocellulose pretreatment

Pretreatment	Biomass	Sugar yield	Advantage	Disadvantage	Refs.
Ionic liquid	Switchgrass	70%	Decrease the cost of ionic liquid pretreatment	The existence of water will hamper the efficiency of pretreatment	Shi et al. (2014a)
Grinding	Wheat straw, corn stover etc.	NA	NA	Result in additional energy requirement and higher final particle size	Barakat et al. (2015)
Supercritical CO_2	Southern yellow pine	84.7%	The increase of initial moisture content will obtain higher final sugar yield	NA	Kim and Hong (2001)
Steam explosion	Corn stover	90%	Improve the sugar yield	Not benefit for the quickly heating of biomass during pretreatment	Sui and Chen (2015)
Ammonia fiber steam explosion	Corn stalk	80%	Increase of moisture content does not hamper the enzymatic hydrolysis of lignocellulose	Ditto	Moniruzzaman et al. (1997)
Steam explosion	Softwood	60–90%	Prompt the enzymatic hydrolysis and reduce the hydrolysis of hemicelluloses content	Ditto	Cullis et al. (2004)
Acid steam explosion	Corn stover	> 90% (Xylose)	High soluble sugar yield	Ditto	Emmel et al. (2003)

NA not available

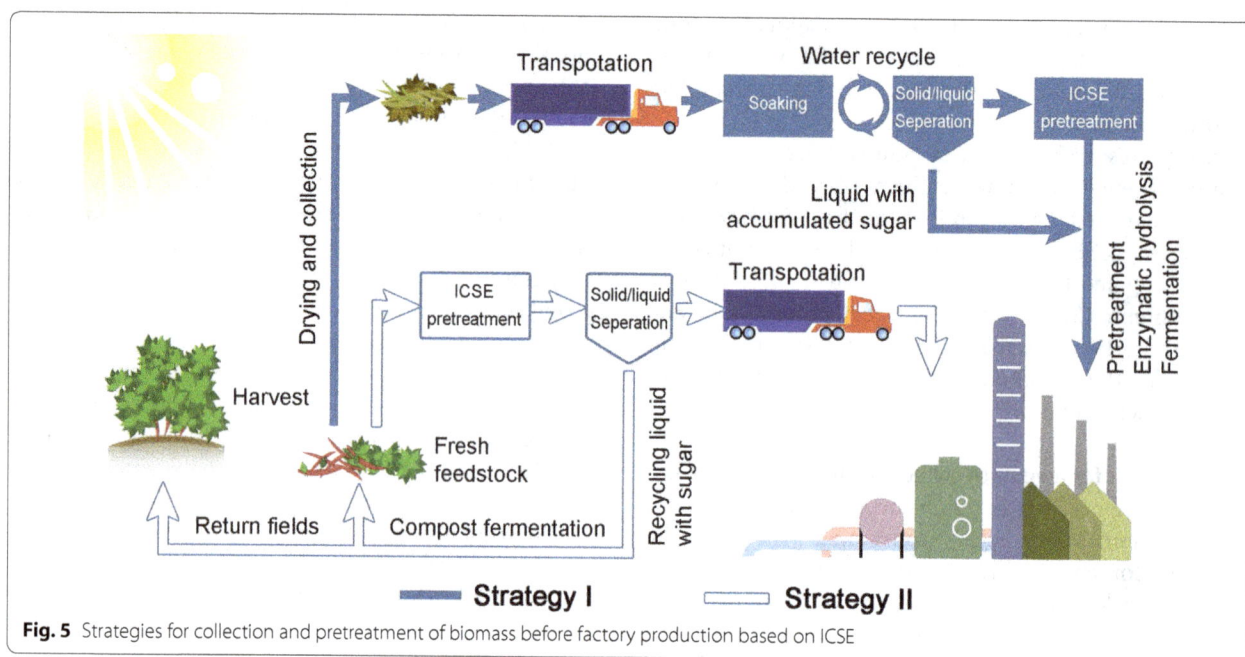

Fig. 5 Strategies for collection and pretreatment of biomass before factory production based on ICSE

transportation of the onsite ICSE-pretreated solid to the biofuel producing plant can also save the time and cost, comparing with onsite drying or transporting fresh biomass (Fig. 5).

Conclusion

The present study focused on developing a low-cost strategy for the pretreatment of feedstock to biofuel-producing industry. The moisture content of lignocellulosic biomass can be utilized to enhance the enzymatic hydrolysis via ICSE, instead of drying. Moreover, ICSE pretreatment raises the carbohydrate content up to 1.43-folds with concomitant lowering of the inhibitors in the hydrolysate. The sugar yield of ICSE-pretreated fresh feedstock was improved by 2.5-folds and reached to 88.05%, which was like soaking the biomass for 60 h, suggesting that the use of fresh biomass would be the best way to run ICSE.

Abbreviations
ICSE: instant catapult steams explosion; FF: fresh feedstock; NDF: naturally dried feedstock; MDF: manually dried feedstock in oven; SF-2: soaking feedstock for 2 h; SF-30: soaking feedstock for 30 h; SF-60: soaking feedstock for 60 h; SEM: scanning electron microscopy; TGA: thermal gravimetric analysis; DTG: differential thermal gravimetric analysis; 5-HMF: 5-hydroxymethyl furfural.

Authors' contributions
LL carried out this experiment, data collection, data analysis, and manuscript preparation. CL conducted the research, investigation process, artwork, and manuscript preparation. JQ and KL made equal contribution to prepare initial manuscript. MAM revised the manuscript to its final form. All authors read and approved the final manuscript.

Author details
[1] State Key Laboratory of Microbial Metabolism, School of Life Sciences and Biotechnology, Shanghai Jiao Tong University, Shanghai 200240, China.
[2] Department of Wood Science, University of British Columbia, Vancouver V6T 1Z4, Canada. [3] School of Life Science and Biotechnology, Dalian University of Technology, Dalian, Liaoning 116023, China. [4] Department of Bioinformatics and Biotechnology, Government College University Faisalabad, Faisalabad 38000, Pakistan.

Acknowledgements
We appreciate the kind support of Prof. Feng-Wu Bai. We also would like to thank Xue-Mi Hao, Bo-Yu Geng, and Bo-Wen Jin for technical assistance and valuable discussion.

Competing interests
The authors declare that they have no competing interests.

Funding
This work was supported by the National Natural Science Foundation of China [Grant Numbers 51561145014, 21536006, 21406030].

References
Agbor VB, Cicek N, Sparling R et al (2011) Biomass pretreatment: fundamentals toward application. Biotechnol Adv 29:675–685. https://doi.org/10.1016/j.biotechadv.2011.05.005

Axelsson L, Franzén M, Ostwald M et al (2012) Perspective: jatropha cultivation in southern India: Assessing farmers' experiences. Biofuels Bioprod Biore fining 6:246–256. https://doi.org/10.1002/bbb

Barakat A, Monlau F, Solhy A, Carrere H (2015) Mechanical dissociation and fragmentation of lignocellulosic biomass: effect of initial moisture, bio chemical and structural proprieties on energy requirement. Appl Energy 142:240–246. https://doi.org/10.1016/j.apenergy.2014.12.076

Borrega M, Kärenlampi PP (2010) Hygroscopicity of heat-treated Norway spruce (Picea abies) wood. Eur J Wood Wood Prod 68:233–235. https://doi.org/10.1007/s00107-009-0371-8

Carrier M, Loppinet-Serani A, Denux D et al (2011) Thermogravimetric analysis as a new method to determine the lignocellulosic composition of biomass. Biomass Bioenergy 35:298–307. https://doi.org/10.1016/j.biombioe.2010.08.067

Cullis IF, Saddler JN, Mansfield SD (2004) Effect of initial moisture content and chip size on the bioconversion efficiency of softwood Lignocellulosics. - Biotechnol Bioeng 85:413–421. https://doi.org/10.1002/bit.10905

Dhakal HN, Zhang ZY, Richardson MOW (2007) Effect of water absorption on the mechanical properties of hemp fibre reinforced unsaturated - polyester composites. Compos Sci Technol 67:1674–1683. https://doi.org/10.1016/j.compscitech.2006.06.019

Emmel A, Mathias AL, Wypych F, Ramos LP (2003) Fractionation of *Eucalyptus grandis* chips by dilute acid-catalysed steam explosion. Bioresour Technol 86:105–115. https://doi.org/10.1016/S0960-8524(02)00165-7

Fao U (2016) FAOSTAT_data_7-19-2017-3. In: Food Agric Organ United Nations. http://www.fao.org/faostat/en/#data/QC

Fernandes DJ, Gil MH, Castro JAAM (2004) Hornification—its origin and interpretation in wood pulps. Wood Sci Technol 37:489–494. https://doi.org/10.1007/s00226-003-0216-2

Frederick N, Zhang N, Ge X et al (2014) Poplar (*Populus deltoides* L.): the effect of washing pretreated biomass on enzymatic hydrolysis and fermentation to ethanol. ACS Sustain Chem Eng 2:1835–1842. https://doi.org/10.1021/sc500188s

Galbe M, Zacchi G (2012) Pretreatment: the key to efficient utilization of lignocellulosic materials. Biomass Bioenergy 46:70–78. https://doi.org/10.1016/j.biombioe.2012.03.026

Gong L, Huang L, Zhang Y (2012) Effect of steam explosion treatment on barley bran phenolic compounds and antioxidant capacity. J Agric Food Chem 60:7177–7184. https://doi.org/10.1021/jf301599a

Jibril H, Perez-Maldonado RA, Mannion PF, Farrell DJ (1999) The nutritive value of sweet potato vines for broilers. In: Australian poultry science symposium, pp 158–161

Jönsson LJ, Alriksson B, Nilvebrant N-O (2013) Bioconversion of lignocellulose: inhibitors and detoxification. Biotechnol Biofuels 6:16. https://doi.org/10.1186/1754-6834-6-16

Kim KH, Hong J (2001) Supercritical CO2 pretreatment of lignocellulose enhances enzymatic cellulose hydrolysis. Bioresour Technol 77:139–144. https://doi.org/10.1016/S0960-8524(00)00147-4

Kudakasseril KJ, Raveendran NG, Hussain A, Vijaya GS (2013) Feedstocks, logistics and pre-treatment processes for sustainable lignocellulosic biorefineries: a comprehensive review. Renew Sustain Energy Rev 25:205–219. https://doi.org/10.1016/j.rser.2013.04.019

Liu CG, Liu LY, Zi LH et al (2014a) Assessment and regression analysis on instant catapult steam explosion pretreatment of corn stover. Bioresour Technol 166:368–372. https://doi.org/10.1016/j.biortech.2014.05.069

Liu LY, Hao XM, Liu CG, Bai FW (2014b) Evaluation of instant catapult steam explosion combined with chemical pretreatments on corn stalk by components and enzymatic hydrolysis analysis. Huagong Xuebao 65:4557–4563

Moniruzzaman M, Dale BE, Hespell RB, Bothast RJ (1997) Enzymatic hydrolysis of high-moisture corn fiber pretreated by AFEX and recovery and recycling of the enzyme complex. Appl Biochem Biotechnol 67(1):113–126

Mosier N, Wyman C, Dale B et al (2005) Features of promising technologies for pretreatment of lignocellulosic biomass. Bioresour Technol 96:673–686. https://doi.org/10.1016/j.biortech.2004.06.025

Qu C, Li B, Wu H, Giesy JP (2012) Controlling air pollution from straw burning in china calls for efficient recycling. Environ Sci Technol 46:7934–7936. https://doi.org/10.1021/es302666s

Rasmussen PE, Allmaras RR, Rohde CR, Roager NC (1980) Crop residue influences on soil carbon and nitrogen in a wheat-fallow system. Soil Sci Soc Am J 44:596–600

Rocha GJM, Gonçalves AR, Oliveira BR et al (2012) Steam explosion pretreatment reproduction and alkaline delignification reactions performed on a pilot scale with sugarcane bagasse for bioethanol production. Ind Crops Prod 35:274–279. https://doi.org/10.1016/j.indcrop.2011.07.010

Sanchez-Silva L, López-González D, Villaseñor J et al (2012) Thermogravimetric-mass spectrometric analysis of lignocellulosic and marine biomass pyrolysis. Bioresour Technol 109:163–172. https://doi.org/10.1016/j.biortech.2012.01.001

Shi J, Balamurugan K, Parthasarathi R et al (2014a) Understanding the role of water during ionic liquid pretreatment of lignocellulose: co-solvent or anti-solvent? Green Chem 16:3830–3840. https://doi.org/10.1039/c4gc00373j

Shi T, Liu Y, Zhang L et al (2014b) Burning in agricultural landscapes: an emerging natural and human issue in China. Landsc Ecol 29:1785–1798

Sluiter A, Hames B, Ruiz R et al (2012) Determination of structural carbohydrates and lignin in biomass. Lab Anal Proced 1617:1–16

Sui W, Chen H (2015) Water transfer in steam explosion process of corn stalk. Ind Crops Prod 76:977–986. https://doi.org/10.1016/j.indcrop.2015.08.001

Tan T, Yu J, Lu J, Zhang T (2010) Biofuels in China. Springer, Berlin

Tian Y, Zhao L, Meng H et al (2009) Estimation of un-used land potential for biofuels development in China. Appl Energy 86:S77–S85. https://doi.org/10.1016/j.apenergy.2009.06.007

Vander PEC, Bakker RR, Baets P, Eggink G (2014) By-products resulting from lignocellulose pretreatment and their inhibitory effect on fermentations for (bio)chemicals and fuels. Appl Microbiol Biotechnol 98:9579–9593. https://doi.org/10.1007/s00253-014-6158-9

Vidal BC, Dien BS, Ting KC, Singh V (2011) Influence of feedstock particle size on lignocellulose conversion—a review. Appl Biochem Biotechnol 164:1405–1421. https://doi.org/10.1007/s12010-011-9221-3

Wang W, Sain M, Cooper PA (2006) Study of moisture absorption in natural fiber plastic composites. Compos Sci Technol 66:379–386. https://doi.org/10.1016/j.compscitech.2005.07.027

Xia L, Wei W, Jiang Y, et al (2013) Comparative analysis of chemical composition of sweet potato vines from different cultivars. 34:238–240. https://doi.org/10.7506/spkx1002-6630-201310052

Zhan J, Lin H, Shen Q et al (2013) Potential utilization of waste sweetpotato vines hydrolysate as a new source for single cell oils production by Trichosporon fermentans. Bioresour Technol 135:622–629. https://doi.org/10.1016/j.biortech.2012.08.068

13

Reconstruction of genome-scale metabolic model of *Yarrowia lipolytica* and its application in overproduction of triacylglycerol

Songsong Wei[1], Xingxing Jian[1], Jun Chen[1], Cheng Zhang[1,2]* and Qiang Hua[1,3]*

Abstract

Background: *Yarrowia lipolytica* is widely studied as a non-conventional model yeast owing to the high level of lipid accumulation. Therein, triacylglycerol (TAG) is a major component of liposome. In order to investigate the TAG biosynthesis mechanism at a systematic level, a novel genome-scale metabolic model of *Y. lipolytica* was reconstructed based on a previous model iYL619_PCP published by our lab and another model iYali4 published by Kerkhoven et al.

Results: The novel model iYL_2.0 contains 645 genes, 1083 metabolites, and 1471 reactions, which was validated more effective on simulations of specific growth rate. The precision of 29 carbon sources utilities reached up to 96.6% when simulated by iYL_2.0. In minimal growth medium, 111 genes were identified as essential for cell growth, whereas 66 essential genes were identified in yeast extract medium, which were verified by database of essential genes, suggesting a better prediction ability of iYL_2.0 in comparison with other existing models. In addition, potential metabolic engineering targets of improving TAG production were predicted by three in silico methods developed in-house, and the effects of amino acids supplementation were investigated based on model iYL_2.0.

Conclusions: The reconstructed model iYL_2.0 is a powerful platform for efficiently optimizing the metabolism of TAG and systematically understanding the physiological mechanism of *Y. lipolytica*.

Keywords: *Yarrowia lipolytica*, Genome-scale metabolic model, iYL_2.0, Triacylglycerol, Gene-level prediction targets

Background

Yarrowia lipolytica is a non-conventional model oleaginous yeast, which is capable of accumulating substantial amounts of neutral lipids, even above than 50% of its dry cell weight (DCW) (Blazeck et al. 2014; Papanikolaou et al. 2002). Today, the chemical industry has been widely developing technology to synthesize multiple kinds of products using oils and fats, such as surfactants, polymers, and biofuels (Ledesma-Amaro and Nicaud 2016). Triacylglycerol (TAG) was discovered as the major constitute of liposome and predominant storage of metabolic energy in microorganism cells, which can also be widely used as valuable resources for dietary consumption and biofuels in industrial application (Yen et al. 2008).

In recent years, great potential of lipid synthesis in *Y. lipolytica* attracts much attention to study its physiological mechanism and explore its application in biofuels. Many efforts, including fermentation and genetic manipulation strategies, have been made for improving TAG production in *Y. lipolytica* (Papanikolaou and Aggelis 2010). It has been reported that the yield of TAG could reach up to 40% of dry cell weight when using glycerol or industrial fats as carbon source (Papanikolaou and Aggelis 2002). In *Y. lipolytica*, TAG can be biosynthesized followed by the Kennedy pathway, and the delta-9 stearoyl-CoA desaturase (SCD) was identified as a rate-limiting enzyme in biosynthesis pathway (Dulermo and Nicaud 2011; Kennedy 1957). Besides, genes YALI0E32769g (DGA1) and YALI0E16797g (LRO1), encoding major triacylglycerol synthases, play important

*Correspondence: cheng.zhang@scilifelab.se; qhua@ecust.edu.cn
[1] State Key Laboratory of Bioreactor Engineering, East China University of Science and Technology, 130 Meilong Road, Shanghai 200237, People's Republic of China
Full list of author information is available at the end of the article

roles in the biosynthesis pathway (Athenstaedt 2011). Additionally, the repression of TGL3 and TGL4, encoding triacylglycerol lipases involved in the degradation of TAG, could lead to high levels of TAG accumulation (Morin et al. 2011). Although much improvement of TAG production and lipid accumulation in *Y. lipolytica* has been achieved, it is still necessary to rationally discover possible limiting factors involved in TAG biosynthesis at a systems level, and in silico analyze its metabolic characteristics under a wide range of genotypic and environmental conditions.

A genome-scale metabolic model (GSMM) is a powerful platform for investigation of microorganism's metabolic processes and rational strain design, as a GSMM may represent the metabolic genotype–phenotype relationships in the microorganism (Feist et al. 2009). GSMMs have been widely used in guidance of strain design, analysis of complex biological phenomena, and contextualization of high-throughput data (Jian et al. 2016a; Loira et al. 2012; Oberhardt et al. 2009; Zhang and Hua 2015). So far, there are three versions of genome-scale metabolic network models of *Y. lipolytica*, namely iYL619_PCP (Pan and Hua 2012), iNL895 (Loira et al. 2012), and iYali4 (Kerkhoven et al. 2016), where yeast consensus network was used as a reference model when iNL895 and iYali4 were reconstructed. Besides, the model iNL895 was integrated automatically, which may cause a series of network gaps when simulating cell growth or the production of several compounds. However, the knowledge in *Y. lipolytica* physiology and metabolism is being updated continuously over time, and an up-to-date metabolic model is most likely required to improve the accuracy of prediction. On the other hand, large quantities of in silico optimization methods have been developed and both gene-level and reaction-level strategies can be identified based on constraint-based flux analysis. Normally, gene-level targets are more feasible and efficient than reaction-level ones according to the complex associations between genes and reactions in metabolic model. For instance, our previously published algorithms, OptGeneKnock (Zhang et al. 2015a), IdealKnock (Gu et al. 2016), and APGC (Jian et al. 2016b), which could provide potential gene-level strategies based on logic transformation of model (LTM) (Zhang et al. 2015b), are more feasible for in vivo implementation.

In this study, to systematically investigate the metabolism of *Y. lipolytica* and gain novel insight into TAG biosynthesis of this strain, a new genome-scale metabolic model of *Y. lipolytica*, named iYL_2.0, was reconstructed. The utilizations of different carbon sources were simulated based on this newly developed metabolic model followed by the validation by experimental data. Also, analysis of gene essentiality of *Y. lipolytica* cells growth

in minimal growth (MG) medium or yeast extract (YE) medium was conducted, and the essential genes predicted based on four *Y. lipolytica* models in YE medium were further verified by database of essential genes (DEG). Additionally, potential genetic strategies were identified by three recent in silico simulation algorithms in order to enhance the biosynthesis of TAG in *Y. lipolytica*, and furthermore, the effects of amino acid supplementation on TAG production were investigated based on the novel model of iYL_2.0.

Results and discussions
Characteristics of the updated genome-scale metabolic model of *Y. lipolytica*

The reconstructed genome-scale metabolic model iYL_2.0 (Additional file 1) included 645 genes, 1083 metabolites, and 1471 reactions (1008 metabolic reactions, 326 transport reactions, and 137 exchange reactions) (Table 1). Compared with iYL619_PCP, the model size of iYL_2.0 was more extensive, but less than iNL895 and iYali4 for the reason that models iNL895 and iYali4 were constructed automatically and several genes, metabolites, and reactions in iNL895 and iYali4 were quoted directly from the metabolic model of *Saccharomyces cerevisiae* (Heavner et al. 2012). The directionalities or reversibility of all metabolic reactions were defined based on thermodynamics, therefore, there exist 266 reversible and 741 irreversible enzymatic reactions in the model of iYL_2.0. Besides, two additional compartments (peroxisome and endoplasmic reticulum) were introduced in iYL_2.0 compared with its previous version of iYL619_PCP. The peroxisome compartment is mainly associated to glyoxylate and dicarboxylate metabolism,

Table 1 Features of four genome-scale metabolic models of *Y. lipolytica*

Features	iYL619_PCP	iNL895	iYali4	iYL_2.0
Compartment	4	16	16	6
Gene	619	895	901	645
Metabolite	843	1847	1683	1083
Reaction	1142	2002	1985	1471
Metabolism reaction	781	1406	1236	1008
Reversible reaction	563	1113	268	266
Irreversible reaction	218	293	968	742
Transport reaction	236	425	595	326
Exchange reaction	125	171	154	137
Biomass	*Y. lipolytica*	*S. cerevisiae*	*S. cerevisiae*	*Y. lipolytica*

while endoplasmic reticulum compartment is related to glycerolipid and glycerophospholipid metabolism.

In addition, 1471 reactions were classified into nine different subsystems based on KEGG pathway database. These subsystems include carbohydrate metabolism, energy metabolism, lipid metabolism, nucleotide metabolism, amino acid metabolism, cofactors and vitamins metabolism, exchange reactions, transport reactions, and other metabolisms (terpenoid backbone biosynthesis and N-glycan biosynthesis). Compared to model iYL619_PCP, the number of reactions in each subsystem of iYL_2.0 was increased (Fig. 1a). The additional reactions in iYL_2.0 were mainly associated with lipid metabolism (79 reactions), amino acid metabolism (54 reactions), and transport reactions (91 reactions). Specifically, lipid metabolism, the largest subsystem in iYL_2.0, accounts for 18.6% of the total reactions (Fig. 1b), in comparison with 16.7% in iYL619_PCP and 14.2% in iYali4.

Analysis and verification of model iYL_2.0

In order to quantitatively demonstrate the advantage of iYL_2.0, we firstly simulate the in silico cell growth characteristics in minimal growth medium based on three previously constructed models and the new model iYL_2.0 using flux balance analysis (FBA) (Orth et al.

2010; Toya and Shimizu 2013). The maximum specific growth rate obtained from iYL_2.0 was 0.0378 h^{-1}, which agreed very well with the experimental value (0.0352 h^{-1}) (Oberhardt et al. 2009), better than the results predicted by iYL619_PCP (0.0439 h^{-1}), iNL895 (3.4367 h^{-1}), and iYali4 (1.7454 h^{-1}) (Table 2). The inaccuracy of simulation result by model iNL895 was largely due to the biomass equation used in iNL895 which was integrated automatically from the yeast consensus model version 4.36 (Herrgard et al. 2008) and might lack of essential gap filling. Similarly, the inconsistency between iYali4 and experimental data might also attribute to the fact that the biomass equation used in iYali4 was quoted directly from the yeast consensus network version 5 (Heavner et al. 2012).

In addition, these models were employed to investigate in silico cell growth on 29 different carbon sources by FBA, in which each substance was used as a sole carbon source and cell growth was optimized. The utilization of different substrates by *Y. lipolytica* has previously been reported (Sitepu et al. 2014; Wehrspann and Fullbrandt 1985). According to the above literatures, *Y. lipolytica* could grow on 11 carbon sources, while cell grown was not detected when other 18 compounds were used as carbon sources (Table 3). However, the utilization of 5, 9, and 7 out of 29 carbon sources were not correctly predicted by using models iYL619_PCP, iNL895, and iYali4, respectively. In comparison, the newly developed model iYL_2.0 could be employed to simulate the growth characteristics on 28 carbon sources (expected for hexadecane). Models iYL619_PCP, iNL895, and iYali4 showed false positive growth on trehalose, while iYL_2.0, with trehalose glucohydrolase removed after careful consideration, showed no growth and were consistent with the experimental data. Moreover, iYL619_PCP, the earlier version of iYL_2.0, failed to show the positive growth on D-mannitol and D-glucitol, which was then fixed by adding two reactions (D-mannitol:NAD$^+$ 2-oxidoreductase and D-glucitol:NAD$^+$ 2-oxidoreductase) enabling the conversion of D-mannitol or D-sorbitol to beta-D-fructose (Scolnick and Lin 1962). On the other hand, both iNL895 and iYali4 exhibited predictions of positive

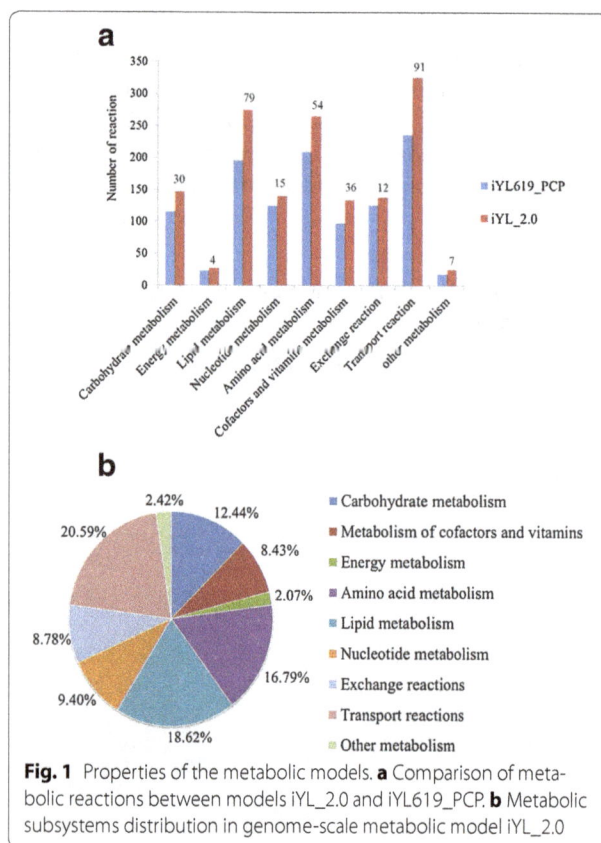

Fig. 1 Properties of the metabolic models. **a** Comparison of metabolic reactions between models iYL_2.0 and iYL619_PCP. **b** Metabolic subsystems distribution in genome-scale metabolic model iYL_2.0

Table 2 Comparison of in silico and in vivo biomass growth rates of *Y. lipolytica* grown in MG medium

Medium condition (mmol/gDCW/h)	In vivo	Growth rate (h^{-1})			
		In silico			
		iYL619_PCP	iNL895	iYali4	iYL_2.0
Glucose ($v = 20$)	0.0352	0.0439	3.4367	1.7454	0.0378

In vivo, experimental results; in silico, simulation results. Biomass reaction in GSMM *Y. lipolytica* was set as the objective function

Table 3 Comparison between simulated results by four models and experimental results on the utilization of different carbon sources in *Y. lipolytica*

Substrate	in vivo	in silico			
		iYL619_PCP	iNL895	iYali4	iYL_2.0
D-glucose	+	🟩	🟩	🟩	🟩
Ethanol	+	🟩	🟩	🟩	🟩
Glycerol	+	🟩	🟩	🟩	🟩
N-acetyl-D-glucosamine	+	🟩	⬛	⬛	🟩
Citrate	+	🟩	🟩	🟩	🟩
D-lactate	+	🟩	🟩	🟩	🟩
L-lactate	+	🟩	🟩	🟩	🟩
Succinate	+	⬛	🟩	🟩	🟩
Hexadecane	+	⬛	⬛	⬛	⬛
D-mannitol	+	⬛	⬛	🟩	🟩
D-sorbitol	+	⬛	🟩	🟩	🟩
Sucrose	-	🟩	⬛	⬛	🟩
Cellobiose	-	🟩	⬛	⬛	🟩
D-glucosamine	-	🟩	⬛	⬛	🟩
Lactose	-	🟩	🟩	🟩	🟩
Inositol	-	🟩	🟩	🟩	🟩
Trehalose	-	⬛	⬛	⬛	🟩
Methanol	-	🟩	🟩	🟩	🟩
L-rhamnose	-	🟩	🟩	🟩	🟩
Galactitol	-	🟩	🟩	🟩	🟩
Maltose	-	🟩	⬛	🟩	🟩
Melibiose	-	🟩	⬛	🟩	🟩
Salicin	-	🟩	🟩	🟩	🟩
D-xylose	-	🟩	⬛	⬛	🟩
L-arabinose	-	🟩	🟩	🟩	🟩
D-arabinose	-	🟩	🟩	🟩	🟩
Raffinose	-	🟩	🟩	🟩	🟩
Mehyl-D-glucoside	-	🟩	🟩	🟩	🟩
Inulin	-	🟩	🟩	🟩	🟩

In vivo: experimental results, In silico: simulation results. '+' represents that the carbon source could be utilized and '−' represents not growth on that carbon source. The green color indicates that in silico result was consistent with in vivo data, whereas red color for the inconsistency. The uptake rate of each carbon source was set at 20 mmol/gDCW/h, and uptake rates of sulfate, phosphate, ammonia, O_2, H_2O, CO_2 and H^+ were set as 1000 mmol/gDCW/h

growth of *Y. lipolytica* on sucrose, cellobiose, D-glucosamine, and D-xylose, and iNL895 also showed positive growth of the strain on maltose and melibiose, which were, however, inconsistent with the in vivo experimental result. The false positive growths on sucrose, cellobiose and D-glucosamine probably attributed to a direct quotation of yeast reactions in iNL895 and iYali4, while the false positive growths on D-xylose, melibiose, and maltose may be activated by intention for simulation purpose (Forster et al. 2007; Wang et al. 2014). Interestingly, all four models considered in this study showed false positive growth on hexadecane. We found that when the annotated genes encoding the required enzymes were added to four models, hexadecane could not be utilized as a sole carbon source, which might be due to the absence of the corresponding regulatory mechanism in the four models.

In addition, the essentialities of individual gene of *Y. lipolytica* grown in both minimal growth medium and yeast extract medium were investigated based on the four above models. All genes involved in iYL_2.0 were classified into three categories: essential genes, partially essential genes, and non-essential genes. As concluded in Additional file 2, 111 (17.21% of the total) and 66 (10.23% of the total) genes were found as essential cells grown in MG medium and YE medium, respectively. The simulated 66 essential genes in YE medium were further examined via BLASTp with the database of essential genes (DEG) (Gao et al. 2015), where 35 genes were found to have a high homology with *S. cerevisiae* (Giaever et al. 2002). For instance, UMP phosphotransferase (YALI0F09339g) is involved in the synthesis of CMP and UMP, which was essential to cell growth. Similarly, thioredoxin reductase (YALI0D27126g) is essential in maintaining the balance of intracellular $NADP^+$/NADPH, which is vital to the whole metabolic process of *Y. lipolytica*. In comparison, only 31 of 71, 25 of 143, and 29 of 112 of essential genes obtained from the simulation of iYL619_PCP, iNL895, and iYali4 were confirmed by DEG, respectively, indicating a better prediction performance of essential genes by iYL_2.0. Further, the essentialities of genes involved in eight subsystems of iYL_2.0 are shown in Fig. 2. Most essential genes were found distributed in amino acid metabolism, lipid metabolism, and nucleotide metabolism in case of MG medium. When analyzing gene essentiality for cells grown in YE medium, only approximately 3% genes categorized in amino acid metabolism were considered essential for cell growth due to the fact that yeast extract contains a mix of peptides, several free amino acids, purine, and pyrimidine bases. For transport reaction, the essential genes associated

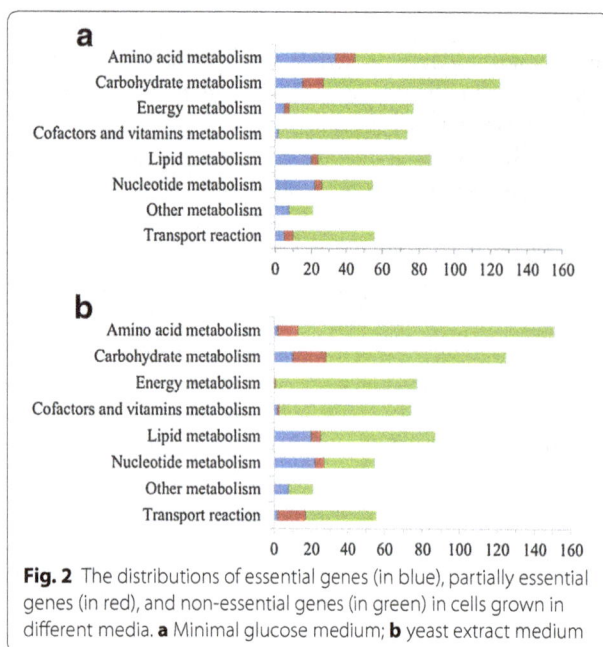

Fig. 2 The distributions of essential genes (in blue), partially essential genes (in red), and non-essential genes (in green) in cells grown in different media. **a** Minimal glucose medium; **b** yeast extract medium

with amino acids metabolism were mainly identified in cells grown in MG medium as the uptake of amino acids needs transport reactions. In addition, for genes categorized in other metabolism, the proportions of essential genes in this category were similar when cells were cultivated in both MG medium and YE medium, implying fewer alternatives routes for biosynthesis of some terpenoids and *N*-glycan in cell metabolic network.

Prediction of metabolic engineering strategies for improving triacylglycerol production

Yarrowia lipolytica is capable of producing TAG by the Kennedy pathway (Fig. 3). As an acyl donor, fatty acyl-CoA is mostly supplied by fatty acid synthase system (FAS) (Schweizer et al. 1988; Wang et al. 2011). In addition, acetyl-CoA and malonyl CoA are essential carbon donor molecules for TAG biosynthesis. In model iYL_2.0, there are multiple pathways that could supply the above essential molecules, including citrate shuttle and conversion from the TCA cycle, amino acids degradation, and other acetyl-CoA generation process (Beopoulos et al. 2011). In addition to TAG production, acetyl-CoA and malonyl-CoA are also used for fatty acid synthesis, amino acid synthesis, and glycerophospholipid synthesis.

To gain additional insight into TAG biosynthesis in *Y. lipolytica*, our previously published simulation methods OptGeneKnock (Zhang et al. 2015a), IdealKnock (Gu et al. 2016), and APGC (Jian et al. 2016b), were employed to identify metabolic engineering strategies to enhance TAG production in cells grown MG medium. The OptGeneKnock algorithm was modified from OptKnock

algorithm through LTM (Zhang et al. 2015), where OptKnock is able to screen the best potential reaction knockout strategies based on a bilevel programming framework. However, IdealKnock algorithm is a top-down framework that scans for the mutants of interest firstly, followed by the identification of the knockout target by iterative searching process. The best single deletion strategy identified by OptGeneKnock was the knockout of carbonate hydrolyase (YALI0F21406g) from the metabolic network, leading to an increase of specific TAG production rate from 0.0126 to 0.0810 mmol/gDCW/h together with the optimal cell growth. The carbonate hydrolyase associated with enzymatic reaction R0450 is involved in energy metabolism and nitrogen metabolism of *Y. lipolytica*. By comparison of the in silico metabolic flux distribution before and after the deletion of YALI0F21406g, we observed that more carbon fluxes were distributed into acetyl-CoA and malonyl-CoA metabolism while fluxes of amino acid synthesis pathway and synthesis pathways of other compounds decreased. Moreover, prediction of double deletion strategy with OptGeneKnock indicated that in addition to YALI0F21406g, further deletion of gene YALI0F15587g (D-glyceraldehyde-3-phosphate glyceronetransferase, R0487) could improve TAG production rate. Actually, TAG synthesis was reported to be indirectly enhanced together with the decrease in pentose phosphate pathway flux and increase in glycerol formation when R0487 was removed from *Y. lipolytica* (Tai and Stephanopoulos 2013).

Meanwhile, another quadruple knockout target (YALI0F19514g, YALI0B19382g, YALI0A19206g, and YALI0D06325g) was identified by IdealKnock program of the COBRA Toolbox v2.0 (Additional file 3). O-Acyltransferase (YALI0F19514g) is a significant enzyme responsible for the synthesis of phosphatidylcholine from 1-acyl-sn-glycero-3-phosphocholine, which was essential in the last step of TAG biosynthesis (Makri et al. 2010). The deficiency of O-acyltransferase might decrease the flux of phosphatidylcholine to glycerophospholipid synthesis and therefore improve the TAG production directly. The gene YALI0B19382g is associated with three reactions (R0161, R0164, and R0176) in the FAS pathway. The in silico deletion of YALI0B19382g could suppress the fatty acid biosynthesis from fatty acyl-CoA, and the TAG synthesis was enhanced accordingly. Furthermore, the removal of YALI0A19206g and YALI0D06325g responsible for glutamate degradation might contribute to the activation of acetyl-CoA carboxylase and the formation of malonyl-CoA due to increased availability of intracellular glutamate (Yu et al. 2003).

To further investigate gene targets of overexpression for enhanced TAG production in MG medium, we

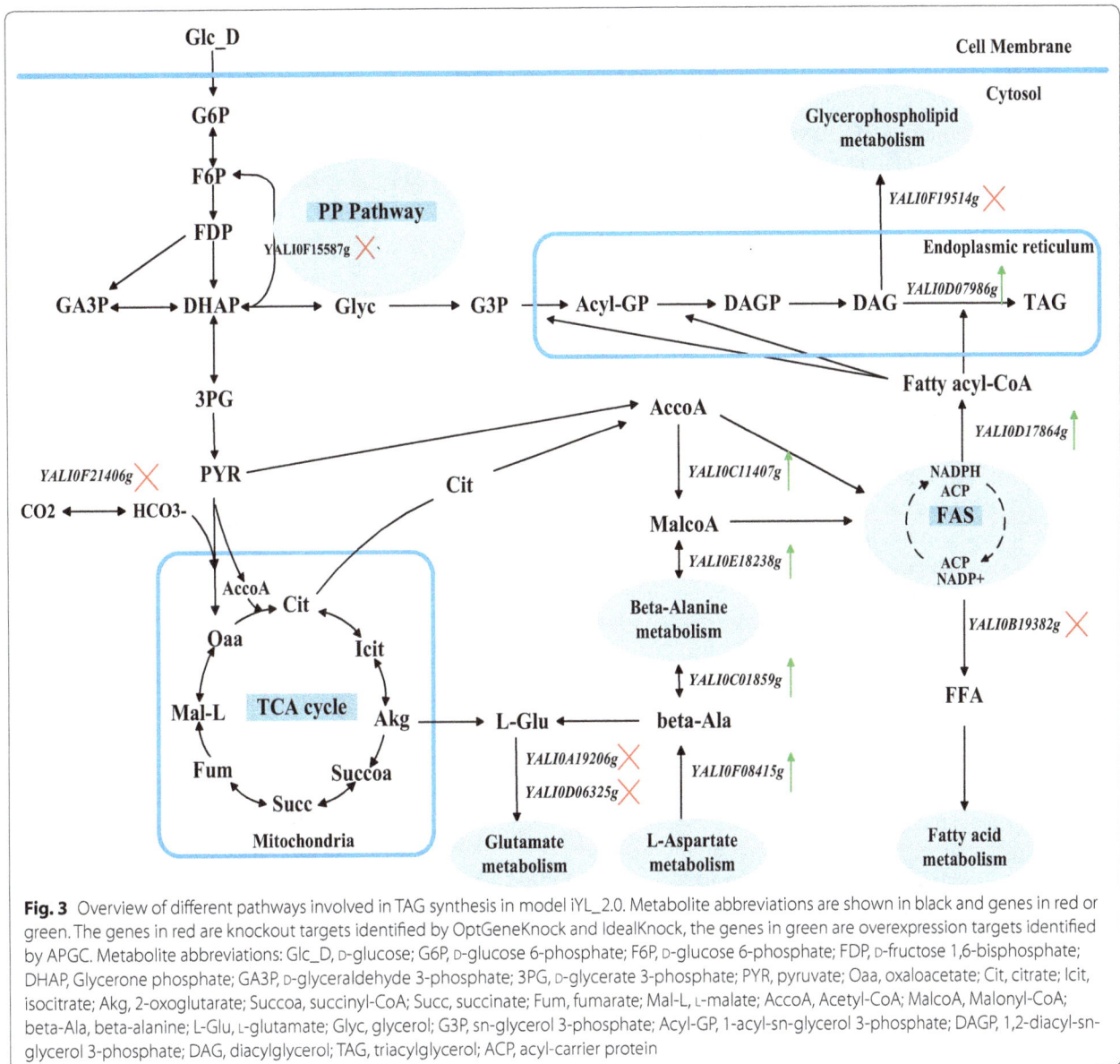

Fig. 3 Overview of different pathways involved in TAG synthesis in model iYL_2.0. Metabolite abbreviations are shown in black and genes in red or green. The genes in red are knockout targets identified by OptGeneKnock and IdealKnock, the genes in green are overexpression targets identified by APGC. Metabolite abbreviations: Glc_D, D-glucose; G6P, D-glucose 6-phosphate; F6P, D-glucose 6-phosphate; FDP, D-fructose 1,6-bisphosphate; DHAP, Glycerone phosphate; GA3P, D-glyceraldehyde 3-phosphate; 3PG, D-glycerate 3-phosphate; PYR, pyruvate; Oaa, oxaloacetate; Cit, citrate; Icit, isocitrate; Akg, 2-oxoglutarate; Succoa, succinyl-CoA; Succ, succinate; Fum, fumarate; Mal-L, L-malate; AccoA, Acetyl-CoA; MalcoA, Malonyl-CoA; beta-Ala, beta-alanine; L-Glu, L-glutamate; Glyc, glycerol; G3P, sn-glycerol 3-phosphate; Acyl-GP, 1-acyl-sn-glycerol 3-phosphate; DAGP, 1,2-diacyl-sn-glycerol 3-phosphate; DAG, diacylglycerol; TAG, triacylglycerol; ACP, acyl-carrier protein

performed APGC algorithm with the use of newly constructed model iYL_2.0. The prediction results showed that 6 genes were identified as potential targets for overexpression (Additional file 3). All these gene targets play important roles in TAG synthetic pathway and these genes are involved in amino acids metabolism, degradation of fatty acids, glycerolipid metabolism, and the growth of biomass. For example, the overexpression of acetyl-CoA:carbon-dioxide ligase (YALI0C11407g) led to an increase in malonyl-CoA synthesis, which hence improved the availability of acetyl-CoA for TAG biosynthesis (Qiao et al. 2015). Three potential gene targets (YALI0F08415g, YALI0C01859g, and YALI0E18238g)

categorized in amino acids metabolism are responsible for the conversion of amino acids into acetyl-CoA and malonyl-CoA. In addition, reaction R0310 (YALI0D07986g) is the last step of TAG biosynthetic pathway, and overexpressing this step might contribute directly to the production enhancement of TAG, which was reported by Yen et al. (2005). The simulation of the overexpression of the above 6 genes showed a 34.1% improvement of TAG production rate compared to the wide-type strain (from 0.0126 to 0.0169 mmol/gDCW/h).

Amino acid metabolism always plays a significant role in lipid metabolism. During the biosynthesis of TAG, additional supplementation of amino acids might be able to promote the biomass growth and TAG production. For this purpose, the effects of individual amino acid additions in MG medium on TAG production were simulated (Additional file 4). The uptake rate of each amino acid was set at 20 mmol/gDCW/h. As shown in Fig. 4, the supplementation of L-glycine, L-alanine, L-cysteine, L-serine, L-threonine, and L-aspartate improved TAG production through enhancing the supply of acetyl-CoA, similar to the results reported for *Schizochytrium limacinum* (Ye et al. 2015a). The TAG production rate was increased by 23.3, 27.7, 27.8, 42.0, 55.5, and 55.5%, respectively. The uptake of above-mentioned amino acids might contribute to the enhanced supply of acetyl-CoA and malonyl-CoA, which therefore increased the production rate of TAG. One example is that additional L-threonine in the medium might be converted to intracellular acetyl-CoA via acetaldehyde:NAD^+/$NADP^+$ oxidoreductase (R0361/R0362) and acetate:CoA ligase (R0358). In addition, via L-aspartate ligase (R0887) and 2-(Nomega-L-arginino) succinate arginine-lyase (R0049), L-aspartate could be converted into either fumarate to enhance the TCA cycle or malonyl-CoA for TAG biosynthesis.

Conclusions

In this study, a novel GSMM of *Y. lipolytica*, named iYL_2.0, was reconstructed based on two previous models and metabolic information from updated databases and recent literatures. The new model contains 645 genes, 1083 metabolites, and 1471 reactions across different compartments. This model was found superior to the two previous models when employed to simulate cell growth characteristics on glucose or other carbon sources. Gene essentiality analysis resulted in 111 essential genes for cells grown in MG medium and 66 essential genes when cells cultivated in YE medium. Using model iYL_2.0, double and quadruple gene knockout targets that could facilitate TAG production were identified by OptGeneKnock and IdealKnock, respectively. In addition, 6 gene amplification targets were predicted by APGC framework, which could lead to a 34.1% improvement of TAG production for increasing the availability of important precursors of acetyl-CoA and malonyl-CoA. Further simulation with iYL_2.0 indicated that the additional supplementation of six amino acids, namely L-glycine, L-alanine, L-cysteine, L-serine, L-threonine, and L-aspartate, might largely contribute to the enhancement of TAG production, where the addition of either L-threonine or L-aspartate resulted in a 55.5% increase in TAG production rate. In summary, the newly developed genome-scale metabolic model iYL_2.0 may serve as a powerful platform for better understanding metabolic viability and guiding rational metabolic engineering strategies for *Y. lipolytica*.

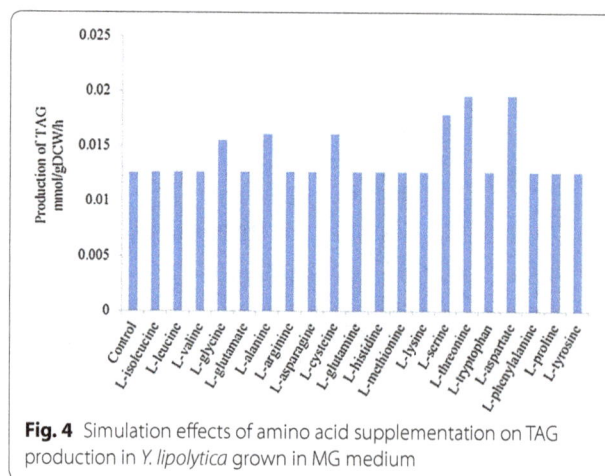

Fig. 4 Simulation effects of amino acid supplementation on TAG production in *Y. lipolytica* grown in MG medium

Methods

Reconstruction of *Y. lipolytica* GSMM

The draft model was reconstructed by amassing reactions from two reference models of *Y. lipolytica*, iYL619_PCP (Pan and Hua 2012), and iYali4 (Kerkhoven et al. 2016). To update and expand the draft model, the absent metabolic reactions were added from KEGG database (Kanehisa et al. 2008). Three databases UniProt (Apweiler et al. 2004), CHEBI (Degtyarenko et al. 2008), and PubChem (Wang et al. 2009) were used to characterize the annotation of metabolites based on their chemical formulas and neutral charges. Additionally, the reaction reversibilities were judged by BioPath (Schreiber 2002) database and thermodynamic data of metabolites. Compared with iYL619_PCP, two new compartments, the peroxisome and endoplasmic reticulum, were determined according to the other reference model iYali4 and subcellular localization prediction tool CELLO (Yu et al. 2004). Transport reactions and exchange reactions were added for metabolic network connectivity. Moreover, to refine the draft model, metabolic gaps were identified by literature data and the GapFind (Satish Kumar et al. 2007) program in COBRA Toolbox v2.0 (Schellenberger et al. 2011).

Simulation and constraint-based flux analysis

The verification and analysis included optimal cell growth in minimal growth (MG) medium (Lanciotti et al. 2005) and utilization of different carbon sources. Meanwhile, essential genes were analyzed by *singlegenedeletion* function in COBRA Toolbox v2.0 in MG medium and YE medium (Ye et al. 2015b). The MG medium only contains basic elements, such as C, H, O, N, P, and S. The glucose uptake rate was set as 20 mmol/gDCW/h, and uptake rates of sulfate, O_2, phosphate, ammonia, H_2O,

CO_2, and H^+ were set as 1000 mmol/gDCW/h. The YE medium contains not only basic elements, but also 20 amino acids, and the uptake rates of amino acids were set as 5 mmol/gDCW/h. Biomass equation was set as the objective function for the different conditions, and optimization was performed by GLPK solver.

The in silico simulations were performed by flux balance analysis (FBA), which is a constraint-based flux analysis method (Orth et al. 2010; Toya and Shimizu 2013). Meanwhile, under a pseudo-steady-state metabolic model, FBA determines metabolic flux distribution based on the assumption that organisms could endeavor to optimize growth or production. Mathematically, the optimization of objective functions subjected to stoichiometric and capacity constraints can be formulated as follows:

$$\max \quad Z = \sum_j c_j v_j$$

$$S.t \quad \sum_j S_{ij} v_j = 0; \quad \forall \text{ metabolite } i \tag{1}$$

$$v_j^{min} \le v_j \le v_j^{max}; \quad \forall \text{ reaction } j.$$

Z represents the cellular objective function of all the metabolic reactions where the relative weights are determined by the coefficient c_j. S_{ij} corresponds to the stoichiometric coefficient of the metabolite i involved in reaction j. v_j denotes the flux or specific rate of metabolic reaction j. v_j^{min} and v_j^{max} refer to the lower and upper bounds of the flux of reaction j, respectively.

Identification of potential targets through gene-level algorithms

For analysis of TAG production, two novel knockout algorithms OptGeneKnock (Zhang et al. 2015a) and IdealKnock (Gu et al. 2016) and one overexpression algorithm, analysis of production and growth coupling (APGC) (Jian et al. 2016b) were carried out for cells grown in MG medium. The APGC program was applied to identify amplification targets for improving production, and OptGeneKnock and IdealKnock were used to identify knockout targets. The OptGeneKnock algorithm was performed by applying logic transformation of model (LTM) (Zhang et al. 2015a) to OptKnock (Burgard et al. 2003). GLPK solver and Gurobi solver were used for linear programming and quadratic programming, respectively. The synthesis reaction of TAG, R0310, was set as objective function to identify the potential targets, and the minimum specific growth rate of biomass was set as 0.0378 h^{-1} in MG medium.

Additional files

Additional file 1. Detailed information about metabolic model iYL_2.0.

Additional file 2. Gene essentiality analysis in minimal growth medium and yeast extract medium by flux balance analysis.

Additional file 3. Potential gene-level targets identified by OptGeneKnock, Idealknock and APGC to improve TAG production.

Additional file 4. Simulation results of amino acid supplementation for TAG production.

Abbreviations

TAG: triacylglycerol; DCW: dry cell weight; SCD: delta-9 stearoyl-CoA desaturase; GSMM: genome-scale metabolic model; MG: minimal growth; YE: yeast extract; FBA: flux balance analysis; DEG: database of essential genes; FAS: fatty acid synthase system; LTM: logic transformation of model; APGC: analysis of production and growth coupling.

Authors' contributions

QH and CZ developed and designed the study. SW carried out the reconstruction of iYL_2.0, performed the statistical analysis, and prepared the manuscript. XJ provided help for implementation of the in silico algorithms. QH and CZ revised the manuscript. All authors read and approved the final manuscript.

Author details

[1] State Key Laboratory of Bioreactor Engineering, East China University of Science and Technology, 130 Meilong Road, Shanghai 200237, People's Republic of China. [2] Science for Life Laboratory, KTH-Royal Institute of Technology, Stockholm SE-171 21, Sweden. [3] Shanghai Collaborative Innovation Center for Biomanufacturing Technology, 130 Meilong Road, Shanghai 200237, China.

Acknowledgements

We thank Prof. Jens Nielsen for providing the Y. lipolytica model of iYali4 and Prof. David James Sherman for providing the Y. lipolytica model of iNL895.

Competing interests

The authors declare that they have no competing interests.

Funding

This study was financially supported by National Natural Science Foundation of China (21776081, 21576089).

References

Apweiler R, Bairoch A, Wu CH, Barker WC, Boeckmann B, Ferro S, Gasteiger E, Huang H, Lopez R, Magrane M et al (2004) UniProt: the universal protein knowledgebase. Nucleic Acids Res 32:D115–D119

Athenstaedt K (2011) YALI0E32769g (DGA1) and YALI0E16797g (LRO1) encode major triacylglycerol synthases of the oleaginous yeast Yarrowia lipolytica. Biochem Biophys Acta 1811:587–596

Beopoulos A, Nicaud JM, Gaillardin C (2011) An overview of lipid metabolism in yeasts and its impact on biotechnological processes. Appl Microbiol Biotechnol 90:1193–1206

Blazeck J, Hill A, Liu LQ, Knight R, Miller J, Pan A, Otoupal P, Alper HS (2014) Harnessing Yarrowia lipolytica lipogenesis to create a platform for lipid and biofuel production. Nat Commun 5:3131

Burgard AP, Pharkya P, Maranas CD (2003) Optknock: a bilevel programming framework for identifying gene knockout strategies for microbial strain optimization. Biotechnol Bioeng 84:647–657

Degtyarenko K, de Matos P, Ennis M, Hastings J, Zbinden M, McNaught A, Alcantara R, Darsow M, Guedj M, Ashburner M (2008) ChEBI: a database and ontology for chemical entities of biological interest. Nucleic Acids Res 36:D344–D350

Dulermo T, Nicaud JM (2011) Involvement of the G3P shuttle and beta-oxidation pathway in the control of TAG synthesis and lipid accumulation in *Yarrowia lipolytica*. Metab Eng 13:482–491

Feist AM, Herrgard MJ, Thiele I, Reed JL, Palsson BO (2009) Reconstruction of biochemical networks in microorganisms. Nat Rev Microbiol 7:129–143

Forster A, Aurich A, Mauersberger S, Barth G (2007) Citric acid production from sucrose using a recombinant strain of the yeast *Yarrowia lipolytica*. Appl Microbiol Biotechnol 75:1409–1417

Gao F, Luo H, Zhang CT, Zhang R (2015) Gene essentiality analysis based on DEG 10, an updated database of essential genes. Methods Mol Biol 1279:219–233

Giaever G, Chu AM, Ni L, Connelly C, Riles L, Veronneau S, Dow S, Lucau-Danila A, Anderson K, Andre B et al (2002) Functional profiling of the *Saccharomyces cerevisiae* genome. Nature 418:387–391

Gu D, Zhang C, Zhou S, Wei L, Hua Q (2016) IdealKnock: a framework for efficiently identifying knockout strategies leading to targeted overproduction. Comput Biol Chem 61:229–237

Heavner BD, Smallbone K, Barker B, Mendes P, Walker LP (2012) Yeast 5—an expanded reconstruction of the *Saccharomyces cerevisiae* metabolic network. BMC Syst Biol 6:55

Herrgard MJ, Swainston N, Dobson P, Dunn WB, Arga KY, Arvas M, Bluthgen N, Borger S, Costenoble R, Heinemann M et al (2008) A consensus yeast metabolic network reconstruction obtained from a community approach to systems biology. Nat Biotechnol 26:1155–1160

Jian X, Li N, Zhang C, Hua Q (2016a) In silico profiling of cell growth and succinate production in *Escherichia coli* NZN111. Bioresour Bioprocess 3:48

Jian X, Zhou S, Zhang C, Hua Q (2016b) In silico identification of gene amplification targets based on analysis of production and growth coupling. Bio Syst 145:1–8

Kanehisa M, Araki M, Goto S, Hattori M, Hirakawa M, Itoh M, Katayama T, Kawashima S, Okuda S, Tokimatsu T et al (2008) KEGG for linking genomes to life and the environment. Nucleic Acids Res 36:D480–D484

Kennedy EP (1957) Metabolism of lipides. Annu Rev Biochem 26:119–148

Kerkhoven EJ, Pomraning KR, Baker SE, Nielsen J (2016) Regulation of amino-acid metabolism controls flux to lipid accumulation in *Yarrowia lipolytica*. NPJ Syst Biol Appl 2:16005

Lanciotti R, Gianotti A, Baldi D, Angrisani R, Suzzi G, Mastrocola D, Guerzoni ME (2005) Use of *Yarrowia lipolytica* strains for the treatment of olive mill wastewater. Bioresour Technol 96:317–322

Ledesma-Amaro R, Nicaud JM (2016) *Yarrowia lipolytica* as a biotechnological chassis to produce usual and unusual fatty acids. Prog Lipid Res 61:40–50

Loira N, Dulermo T, Nicaud JM, Sherman DJ (2012) A genome-scale metabolic model of the lipid-accumulating yeast *Yarrowia lipolytica*. BMC Syst Biol 6:35

Makri A, Fakas S, Aggelis G (2010) Metabolic activities of biotechnological interest in *Yarrowia lipolytica* grown on glycerol in repeated batch cultures. Bioresour Technol 101:2351–2358

Morin N, Cescut J, Beopoulos A, Lelandais G, Le Berre V, Uribelarrea JL, Molina-Jouve C, Nicaud JM (2011) Transcriptomic analyses during the transition from biomass production to lipid accumulation in the oleaginous yeast *Yarrowia lipolytica*. PLoS ONE 6:e27966

Oberhardt MA, Palsson BO, Papin JA (2009) Applications of genome-scale metabolic reconstructions. Mol Syst Biol 5:320

Orth JD, Thiele I, Palsson BO (2010) What is flux balance analysis? Nat Biotechnol 28:245–248

Pan P, Hua Q (2012) Reconstruction and in silico analysis of metabolic network for an oleaginous yeast, *Yarrowia lipolytica*. PloS ONE 7:e51535

Papanikolaou S, Aggelis G (2002) Lipid production by *Yarrowia lipolytica* growing on industrial glycerol in a single-stage continuous culture. Bioresour Technol 82:43–49

Papanikolaou S, Aggelis G (2010) *Yarrowia lipolytica*: a model microorganism used for the production of tailor-made lipids. Eur J Lipid Sci Technol 112:639–654

Papanikolaou S, Chevalot I, Komaitis M, Marc I, Aggelis G (2002) Single cell oil production by *Yarrowia lipolytica* growing on an industrial derivative of animal fat in batch cultures. Appl Microbiol Biotechnol 58:308–312

Qiao K, Imam Abidi SH, Liu H, Zhang H, Chakraborty S, Watson N, Kumaran Ajikumar P, Stephanopoulos G (2015) Engineering lipid overproduction in the oleaginous yeast *Yarrowia lipolytica*. Metab Eng 29:56–65

Satish Kumar V, Dasika MS, Maranas CD (2007) Optimization based automated curation of metabolic reconstructions. BMC Bioinform 8:212

Schellenberger J, Que R, Fleming RM, Thiele I, Orth JD, Feist AM, Zielinski DC, Bordbar A, Lewis NE, Rahmanian S et al (2011) Quantitative prediction of cellular metabolism with constraint-based models: the COBRA Toolbox v2.0. Nat Protoc 6:1290–1307

Schreiber F (2002) High quality visualization of biochemical pathways in BioPath. In Silico Biol 2:59–73

Schweizer E, Kottig H, Regler R, Rottner G (1988) Genetic control of *Yarrowia lipolytica* fatty acid synthetase biosynthesis and function. J Basic Microbiol 28:283–292

Scolnick EM, Lin EC (1962) Parallel induction of D-arabitol and D-sorbitol dehydrogenases. J Bacteriol 84:631–637

Sitepu I, Selby T, Lin T, Zhu S, Boundy-Mills K (2014) Carbon source utilization and inhibitor tolerance of 45 oleaginous yeast species. J Ind Microbiol Biotechnol 41:1061–1070

Tai M, Stephanopoulos G (2013) Engineering the push and pull of lipid biosynthesis in oleaginous yeast *Yarrowia lipolytica* for biofuel production. Metab Eng 15:1–9

Toya Y, Shimizu H (2013) Flux analysis and metabolomics for systematic metabolic engineering of microorganisms. Biotechnol Adv 31:818–826

Wang Y, Xiao J, Suzek TO, Zhang J, Wang J, Bryant SH (2009) PubChem: a public information system for analyzing bioactivities of small molecules. Nucleic Acids Res 37:W623–W633

Wang JJ, Zhang BR, Chen SL (2011) Oleaginous yeast *Yarrowia lipolytica* mutants with a disrupted fatty acyl-CoA synthetase gene accumulate saturated fatty acid. Process Biochem 46:1436–1441

Wang W, Wei H, Alahuhta M, Chen X, Hyman D, Johnson DK, Zhang M, Himmel ME (2014) Heterologous expression of xylanase enzymes in lipogenic yeast *Yarrowia lipolytica*. PLoS ONE 9:e111443

Wehrspann P, Fullbrandt U (1985) Report of a case of *Yarrowia lipolytica* (Wickerman et al.) van der Walt & von Arx isolated from a blood culture. Mykosen 28:217–222

Ye C, Qiao W, Yu X, Ji X, Huang H, Collier JL, Liu L (2015a) Reconstruction and analysis of the genome-scale metabolic model of *Schizochytrium limacinum* SR21 for docosahexaenoic acid production. BMC Genomics 16:799

Ye C, Xu N, Chen H, Chen YQ, Chen W, Liu L (2015b) Reconstruction and analysis of a genome-scale metabolic model of the oleaginous fungus *Mortierella alpina*. BMC Syst Biol 9:1

Yen CLE, Monetti M, Burri BJ, Farese RV (2005) The triacylglycerol synthesis enzyme DGAT1 also catalyzes the synthesis of diacylglycerols, waxes, and retinyl esters. J Lipid Res 46:1502–1511

Yen CL, Stone SJ, Koliwad S, Harris C, Farese RV Jr (2008) Thematic review series: glycerolipids. DGAT enzymes and triacylglycerol biosynthesis. J Lipid Res 49:2283–2301

Yu LJ, Qin WM, Lan WZ, Zhou PP, Zhu M (2003) Improved arachidonic acids production from the fungus *Mortierella alpina* by glutamate supplementation. Bioresour Technol 88:265–268

Yu CS, Lin CJ, Hwang JK (2004) Predicting subcellular localization of proteins for Gram-negative bacteria by support vector machines based on n-peptide compositions. Protein Sci 13:1402–1406

Zhang C, Hua Q (2015) Applications of genome-scale metabolic models in biotechnology and systems medicine. Front Physiol 6:413

Zhang C, Ji B, Mardinoglu A, Nielsen J, Hua Q (2015) Logical transformation of genome-scale metabolic models for gene level applications and analysis. Bioinformatics 31:2324–2331

14

Biofabrication of gold nanoparticles by *Shewanella* species

Jhe-Wei Wu and I-Son Ng[*] [iD]

Abstract

Background: *Shewanella oneidensis* MR-1 (MR-1) and *Shewanella xiamenensis* BC01 (SXM) are facultative anaerobic bacteria that exhibit outstanding performance in the dissimilatory reduction of metal ions. *Shewanella* species have been reported to produce metal nanoparticles, but the mechanism and optimization are still not extensively studied and clearly understood. Herein, the effects of pH, biomass, gold ion concentration, and photoinduction are evaluated to optimize gold nanoparticle (Au@NP) production by *Shewanella*.

Results: The highest amount of Au@NPs produced by SXM and MR-1 were 108 and 62 ppm, respectively, at pH 5 when 2.4 g/L biomass was immersed in 300 ppm gold ions and 50 mM lactate under a light intensity of 100 μmol/m^2/s. By scanning election microscopy and zeta potential analysis, the proposed mechanism of Au@NP formation was that *Shewanella* used lactate as electron donors for the Mtr pathway, stimulated by photosensitive proteins resulting in the nucleation of NPs on the cell membrane. Besides, the resting cells retained the ability for biofabrication of nanoparticles for nearly 25 days.

Conclusions: The optimal conditions evaluated for Au@NPs production by *Shewanella* were biomass, pH, ions concentration, and photoinduction. To the best of our knowledge, this is the first attempt to explore a two-step mechanism for Au@NPs formation in *Shewanella*. First, the HAuCl$_4$ solution reacted with sodium lactate to form metallic gold ions. Second, the metallic gold ions were adsorbed onto the outer membrane of cell, and the formation of Au@NPs at the surface was triggered. *Shewanella*-based Au@NPs production could be a potential ecofriendly solution for the recovery of Au ions from secondary resources like industrial waste.

Keywords: *Shewanella*, Gold nanoparticle, Resting cell, Optimization, Photoinduction

Background

Because of global modernization and industrialization, pollution caused by the release of heavy metals into the environment has become a critical issue, and thus concerns surrounding the recovery of such heavy metals have been rising (Dodson et al. 2015). Among all metals commonly seen in wastes, gold is a noble metal and can be used in luxury jewelry, electronics, and medical applications because of its unique physical and chemical properties such as high biocompatibility and long-term stability (Ramesh et al. 2008; Spitzer and Bertazzoli 2004). Gold ions fabricated in nanoscale (gold nanoparticles or Au@NPs) with shape-dependent and optoelectronic properties have broadly applicable physiochemical characteristics and biological functions, and are thus of great interest to scientists (Klaus et al. 2001; Shedbalkar et al. 2014; Suresh et al. 2011).

Many traditional methods of gold recovery, such as cyanide leaching, precipitation and filtration, and electrochemical treatments, have been reported (Mata et al. 2009). However, these methods are challenged by restricted selectivity and large amounts of toxic chemicals involved in the process resulting in secondary pollution (He et al. 2015). Compared with traditional methods, biological methods have great appeal because of their simplicity, elimination of toxic chemicals (Mishra et al. 2014), and capability of controlling nanoparticle size (Bai et al. 2009; Sathishkumar et al. 2010a). Over the

*Correspondence: yswu@mail.ncku.edu.tw
Department of Chemical Engineering, National Cheng Kung University, Tainan 70101, Taiwan

last decade, adsorption and reduction of heavy metals by microorganisms including bacteria, yeast, and fungi have been reported (Farooq et al. 2010; Lo et al. 2014; Tan et al. 2017). The biofabrication of nanoparticles with optimal conditions, such as pH, incubation time, and metal ion concentration (Narayanan and Sakthivel 2010) also showed high potential in commercial applications and large-scale production.

Shewanella species, which are dissimilatory metal-reducing bacteria widely distributed in sediment or seawater, grow optimally between 25 and 40 °C and can reduce metal ions via a special electron pathway called the Mtr pathway (Fredrickson et al. 2008; Shi et al. 2012; Wang et al. 2017). *Shewanella oneidensis* MR-1, isolated from Oneida Lake, New York (Myers and Nealson 1988), has been reported to reduce Au^{3+} into discrete spherical Au@NPs that are well dispersed with homogeneous sizes. The nanoparticle sizes of gold produced by MR-1 were in the range of 2–50 nm and had high potential for use in biomaterials (Suresh et al. 2011). In the MR-1 strain, the outer membrane c-type cytochromes, MtrC and OmcA, were considered as important proteins for metal reduction (Wu et al. 2013a). However, the MR-1 wild type and its mutants *omcA* and *mtrC* were still capable of reducing Au^{3+} into Au@NPs (Wu et al. 2013b), while the particle size of the extracellular nanoparticles were decreased *mtrC* and *omcA* mutants (Ng et al. 2013). On the other hand, *Shewanella xiamenensis*, which is isolated from coastal sediment collected off Xiamen, China, is a close-related strain to *S. oneidensis* based on phylogenetic tree analysis by 16S rRNA and gyrase gene (Huang et al. 2010). It has also been reported to reduce mediators from the medium to nanoparticles (Ng et al. 2015a) and also showed resistance to different kinds of metal ions (Ng et al. 2015b).

The mechanism and conditional optimization of Au@NP production by *Shewanella* have never been reported. The aim of this study is to explore the mechanism and to accomplish optimization by determining the effects of pH, biomass, and gold concentration on Au@NP production by *S. oneidensis* MR-1 and *S. xiamenensis* BC01 (SXM). Finally, the resting cell activity on the biofabrication of Au@NPs and silver nanoparticles (Ag@NPs) is also examined.

Methods
Chemicals
For the reduction of gold and silver ions, chloroauric acid ($HAuCl_4 \cdot 3H_2O$) was purchased from Alfa Aesar, silver nitrate ($AgNO_3$) was purchased from Sigma (209139), and sodium lactate (50% w/w) was purchased from Showa (G1510E). For scanning electron microscopy, formvar solution was purchased from Sigma (09823), and

tert-butanol was purchased from Shimakyu Chemical Co. Ltd.

Bacteria culture
Both *S. xiamenensis* BC01 (SXM) and *S. oneidensis* MR-1 were grown in Luria–Bertani broth containing yeast extract (5 g/L), sodium chloride (10 g/L), and tryptone (10 g/L). The cells were maintained at 4 °C on LB plates, and a single colony was inoculated into 2 mL of LB medium and cultured at 30 °C and 150 rpm for 12 h for preculture. Then, 1% (v/v) of the precultured cells was transferred into a 50-mL flask containing 10 mL of LB medium and grown aerobically at 30 °C and 150 rpm for another 12 h.

Biofabrication and characterization of nanoparticles
Cells were collected by centrifugation at 8000×*g* for 5 min and washed twice with 0.5 mL of distilled water to obtain a final biomass concentration of 2.4 g/L. The precipitate was resuspend in 50 mM sodium lactate as the electron donor with 0.5 mL of 300 ppm Au^{3+} or 100 ppm Ag^+ solution. The samples were incubated at room temperature (approximately 30 °C) with a light intensity of 100 μmol/m^2/s for 24 h.

UV–vis spectroscopy analysis of nanoparticle formation
The nanoparticles were analyzed using UV–vis spectroscopy (Molecular Devices, SpectraMax 340PC^{384}, USA) to measure the surface plasmon resonance at 530 nm for Au@NPs and 410 nm for Ag@NPs. The color of the sample changed from pale yellow to purple to indicate the formation of Au@NPs (Kumar et al. 2008). For Ag@NPs, the color changed from pale yellow to orange.

Scanning electron microscopy (SEM)
The samples were fixed in 2.5% (w/v) glutaraldehyde for 2 h and washed three times with phosphate buffer (0.1 M, pH 7.4). The sample (100 μL) was carefully dropped onto a formvar-coated silicon chip for 1 h and washed three times with phosphate buffer. The cells were dehydrated in a series of ethanol washes with increasing ethanol concentration (30, 50, 70, and 100%). After three final washes in 100% ethanol, the samples were immersed in tert-butanol and dried by lyophilization (KINGMECH, FD3-12P, Taiwan) for 0.5 h. Dehydrated samples were analyzed using SEM (JEOL JSM-6700F, Japan).

Inductively coupled plasma-optical emission spectrometry (ICP-OES)
Cells were centrifuged at 10,000 rpm for 10 min, and the supernatant was filtered using a 0.22-μm filter (Millipore, USA). Au^{3+} concentration was measured using ICP-OES (ULTIMA 2000, Japan). A standard solution

containing 1000 ppm Au^{3+} (High-Purity Standards, USA) or 200 ppm Ag$^+$ was used as the starting solution and diluted in the range from 0 to 100 ppm for ICP-OES analysis.

Zeta potential

Samples for zeta potential measurement were prepared as follows. Cells were added into 1 mL of distilled water in Falcon tubes with a final biomass concentration of 1.2 g/L. Different samples with a volume of 750 μL were introduced into cuvettes, and the zeta potential was measured (Malvern, Zetasizer Nano ZS, UK).

Preparation of resting cells

Strains were grown aerobically at 30 °C and 150 rpm in 250-mL flasks containing 50 mL of Luria–Bertani medium for 12 h. Cells were collected by centrifugation at 8000×g for 10 min and washed twice by distilled water. The precipitate was resuspended by 1 mL of distilled water, and the sample was quickly frozen in liquid nitrogen. The resting cells were generated by overnight lyophilization into a powder and stored at − 20 °C for long-term storage.

Results and discussion

Effect of pH

There have been reports on the biosynthesis of metal nanoparticles using *Shewanella* species (Ng et al. 2013; Suresh et al. 2011), but the physical factors for optimization have rarely been reported. Therefore, it is critical and meaningful to explore the optimal conditions of Au@NP biofabrication by SXM and MR-1. By adjusting the pH to 3, 4, 5, and 6, *Shewanella* produced different quantities of Au@NPs. It was obvious that SXM produced more Au@NPs than MR-1 did, as shown in Table 1. SXM produced 116 and 108 ppm Au@NPs at pH 4 and 5, respectively, while MR-1 only produced 62 ppm Au@NPs at pH 5. The production of Au@NPs by both strains decreased dramatically at pH 6. The value of pH was considered an important factor for gold reduction and nanoparticle formation (Mishra et al. 2012), as the different pH values affect the zeta potential, thus influenced the electric properties on the cell surface to form nanoparticles. The effect of pH was consistent with the results of *S. haliotis*, which has an optimal pH of 5 (Zhu et al. 2016). In addition, neutral conditions were not appropriate for *Shewanella* for the production of Au@NPs.

Effect of biomass and concentration of gold ions

As shown in Fig. 1a, the dark purple colors indicated Au@NPs formation by both *Shewanella*. As shown in Fig. 1b, the gold ions were adsorbed onto the cell surface

Table 1 The effect of pH on *Shewanella* at biomass of a 0.6 g/L to produce Au@NP with 300 ppm Au^{3+} after 24 h

pH	Au@NPs (ppm)	
	S. xiamenensis BC01	*S. oneidensis* MR-1
3	17.0 ± 3.5	33.1 ± 4.2
4	116.7 ± 27.5	44.7 ± 7.9
5	108.0 ± 9.2	62.0 ± 9.1
6	45.9 ± 4.6	1.6 ± 4.7

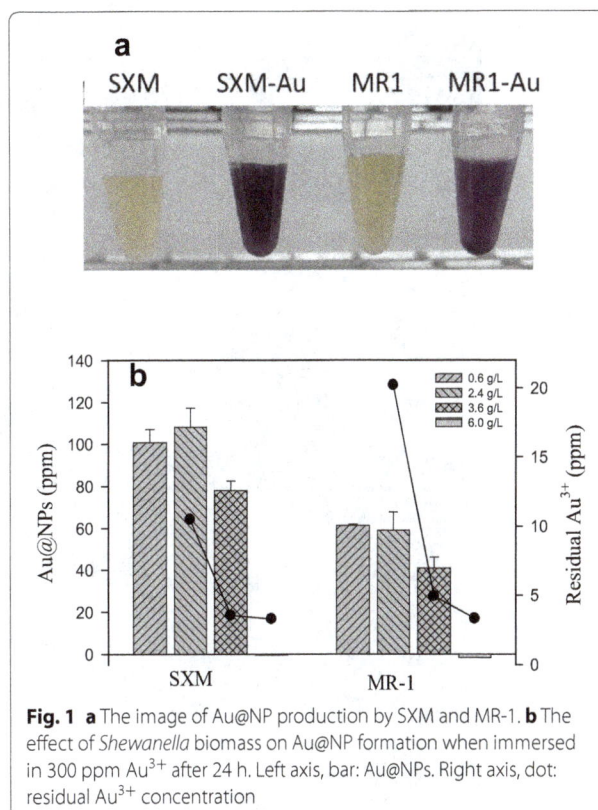

Fig. 1 **a** The image of Au@NP production by SXM and MR-1. **b** The effect of *Shewanella* biomass on Au@NP formation when immersed in 300 ppm Au^{3+} after 24 h. Left axis, bar: Au@NPs. Right axis, dot: residual Au^{3+} concentration

when the biomass increased, and the concentration of gold dropped from 300 to 3 ppm with 6 g/L biomass. As a result, the residual concentration of Au decreased significantly when 3.6 g/L biomass was used, and residual Au^{3+} was relatively low (i.e., < 5 ppm) in both strains. However, the quantity of Au@NPs did not increase when the biomass increased from 2.4 to 6.0 g/L (Fig. 1b, bar graph). Maximal amounts of Au@NPs were generated at 2.4 g/L biomass (108 and 58.9 ppm for SXM and MR-1, respectively). The morphology of *Shewanella* in different biomass concentrations and the amounts of Au@NPs formed could be further analyzed by SEM as shown in Fig. 2. It was evident that Au@NPs were fabricated on the outer membrane of cells at 2.4 and 3.6 g/L biomass. When the biomass increased to 6.0 g/L, the cell

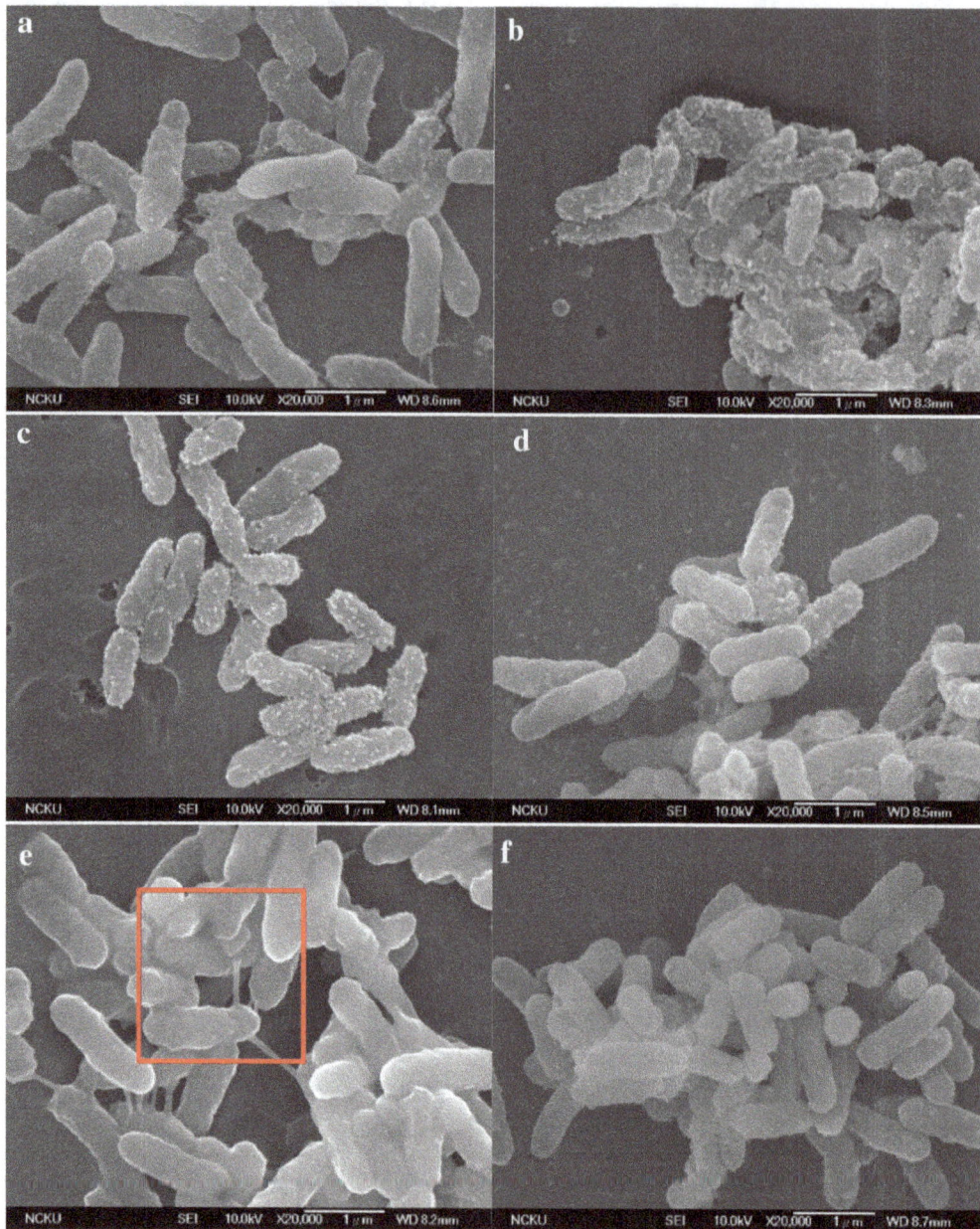

Fig. 2 SEM analysis of SXM at biomass of **a** 0.6 g/L, **c** 2.4 g/L, and **e** 6.0 g/L, and of MR-1 at **b** 0.6 g/L, **d** 2.4 g/L, and **f** 6.0 g/L when immersed in 300 ppm Au^{3+} for 24 h

membrane became thicker and sticky (shown in red rectangle in Fig. 2e). Although gold ions would adsorb on the cell surface, it was not beneficial to generate Au@ NPs through nucleation layer by layer when biomass increased up to 3.6 g/L.

The effect of gold concentration on Au@NP formation was tested within the range of 10–300 ppm Au^{3+}. As shown in Table 2, no Au@NPs were produced at 0.6 and 2.4 g/L biomass when the concentration of gold

ions was 50 ppm. Even when the concentration of gold ions increased to 100 ppm, only approximately 5 ppm Au@NPs were generated in (i.e., 5% conversion). When the gold concentration was greater than 100 ppm, *Shewanella* produced Au@NPs that proportionally increased when the Au^{3+} increased from 100 to 300 ppm. As shown in Table 2, it was obvious that SXM produced more Au@NPs than MR-1 with the increasing amount of Au^{3+}. Moreover, the threshold of gold concentration for

Table 2 The effect of gold ion concentration on *Shewanella* in Au@NP formation after 24 h

Au^{3+} (ppm)	Au@NPs (ppm)			
	S. xiamenensis BC01		*S. oneidensis* MR-1	
	0.6 g/L	2.4 g/L	0.6 g/L	2.4 g/L
50	Nd	Nd	Nd	Nd
100	5.3	Nd	4.9	Nd
200	33.9	48.2	31.9	12.8
300	100.8	108.0	61.2	59.0

Nd not detected

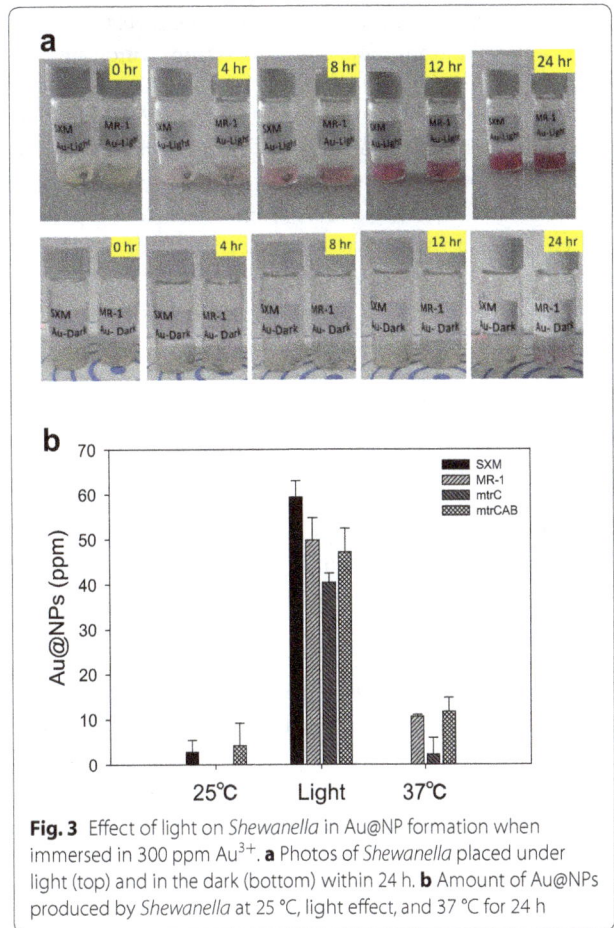

Fig. 3 Effect of light on *Shewanella* in Au@NP formation when immersed in 300 ppm Au^{3+}. **a** Photos of *Shewanella* placed under light (top) and in the dark (bottom) within 24 h. **b** Amount of Au@NPs produced by *Shewanella* at 25 °C, light effect, and 37 °C for 24 h

Shewanella to produce Au@NPs is reported here for the first time.

A comparison of Au@NP formation using other microbes is shown in Table 3. The optimal biomasses of *S. haliotis*, *Aspergillus oryzae* var. *viridis*, and *Sargassum* sp. were 5.3, 10, and 5 g/L, respectively (Binupriya et al. 2010; Sathishkumar et al. 2010b; Zhu et al. 2016). Herein, the optimal biomass of SXM and MR-1 was 2.4 g/L, which was much lower than those of the other bacteria. Moreover, the formation of Au@NPs occurred within 4 h, which was much faster than those shown in previous reports. The absorbance intensities of SXM and MR-1, corresponding to the amounts of Au@NPs, were significantly higher than those of the other species. Because of the lower biomass used, higher reaction rates, and higher productivity, MR-1 and SXM have great advantages in gold reduction to nanoparticles.

Photo effect on Au@NP formation

As shown in Fig. 3a, the color of the solution changed rapidly within 4 h from pale yellow to purple when exposed to light at an intensity of 100 µmol photons/m^2/s, indicating that the reaction rate of *Shewanella* in the presence of light was far greater than that of the solution in darkness. We also compared the wild type *Shewanella* (i.e., MR-1 and SXM) and genetically modified strains (i.e., cells harboring *mtr*C and *mtr*CAB genes). As shown in Fig. 3a, cells placed in light produced more Au@NPs in the time series from 0 to 24 h. Temperature was not a critical factor for Au@NP formation as shown in Fig. 3b. Very low levels of Au@NPs were produced at 25 and 37 °C without light induction. In addition, the MR-1 strains harboring

Table 3 Comparison of different strains in biofabrication of Au@NPs

Microorganisms	pH	Temp (°C)	Biomass (g/L)	Light density (µmol photons/m^2/s)	Color change time (h)	Absorbance intensity after 24 h	References
Shewanella xiamenensis BC01	5	RT	2.4	100	4	2.5	This study
Shewanella oneidensis MR-1	5	RT	2.4	100	4	1.7	This study
Shewanella haliotis	5	30	5.3	X	12	1.0	(Zhu et al. 2016)
Aspergillus oryzae var. *viridis*	7	25	10	X	10	0.4	(Binupriya et al. 2010)
Sargassum sp.	8	RT	5	X	0.5	0.05	(Sathishkumar et al. 2010b)

RT room temperature

*mtr*C or *mtr*CAB genes showed the same levels of nanoparticle production under light induction, which implies that the Mtr pathway proteins are not stimulated by light. It is known that cyanobacteria possess light-sensitive phytochromes to control photosynthesis, phototaxis, and production of pigments (Schmitz et al. 2000; Yeh et al. 1997). A phytochrome is a two-component system, with a membrane-bound sensor protein and an intracellular response regulator protein, which function in sequence in response to an extracellular signal. An *E. coli* strain has been artificially engineered to respond to light by replacing the osmolarity sensing domain EnvZ to a photosensing domain from the cyanobacteria in the native EnvZ–OmpR two-component system (Levskaya et al. 2005). The engineered *E. coli* strain was photosensitive and could turn on or shut down the expression of the reporter gene according to illumination. Therefore, we hypothesize that photoinduction in *Shewanella* which caused the formation of Au@NPs could be regulated by a two-component protein system. In our experiments, the formation of Au@NPs was significantly increased by photoinduction. From the genomic database of MR-1 (Accession Number NC004347), a putative two-component system with a photoreactivation-associated protein (PhrA) and photolyase (PhrB) could be involved in sensing of light. These light-sensitive proteins may be the key factor in stimulating the formation of Au@NPs. The search for key proteins by genetic method, i.e., knock-out *phr*A and *phr*B, is an inevitable further research motive for this study. In the past, the *S. algae* strain BRY was found to reduce Au^{3+} to Au using hydrogen as the electron donor (Kashefi et al. 2001). Alternatively, *Shewanella* species have been considered for the bioremediation of different kinds of metals ions or for use in microbial fuel cells (Chen et al. 2015; Liu and Logan 2004; Xu et al. 2006). This is the first attempt to discover the effects of temperature and light on the formation of Au@NPs.

The proposed mechanism of Au@NP formation

The zeta potential analysis is shown in Fig. 4a. Significant differences in zeta potential were observed at biomasses of 2.4 and 1.2 g/L for SXM and MR-1, respectively. The difference in zeta potential gradually decreased when the biomass was greater than 3.6 g/L. At 6.0 g/L biomass, no difference in zeta potential was detected. This shows the disadvantage of high biomass (6.0 g/L) for Au@NP formation. Recent research shows that the changes in zeta potential of MR-1 not only reflect on the production of Au@NPs, but also the formation of biofilm (Ishiki et al. 2017). Therefore, with higher cell density, increased biofilm formation could affect the nucleation of gold atoms to NPs. On the other hand, the ICP-OES analysis of residual ion concentration after immersion of SXM or

MR-1 in 300 ppm gold solution or 100 ppm silver solution with biomass concentrations ranging from 2.4 to 6 g/L, we found that only 20.8 to 3.29 ppm of residual gold ions or 2.26 to 0.81 ppm of residual silver ions were present for both SXM and MR-1 at lower or higher biomass (Table 4). Our hypothesis for the mechanism of Au@NPs formation by *Shewanella* is shown in Fig. 4b.

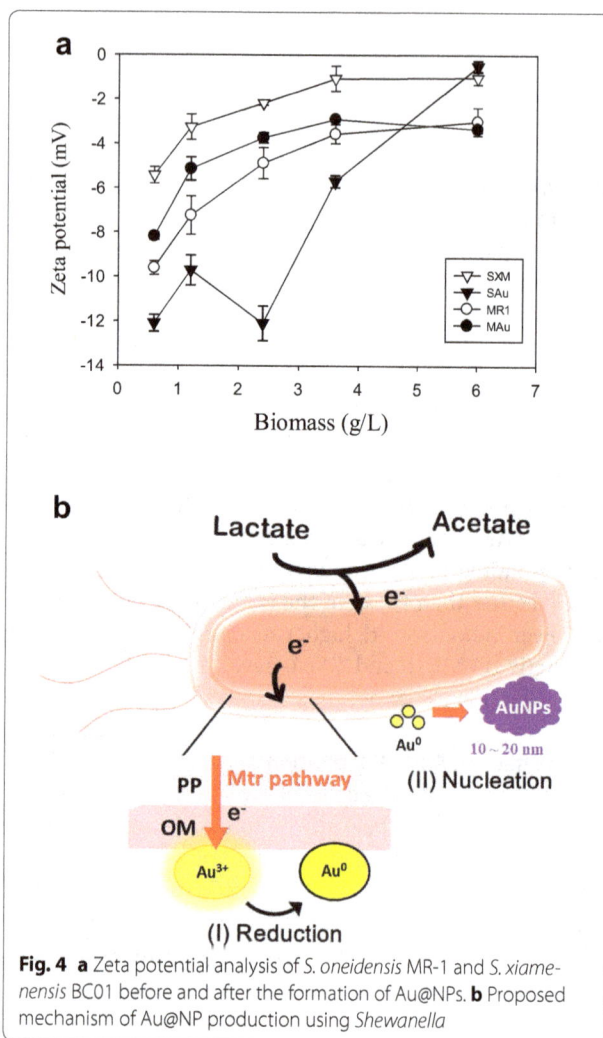

Fig. 4 **a** Zeta potential analysis of *S. oneidensis* MR-1 and *S. xiamenensis* BC01 before and after the formation of Au@NPs. **b** Proposed mechanism of Au@NP production using *Shewanella*

Table 4 ICP-OES analysis of residual ion concentration after immersion in Au^{3+} or Ag^+ by SXM or MR-1 at different biomasses for 24 h

Metal conc.	Biomass conc. (g/L)	Final conc. (ppm)	
		SXM	MR-1
300 ppm Au^{3+}	2.4	11.9 ± 0.12	20.8 ± 0.11
300 ppm Au^{3+}	6.0	3.29 ± 0.08	3.31 ± 0.13
100 ppm Ag^+	2.4	2.21 ± 0.02	2.26 ± 0.05
100 ppm Ag^+	6.0	0.81 ± 0.03	1.17 ± 0.04

Sodium lactate acted as the electron donor at a final concentration of 50 mM. Then, *Shewanella* adsorbed the gold ions, and nucleation was triggered on the cell membrane layer by layer, while higher biomass blocked the nucleation. Because of the distinct electron pathway of *Shewanella*, gold ions were reduced on the outer membrane of the cells, further accomplishing the process of nanoparticle fabrication and resulted in different zeta potentials on cell.

In the formation of Au@NPs, the first step is the reduction of gold ions to metallic gold atoms. This process is supposed to be the common reduction, where the electron donor lactate goes through the Mtr pathway (i.e., *mtr*A, *mtr*B, *mtr*C, and *omc*A) and excess electrons enter the cytoplasmic membrane-anchored tetrahaem c-type cytochrome CymA to accomplish the reduction. The importance of proteins in Mtr pathway and cytochrome c-type has also been demonstrated in DMSO or dinitrotoluene reduction (Coursolle and Gralnick 2010; Liu et al. 2017). Moreover, as Au@NPs formation by *Shewanella* was photoinduced, there is the supposed involvement of a prospective two-component system [photoreactivation-associated protein (PhrA) and photolyase (PhrB)] which was activated by light energy to drive electron transfer and accelerate the reduction (Ng et al. 2000; Sancar 2003). Second, the nucleation of gold atom as a nanoparticle on cell surface should be a "layer by layer" processing, obeying thermodynamics and kinetic behavior. Thus, with higher cell density, the thicker biofilm formed would be a drawback for nanoparticles formation.

The activity of resting cells

Apart from the fresh cells, resting cells were also used for Au@NP formation. Resting SXM and MR-1 cells were prepared based on the method described in "Preparation of resting cells". They were stored at − 20 °C for long-term conservation as shown in Fig. 5a. The ability of resting cells to produce Au@NPs was analyzed after 1, 15, 20, and 25 days. The resting cells had nearly 60% capability for Au@NPs production after 25 days (Fig. 5b), the variations in the activity could be attributed to the nonuniform cell powder. The resting cells were preferable as they could be used at any time without several repeated steps. Although there was a report showing that resting *S. algae* cells generated reduced amounts of platinum nanoparticles (Konishi et al. 2007), this is also the first attempt in showing that resting *Shewanella* cells possess long-term stability for the fabrication of Au@NPs.

The selectivity of gold and silver ions by *Shewanella*

The ability for Ag@NPs formation by SXM and MR-1 was shown in the blue line shown in Fig. 6. The Ag@NPs exhibited surface plasmon resonance at 410 nm. In order to confirm the selectivity of Au^{3+} and Ag^+ ions by *Shewanella*, both strains were examined in a solution containing 100 ppm Au^{3+} and 100 ppm Ag^+ ions (red line in Fig. 6) or 250 ppm Au^{3+} and 100 ppm Ag^+ ions (green line in Fig. 6) for 24 h. The results showed that both SXM (Fig. 6a) and MR-1 (Fig. 6b) reduced silver ions at the same concentration, and reduced Au^{3+} when immersed in 250 ppm Au^{3+} and 100 ppm Ag^+ ions. The selectivity can also be controlled by the concentration of ions inside the system.

Conclusions

Shewanella xiamenensis BC01 and MR-1 reduced Au^{3+} to Au@NPs, which were localized on the surface. By measuring the optimal condition, the highest amounts

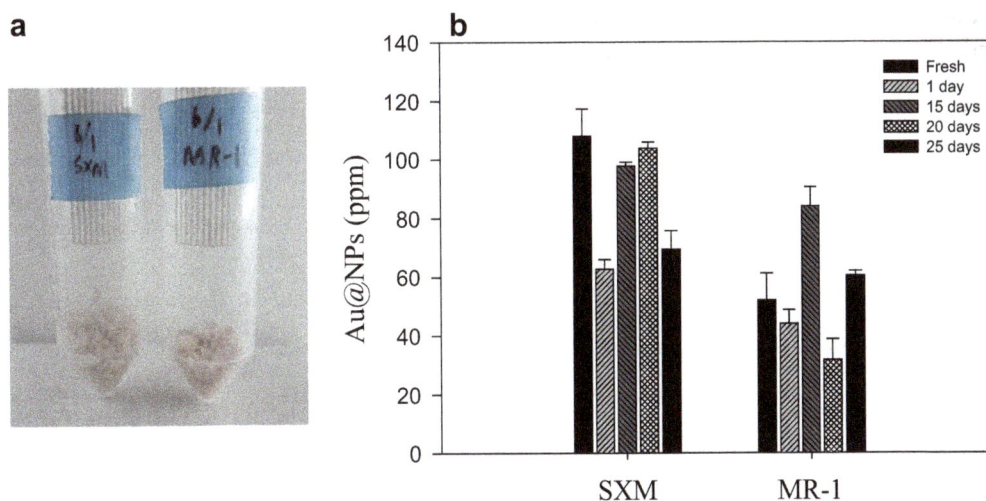

Fig. 5 a Resting *Shewanella* cells. **b** Biofabrication of Au@NPs by resting cells after 1, 15, 20, and 25 days

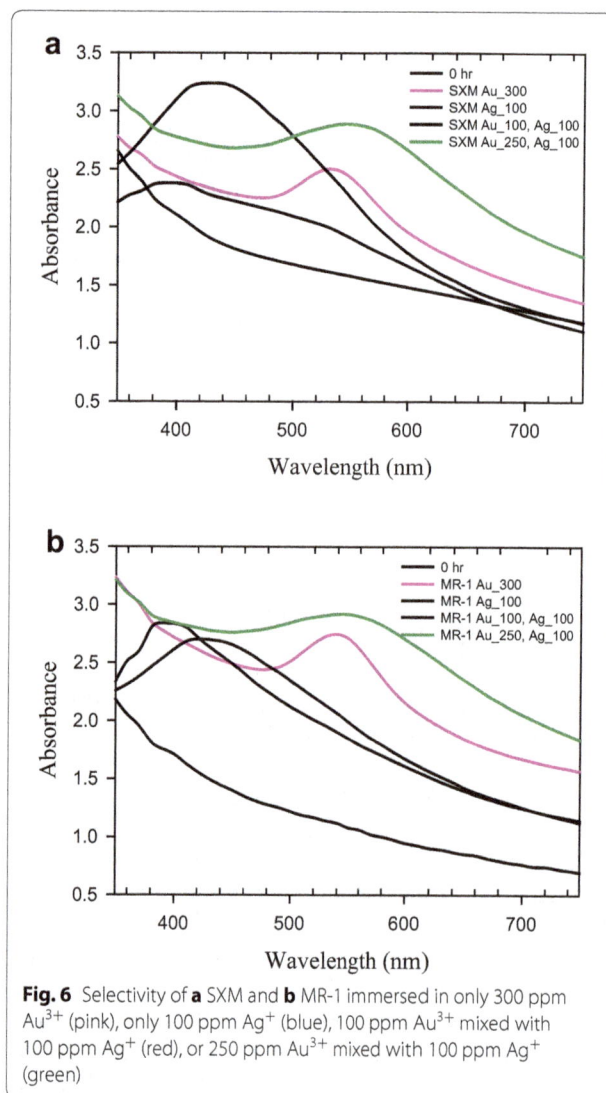

Fig. 6 Selectivity of **a** SXM and **b** MR-1 immersed in only 300 ppm Au³⁺ (pink), only 100 ppm Ag⁺ (blue), 100 ppm Au³⁺ mixed with 100 ppm Ag⁺ (red), or 250 ppm Au³⁺ mixed with 100 ppm Ag⁺ (green)

of Au@NPs were 100 ppm for SXM and 62 ppm for MR-1, respectively. It was found that an increase in biomass resulted in a decrease in Au@NPs. The presence of light dramatically accelerated nanoparticle formation. The mechanism of Au@NP formation and light-induced effect in *Shewanella* have been reported for the first time. The recovery of Au ions from industrial waste via *Shewanella* is a potential bioremediation option.

Abbreviations
Au@NPs: gold nanoparticles; Ag@NPs: silver nanoparticles; SXM: *Shewanella xiamenensis* BC01; MR-1: *Shewanella oneidensis* MR-1; SEM: scanning electron microscopy; ICP-OES: inductively coupled plasma-optical emission spectrometry.

Authors' contributions
ISN designed the experiment and analyzed the data, JWW performed most of experiments. ISN and JWW wrote the manuscript. Both authors read and approved the final manuscript.

Acknowledgements
The authors are grateful for the financial support for this study provided by the Ministry of Science and Technology (MOST 105-2221-E-006-225-MY3 and MOST-105-2621-M-006-012-MY3) in Taiwan.

Competing interests
The authors declare that they have no competing interests.

Funding
This work was supported by the Ministry of Science and Technology (MOST 105-2221-E-006-225-MY3 and MOST 105-2621-M-006-012-MY3) in Taiwan.

References
Bai H, Zhang Z, Guo Y, Jia W (2009) Biological synthesis of size-controlled cadmium sulfide nanoparticles using immobilized *Rhodobacter sphaeroides*. Nanoscale Res Lett 4(7):717–723

Binupriya A, Sathishkumar M, Vijayaraghavan K, Yun SI (2010) Bioreduction of trivalent aurum to nano-crystalline gold particles by active and inactive cells and cell-free extract of *Aspergillus oryzae* var. *viridis*. J Hazard Mater 177(1):539–545

Chen T, Zhou Y, Ng IS, Yang CS, Wang HY (2015) Formation and characterization of extracellular polymeric substance from *Shewanella xiamenensis* BC01 under calcium stimulation. J Taiwan Inst Chem Eng 57:175–181

Coursolle D, Gralnick JA (2010) Modularity of the Mtr respiratory pathway of *Shewanella ondeidensis* strain MR-1. Mol Microbiol 77(4):995–1008

Dodson JR, Parker HL, Muñoz García A, Hicken A, Asemave K, Farmer TJ, He H, Clark JH, Hunt AJ (2015) Bio-derived materials as a green route for precious & critical metal recovery and re-use. Green Chem 17(4):1951–1965

Farooq U, Kozinski JA, Khan MA, Athar M (2010) Biosorption of heavy metal ions using wheat based biosorbents—a review of the recent literature. Bioresour Technol 101(14):5043–5053

Fredrickson JK, Romine MF, Beliaev AS, Auchtung JM, Driscoll ME, Gardner TS, Nealson KH, Osterman AL, Pinchuk G, Reed JL, Rodionov DA, Rodrigues JLM, Saffarini DA, Serres MH, Spormann AM, Zhulin IB, Tiedje JM (2008) Towards environmental systems biology of *Shewanella*. Nat Rev Microbiol 6(8):592–603

He YR, Cheng YY, Wang WK, Yu HQ (2015) A green approach to recover Au(III) in aqueous solution using biologically assembled rGO hydrogels. Chem Eng J 270:476–484

Huang J, Sun B, Zhang X (2010) *Shewanella xiamenensis* sp. nov., isolated from coastal sea sediment. Int J Syst Evol Microbiol 60(7):1585–1589

Ishiki K, Okada K, Le DQ, Shiigi H, Nagaoka T (2017) Investigation concerning the formation process of gold nanoparticles by *Shewanella oneidensis* MR-1. Anal Sci 33(2):129–131

Kashefi K, Tor JM, Nevin KP, Lovley DR (2001) Reductive precipitation of gold by dissimilatory Fe(III)-reducing bacteria and archaea. Appl Environ Microbiol 67(7):3275–3279

Klaus T, Joerger R, Olsson E, Granqvist CG (2001) Bacteria as workers in the living factory: metal-accumulating bacteria and their potential for materials science. Trends Biotechnol 19(1):15–20

Konishi Y, Ohno K, Saitoh N, Nomura T, Nagamine S, Hishida H, Takahashi Y, Uruga T (2007) Bioreductive deposition of platinum nanoparticles on the bacterium *Shewanella algae*. J Biotechnol 128(3):648–653

Kumar SA, Peter YA, Nadeau JL (2008) Facile biosynthesis, separation and conjugation of gold nanoparticles to doxorubicin. Nanotechnology 19(49):495101

Levskaya A, Chevalier AA, Tabor JJ, Simpson ZB, Lavery LA, Levy M, Davidson EA, Scouras A, Ellington AD, Marcotte EM, Voigt CA (2005) Synthetic biology: engineering *Escherichia coli* to see light. Nature 438(7067):441–442

Liu H, Logan BE (2004) Electricity generation using an air-cathode single chamber microbial fuel cell in the presence and absence of a proton exchange membrane. Environ Sci Technol 38(14):4040–4046

Liu DF, Min D, Cheng L, Zhang F, Li DB, Xiao X, Sheng GP, Yu HQ (2017) Anaerobic reduction of 2,6-dinitrotoluene by *Shewanella oneidensis* MR-1: roles of Mtr respiratory pathway and NfnB. Biotechnol Bioeng 114(4):761–768

Lo YC, Cheng CL, Han YL, Chen BY, Chang JS (2014) Recovery of high-value metals from geothermal sites by biosorption and bioaccumulation. Bioresour Technol 160:182–190

Mata YN, Torres E, Blazquez ML, Ballester A, Gonzalez F, Munoz JA (2009) Gold(III) biosorption and bioreduction with the brown alga *Fucus vesiculosus*. J Hazard Mater 166(2–3):612–618

Mishra A, Tripathy SK, Yun SI (2012) Fungus mediated synthesis of gold nanoparticles and their conjugation with genomic DNA isolated from *Escherichia coli* and *Staphylococcus aureus*. Process Biochem 47(5):701–711

Mishra A, Kumari M, Pandey S, Chaudhry V, Gupta KC, Nautiyal CS (2014) Biocatalytic and antimicrobial activities of gold nanoparticles synthesized by *Trichoderma* sp. Bioresour Technol 166:235–242

Myers CR, Nealson KH (1988) Bacterial manganese reduction and growth with manganese oxide as the sole electron acceptor. Science 240(4857):1319

Narayanan KB, Sakthivel N (2010) Biological synthesis of metal nanoparticles by microbes. Adv Colloid Interface Sci 156(1):1–13

Ng WO, Zentella R, Wang Y, Taylor AJS, Pakrasi BH (2000) PhrA, the major photoreactivating factor in the cyanobacterium *Synechocystis* sp. strain PCC 6803 codes for a cyclobutane-pyrimidine-dimer-specific DNA photolyase. Arch Microbiol 173:412–417

Ng CK, Sivakumar K, Liu X, Madhaiyan M, Ji L, Yang L, Tang C, Song H, Kjelleberg S, Cao B (2013) Influence of outer membrane c-type cytochromes on particle size and activity of extracellular nanoparticles produced by *Shewanella oneidensis*. Biotechnol Bioeng 110(7):1831–1837

Ng IS, Xu F, Zhang X, Ye C (2015a) Enzymatic exploration of catalase from a nanoparticle producing and biodecolorizing algae *Shewanella xiamenensis* BC01. Bioresour Technol 184:429–435

Ng IS, Ndive CI, Zhou Y, Wu X (2015b) Cultural optimization and metal effects of *Shewanella xiamenensis* BC01 growth and swarming motility. Bioresour Bioprocess 2(1):28

Ramesh A, Hasegawa H, Sugimoto W, Maki T, Ueda K (2008) Adsorption of gold(III), platinum(IV) and palladium(II) onto glycine modified crosslinked chitosan resin. Bioresour Technol 99(9):3801–3809

Sancar A (2003) Structure and function of DNA photolyase and cryptochrome blue-light photoreceptors. Chem Rev 103:2203–2237

Sathishkumar M, Sneha K, Yun YS (2010a) Immobilization of silver nanoparticles synthesized using *Curcuma longa* tuber powder and extract on cotton cloth for bactericidal activity. Bioresour Technol 101(20):7958–7965

Sathishkumar M, Mahadevan A, Vijayaraghavan K, Pavagadhi S, Balasubramanian R (2010b) Green recovery of gold through biosorption, biocrystallization, and pyro-crystallization. Ind Eng Chem Res 49(16):7129–7135

Schmitz O, Katayama M, Williams BS, Kondo T, Golden SS (2000) CikA, a bacteriophytochrome that resets the cyanobacterial circadian clock. Science 289:765–768

Shedbalkar U, Singh R, Wadhwani S, Gaidhani S, Chopade BA (2014) Microbial synthesis of gold nanoparticles: current status and future prospects. Adv Colloid Interface Sci 209:40–48

Shi L, Rosso KM, Clarke TA, Richardson DJ, Zachara JM, Fredrickson JK (2012) Molecular underpinnings of Fe(III) oxide reduction by *Shewanella oneidensis* MR-1. Front Microbiol 3:50

Spitzer M, Bertazzoli R (2004) Selective electrochemical recovery of gold and silver from cyanide aqueous effluents using titanium and vitreous carbon cathodes. Hydrometallurgy 74(3):233–242

Suresh AK, Pelletier DA, Wang W, Broich ML, Moon JW, Gu B, Allison DP, Joy DC, Phelps TJ, Doktycz MJ (2011) Biofabrication of discrete spherical gold nanoparticles using the metal-reducing bacterium *Shewanella oneidensis*. Acta Biomater 7(5):2148–2152

Tan SI, Ng IS, Yu YY (2017) Heterologous expression of an acidophilic multicopper oxidase in *Escherichia coli* and its applications in biorecovery of gold. Bioresour Bioprocess 4:20

Wang G, Zhang B, Li S, Yang M, Yin C (2017) Simultaneous microbial reduction of vanadium(V) and chromium(VI) by *Shewanella loihica* PV-4. Bioresour Technol 227:353–358

Wu C, Cheng YY, Li BB, Li WW, Li DB, Yu HQ (2013a) Electron acceptor dependence of electron shuttle secretion and extracellular electron transfer by *Shewanella oneidensis* MR-1. Bioresour Technol 136:711–714

Wu R, Cui L, Chen L, Wang C, Cao C, Sheng G, Yu H, Zhao F (2013b) Effects of bio-Au nanoparticles on electrochemical activity of *Shewanella oneidensis* wild type and ΔomcA/mtrC mutant. Sci Rep 3:3307

Xu M, Guo J, Zeng G, Zhong X, Sun G (2006) Decolorization of anthraquinone dye by *Shewanella decolorationis* S12. Appl Microbiol Biotechnol 71(2):246–251

Yeh KC, Wu SH, Murphy JT, Lagarias JC (1997) A cyanobacterial phytochrome two-component light sensory system. Science 277:1505–1508

Zhu N, Cao Y, Shi C, Wu P, Ma H (2016) Biorecovery of gold as nanoparticles and its catalytic activities for p-nitrophenol degradation. Environ Sci Pollut Res Int 23(8):7627–7638

15

Photocontrolled reversible self-assembly of dodecamer nitrilase

Qiao Yu[1], Yong Wang[2], Shengyun Zhao[3] and Yuhong Ren[4*]

Abstract

Background: Naturally photoswitchable proteins act as a powerful tool for the spatial and temporal control of biological processes by inducing the formation of a photodimerizer. In this study, a method for the precise and reversible inducible self-assembly of dodecamer nitrilase in vivo (in *Escherichia coli*) and in vitro (in a cell-free solution) was developed by means of the photoswitch-improved light-inducible dimer (iLID) system which could induce protein–protein dimerization.

Results: Nitrilase was fused with the photoswitch protein AsLOV2-SsrA to achieve the photocontrolled self-assembly of dodecamer nitrilase. The fusion protein self-assembled into a supramolecular assembly when illuminated at 470 nm. Scanning electron microscopy showed that the assembly formed a circular sheet structure. Self-assembly was also induced by light in *E. coli*. Dynamic light scattering and turbidity assay experiments showed that the assemblies formed within a few seconds under 470-nm light and completely disassembled within 5 min in the dark. Assembly and disassembly could be maintained for at least five cycles. Both in vitro and in vivo, the assemblies retained 90% of the initial activity of nitrilase and could be reused at least four times in vitro with 90% activity.

Conclusions: An efficient method was developed for the photocontrolled assembly and disassembly of dodecamer nitrilase and for scaffold-free reversible self-assembly of multiple oligomeric enzymes in vivo and in vitro, providing new ideas and methods for immobilization of enzyme without carrier.

Keywords: Disassembly, iLID, Nitrilase, Photoswitch, Self-assembly

Background

Protein assembly has been extensively studied (King and Lai 2013; Yu et al. 2015), and represents a useful tool for biology and bioengineering processes owing to its potential applications in protein material development nanobiotechnology, and multienzyme biocatalysis (Matsumoto et al. 2016; Smith et al. 2011; Fairman and Åkerfeldt 2005; Holmes 2002). Although various protein assembly strategies have been reported, such as coimmobilization, protein fusion, protein crosslinking, and scaffold-mediated multienzyme colocalization (King and Lai 2013; Schoffelen and van Hest 2012; Brady and Jordaan 2009), those strategies do not have the ability to achieve the precise

control of the assembly process, or realize the reversible disassembly in cell-free solutions (in vitro) or in bacterial cell cultures (in vivo). Including a control mechanism would solve this problem. Optogenetics provides a powerful tool for controlling biological processes through several light receptors such as cryptochromes (Kennedy et al. 2010), phytochrome B, phototropins (LOV), and rhodopsins (Levskaya et al. 2009; Yin and Wu 2013). Thus, we aimed to use light for the highly tunable, nontoxic, simple, and high-resolution spatiotemporal control of protein assembly. We focused on using the photosensitive LOV2 domain from *Avena sativa* (oat) (AsLOV2) (Strickland et al. 2010). AsLOV2 consists of a core Per-Arnt-Sim (PAS) domain that is sensitive to blue light and a C-terminal helical extension (Jα). Upon irradiation with 470-nm blue light, a covalent adduct is generated between a conserved cysteine residue in the Per-Arnt-Sim core of AsLOV2 and flavin mononucleotide (FMN) C(4a), leading to the unfolding of the Jα helix (Halavaty

*Correspondence: yhren@ecust.edu.cn
[4] State Key Laboratory of Bioreactor Engineering, New World Institute of Biotechnology, East China University of Science and Technology, Shanghai, China
Full list of author information is available at the end of the article

and Moffat 2007; Harper et al. 2004; Guntas et al. 2014). Several strategies have been developed to utilize AsLOV2 as a photoswitch, including using tunable light-inducible dimerization tags to control the interaction between the AsLOV2 domain and an engineered PDZ domain, and using the improved light-inducible dimer (iLID) system to control the location of transmembrane proteins (Guntas et al. 2014). The iLID system consists of an AsLOV2 mutant domain fused to the N-terminal of SsrA (called iLID-micro) and a 13-kDa adaptor protein, SspB, that can dimerize spontaneously and bind to SsrA (Guntas et al. 2014). After irradiation with 470-nm blue light, the conformation of the AsLOV2 domain changes, leading to the exposure of SsrA, which subsequently binds to SspB (Guntas et al. 2014; Zimmerman et al. 2016). The iLID presents several advantages, such as possessing strictly monomeric components, being easily expressed in *Escherichia coli*, and having a broad dynamic range, a highly tunable affinity, and a fast reversion rate in the dark (Guntas et al. 2014).

In this study, the dodecamer nitrilase from *Burkholderia cenocepacia* J2315 (Wang et al. 2013) (BCNIT) was selected as the model enzyme, and a light-controllable enzyme assembly method was developed using iLID and BCNIT. Dynamic light scattering (DLS), fluorescence complementation, scanning electron microscopy (SEM), optical density measurements (OD_{600}), and enzyme activity assays were performed to study the mechanisms involved.

Methods

Materials

PrimeSTAR® Max DNA polymerase (Takara Biotech, Dalian, China) was used for PCR amplification. The sequence information for AsLOV2-SsrA and SspB was acquired from the Protein Data Bank (4WF0 and 1YFN) and the genes and primers were synthesized (Generay, Shanghai, China) with codon optimized for expression in *E. coli*. FMN, mandelonitrile, and mandelic acid were purchased from Sigma-Aldrich (Shanghai, China). All other chemicals were purchased from SinopHarm Chemical Reagent Co., Ltd. (Shanghai, China). The BCNIT gene was acquired from our laboratory.

Construction of fusion genes

The plasmids pET28a/BCNIT-AsLOV2-SsrA and pET28a/SspB were constructed by PCR cloning using PrimeSTAR® Max DNA polymerase. The oligonucleotide sequence of the AsLOV2-SsrA domain containing 5′ *Spe*I and 3′ *Xho*I restriction sites was constructed on pET28a, and the flexibility linker of $(GGGGS)_2$ was modified on the N terminal of AsLOV2-SsrA using the primers P3/R3. The BCNIT gene with 5′ *Nde*I and 3′ *Spe*I restriction sites was amplified by PCR based on the BCNIT template using the

primers P2/R2. The genes were sequentially inserted into a modified pET28a plasmid with *Spe*I restriction sites, producing the pET28a/BCNIT-AsLOV2-SsrA (BNAS) domain, and then pET21a/BCNIT-AsLOV2-SsrA was constructed by inserting the BNAS domain into pET21a. The SspB gene was prepared by gene synthesis and constructed on pET21a and a pET21c plasmid modified with a resistance gene against chloromycetin instead of ampicillin. The sequences mN159 and mC160 were inserted into pET21a/BCNIT-AsLOV2-SsrA and pET28a/BCNIT-AsLOV2-SsrA using the primers P4/R4 and P5/R5 to produce the plasmids pET21a/BCNIT-mN159-AsLOV2-SsrA (BNMnAS) and pET28a/BCNIT-mC160-AsLOV2-SsrA (BNMcAS), respectively. The plasmids were transfected into *E. coli* BL21 (DE3) for recombinant protein expression. All the primers used are shown in Table 1. The detail information of construction of fusion genes is described in the supporting information (See Additional file 1).

Protein expression and purification

Recombinant *E. coli* BL21 (DE3) cells were cultured in Luria–Bertani medium at 37 °C until reaching an optical density at 600 nm (OD_{600}) of 0.5–0.7. Protein expression was induced by adding 0.1 mM isopropyl β-D-1-thiogalactopyranoside and incubating the cells at 18 °C for 20 h. The BCNIT, BNAS, and SspB proteins were purified with Ni-nitrilotriacetic acid columns (GE Healthcare, Waukesha, WI, USA). Protein concentrations were determined using the Bradford assay (Beyotime Biotechnology, Shanghai, China).

Optical control of enzyme assembly and disassembly in *E. coli*

The assembly of the fusion proteins in *E. coli* was monitored using the fluorescence complementation assay (Gao et al. 2015). The three plasmids pET21a/BNMnAS, pET28a/BNMcAS, and pET21c/SspB were cotransfected into *E. coli* BL21 (DE3). After centrifugation at 8000×*g*

Table 1 Primers used for cloning and transcript amplification in this study

Primer	Sequence (5′–3′)
P1	GGAATTCCATATGGAATACAAATCCTC
R1	CCGCTCGAGTTATTCATCGTAGATTTCTTCAG
P2	GGAATTCCATATGACCATCAATCACCCG
R2	GACTAGTCGAGCCACCGCCACCAGCGGGTGTGACGCGC
P3	GGACTAGTGGCCGGTGGCGGATCTTTAGCCACTACTTTAGAAAGG
R3	ATAAGAATGCGGCCGCAAAATAATTTTCATCATTAG
P4	AGGTGGCTCTACTAGTGGCGGAGGTGGCTCTGTG
R4	CACCTCCGCCACTAGTCGATCCGCCACCGCCGTC
P5	AGGTGGCTCTACTAGTGGCGGTGGCGGATCTGGC
R5	CACCTCCGCCACTAGTAGAGCCACCTCCGCCGCT

for 10 min and resuspension of the cell pellets in 5 mL 20 mM phosphate-buffered saline (PBS; pH 7.4), 10 μM FMN was added and the cells were incubated for 10 min at 37 °C under a 470-nm blue light-emitting diode array (Guntas et al. 2014). Meanwhile, an equivalent number of cells coexpressing BNMnAS, BNMcAS, and SspB in the dark, and cells coexpressing mN159, mC160, and SspB under 470-nm light were used as controls and assayed in the same manner. We monitored the differences between the samples placed under the blue light and those kept in the dark by measuring epifluorescence (excitation: 562–640 nm, emission: 590–650 nm).

Optical control of enzyme assembly and disassembly in vitro

The BNAS fusion proteins were assembled on SspB by triggering the affinity between AsLOV2-SsrA and SspB under illumination at 470 nm. Self-assembly via photopolymerization occurred in PBS buffer (pH 7.4) after mixing the freshly purified BNAS and SspB proteins at an equimolar ratio under freezing conditions and adding 10 μM FMN to induce a stable supramolecular polymerization. The disassembly measurement was performed in the dark after 470-nm blue light illumination. The optical density at 600 nm (OD_{600}) was used to characterize the turbidity of the supramolecular assembly (Kanekura et al. 2016). The reactions before measured the OD_{600} were carried out in an ice bath. Moreover, the assemblies were visualized using scanning electron microscopy. BNAS or BCNIT and SspB were mixed at a 1:1 molar ratio.

Dynamic light scattering (DLS) assay

To explore the assembly and disassembly before and after illumination, we used a DynaPro NanoStar® instrument (Wyatt Technology, Santa Barbara, CA, USA) to analyze the changes in the particle size of the BNAS–SspB (BNASS) supramolecular complex. Samples were filtered with a pore size of 0.22 μm prior to analysis, and each measurement was repeated three times at 18 °C.

Field-emission scanning electron (FESEM)

The sample was repeatedly washed with deionized water and resuspended, dried under air, and applied onto a slide. Images were collected on an S4800 scanning electron microscope (Hitachi, Tokyo, Japan) operated at 15 kV. In order to more clearly observe morphology and structure of BNASS supramolecular assemblies, we collected images with different resolutions.

Enzymatic activity assay in vitro and in vivo

Nitrilase activity was measured using reverse-phase high-performance liquid chromatography by monitoring the decrease of mandelonitrile (substrate) or the increase of mandelic acid (product) at 210 nm. The standard assay mixtures contained 20 mM mandelonitrile, 10 μM FMN, 100 mM PBS (pH 7.4), and 6 μM pure enzyme (in vitro) or 10 mg/mL of E. coli cells expressing nitrilase (in vivo). After reaction at 30 °C with agitation at 200 rpm for 20 min, 100 μL of 1 M HCl was added to stop the reaction, and centrifugation was performed at $13,000 \times g$ for 10 min, before removing 500 μL of sample for reverse-phase high-performance liquid chromatography analysis using a Zorbax® SB-Aq column (250 × 4.6 mm, 5 μm; Agilent Technologies, USA) at a detection wavelength of 210 nm (Ni et al. 2013). The detail information of enzymatic activity assay is described in the supporting information (See Additional file 1).

Results and discussion

The strategy for the light-controlled self-assembly of dodecamer nitrilase in vitro and in vivo is outlined in Fig. 1. The fusion protein BNAS could be assembled under 470-nm light, and the resulting assemblies could be disassembled in the dark.

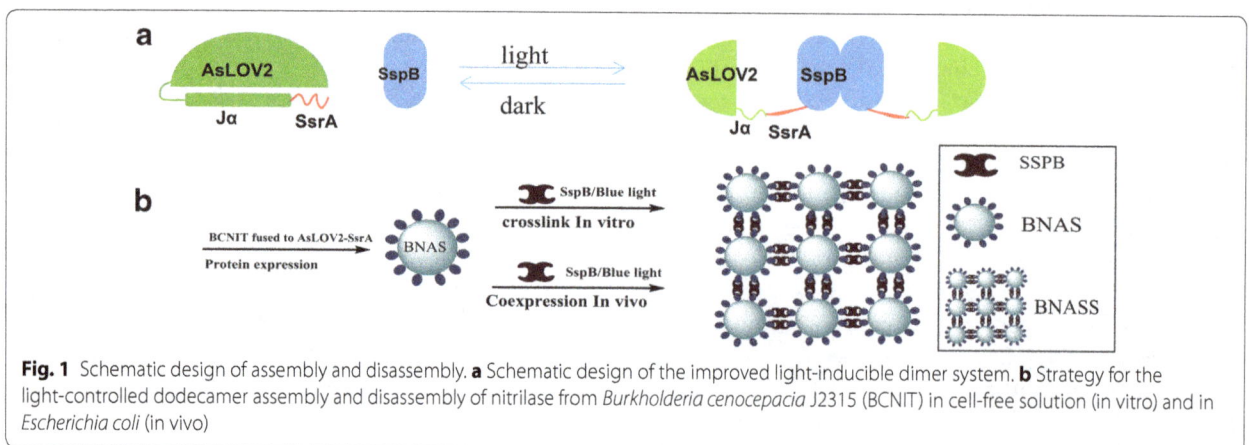

Fig. 1 Schematic design of assembly and disassembly. **a** Schematic design of the improved light-inducible dimer system. **b** Strategy for the light-controlled dodecamer assembly and disassembly of nitrilase from *Burkholderia cenocepacia* J2315 (BCNIT) in cell-free solution (in vitro) and in *Escherichia coli* (in vivo)

Light-controlled assembly and disassembly in vitro

The in vitro light-controlled assembly was performed by expressing and purifying SspB and BNAS in *E. coli* BL21 (DE3). As shown in the Fig. 2a, the purified proteins were observed with the molecular mass of 13 and 58 kDa for SspB and BNAS, respectively. BNAS retained about 97% of the specific activity of BCNIT (Fig. 2b), indicating that the protein fused with AsLOV2-SsrA did not suffer any drastic structural change in vitro. The light-controlled assembly was triggered by mixing the purified SspB and BNAS proteins with FMN under illumination at 470 nm. Following illumination, the solution changed rapidly from clear to turbid (Fig. 3A), and the DLS measurements indicated that BNAS was converted into a BNASS supramolecular polymer (Fig. 3B). The hydrodynamic diameter of pure BNAS is about 30 nm. The hydrodynamic diameter of the proteins decreased from 1000 to 100 nm after the light was removed, suggesting that the supramolecular polymer was disassembled within 5 min. However, the assembly could not disassembled completely, which might result from a remnant weak affinity between SsrA and SspB in the dark.

The morphology and structure of BNASS supramolecular polymers were observed by scanning electron microscopy (SEM), and it was found that the assembly formed a circular sheet structure with a hydrodynamic diameter about 1 μm (Fig. 3C). When the blue light was removed for 10 min, the hydrodynamic diameter of the assembly was found to vary from 30 to 200 nm by SEM (Fig. 3D). The results confirmed that SspB and AsLOV2-SsrA could mediate the formation of a supramolecular polymer under illumination at 470 nm and the polymer would be disassembled in the dark. Figure 4a shows the results obtained when monitoring the disassembly of the

supramolecular polymer in the dark. The change of OD_{600} of the supramolecular polymer suggested that the supramolecular polymer disassembled dramatically, confirming the DLS results. An experiment was performed to examine whether different illumination intensities could influence the assembly under blue light and the disassembly in the dark. Figure 4b shows that the OD_{600} increased gradually with the increasing illumination intensities. This result indicated that the light intensity has a great effect on the enzyme assembly, possibly because the light intensity affected the responding time of the AsLOV2-SsrA to the blue light, resulting in different degrees of assembly at the same time. However, when the light was removed, assemblies with different OD_{600} values showed varying degrees of disassembly (Fig. 4c). This observation could be explained by larger assemblies exhibiting a stronger clustering effect (Pieters 2009; Sengupta et al. 2011; Jaenicke 2000), thereby weakening the degree of disassembly. The disassembly efficiency is optimal when the OD_{600} is 0.7–0.8. Therefore, we employed the assembly with an OD_{600} value of 0.7 in the following experiments.

The recyclability of multienzyme polymers is an important factor when it comes to applications in synthetic biology and biocatalysis, especially for light-controlled enzyme assembly processes. The recyclability of the light-controlled assembly was evaluated by examining the change in the OD_{600} value of the mixture of BNAS and SspB when the mixture solution was repeatedly transferred between blue light illumination and the dark conditions. As shown in Fig. 4d, the OD_{600} value decreased from 0.7 to 0.1 within 5 min in the dark, then increased to 0.7 within a few seconds under illumination at 470 nm, which is in accordance with the results displayed in Fig. 3B. Furthermore, neither the assembly nor

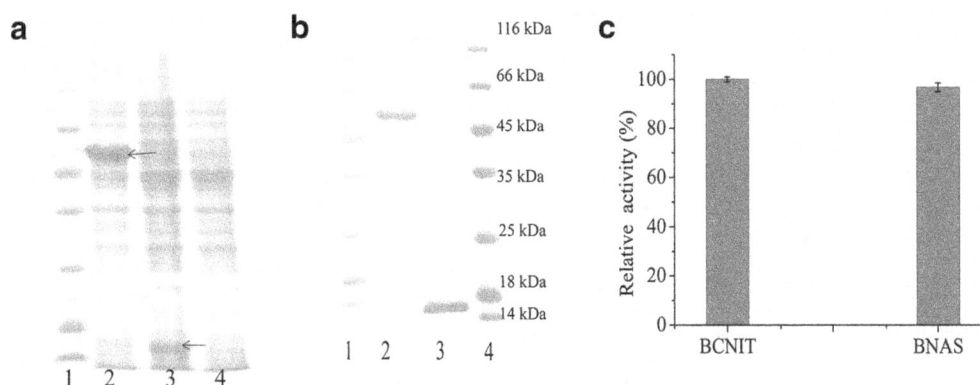

Fig. 2 Expressions of BNAS and SspB proteins in *E. coli* BL21 (DE3) cells. **a** SDS-PAGE image of the cells expressing BNAS and SspB at 20 h after IPTG addition. *Lanes 1* marker; *lane 2* BNAS; *lane 3* SspB; *lane 4* control. The *arrow* shows the destination band. **b** SDS-PAGE analyses of purified BNAS and SspB in vitro. *Lanes 1* and *4* marker; *lane 2* purified BNAS; *lane 3* purified SspB. **c** Comparison of the enzyme activities of BCNIT and BNAS in vitro. The determination of nitrilase activity was performed at 30 °C for 30 min with 10 μM FMN under 470-nm blue light illumination. Under the experimental conditions, the relative activity was expressed as a percentage of the maximum activity. *Error bars* represent the standard error of three replicates

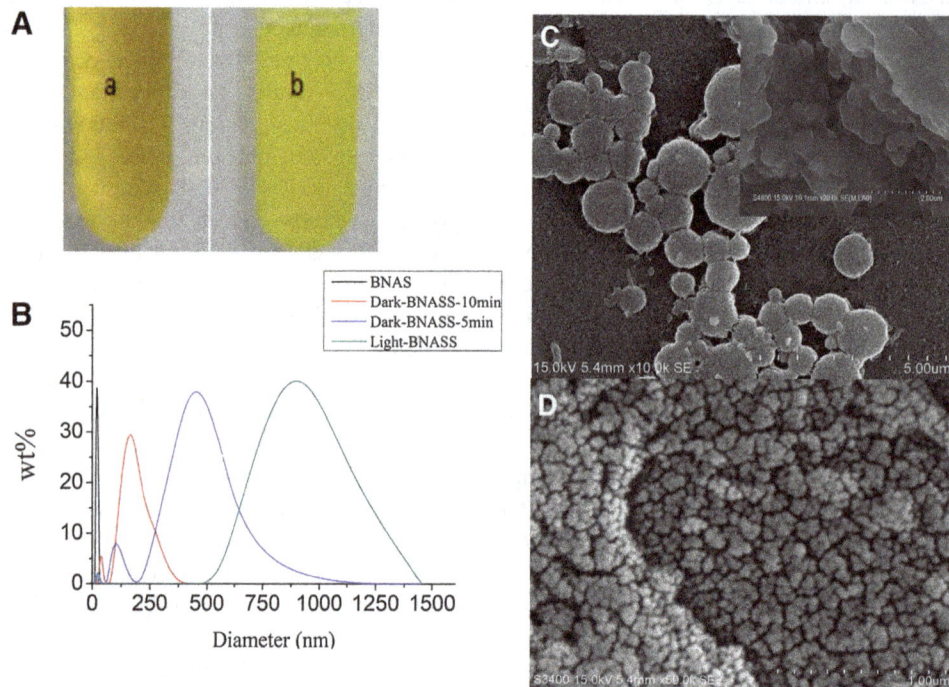

Fig. 3 The light-controlled assembly, disassembly, and morphology of the BCNIT-AsLOV2-SsrA-SspB (BNASS) complex in vitro. **A** The solution states of the mixture of BNAS, SspB, and FMN either (*a*) kept in the dark or (*b*) exposed to light. **B** Dynamic light scattering analysis of the hydrodynamic diameter of pure BCNIT-AsLOV2-SsrA (BNAS) and the mixture of SspB and BNAS. **C** Structure of the BNASS complex in vitro visualized by FESEM and SEM after illumination with blue light. **D** Structure of the BNASS complex in vitro visualized by SEM when the light is removed for 10 min. The reactions were carried out in an ice bath

the disassembly was affected by repeating the light/dark cycling up to five times.

To further determine whether the catalytic activity of the assemblies is affected by repeated assembly–disassembly cycles, we analyzed the activity of the supramolecular assembly through seven cycles of illumination at 470 nm followed by incubation in the dark. Figure 5 shows that the light-controlled assembly retained a high level of activity as the number of cycles increased, and still retained 90% of its initial activity after four cycles. The dramatic decrease of enzyme activity after five or more cycles may be caused by the disruption of the enzyme assembles after the violent resuspension, the loss of some smaller enzyme assembles during centrifugal isolation, and the toxicity of nitrile substrate after multicycles.

Light-controlled assembly in vivo

To assess whether the light-controlled assembly and disassembly of the fusion proteins can occur within *E. coli* cells, BNAS and SspB were coexpressed in *E. coli* BL21 (DE3), and both of them could be expressed in their soluble forms. To monitor the light-controlled formation of the assemblies in vivo, a fluorescence

complementation assay was also performed. The red fluorescent protein mCheery was divided into two parts, mN159 and mC160, and was fused to the site between BCNIT and AsLOV2-SsrA, thus producing the BCNIT-mN159-AsLOV2 (BNMnAS) and BCNIT-mC160-AsLOV2 (BNMcAS), respectively. Little fluorescence was observed when mN159, mC160, and SspB were coexpressed in *E. coli* BL21 (DE3) and the cells were illuminated under blue light (Fig. 6a). However, the fluorescence intensity increased markedly when BNMnAS and BNMcAS were coexpressed with SspB under illumination at 470 nm (Fig. 6b), indicating that the strong fluorescence is due to the presence of AsLOV2-SsrA. Concordant with the increase of fluorescence intensity resulting from induction by blue light illumination, the strains coexpressing BNMnAS, BNMcAS, and SspB showed significantly lower fluorescence intensities in the dark (Fig. 6c). Although the core domain of the AsLOV2 covered the binding site of the interaction between SsrA and SspB in darkness, the SsrA domain was not completely inactivated, and hence SsrA and SspB kept a weaker affinity in darkness (Guntas et al. 2014). The weaker affinity led to the assembly of fewer nitrilases, and thus there were weak fluorescent signals

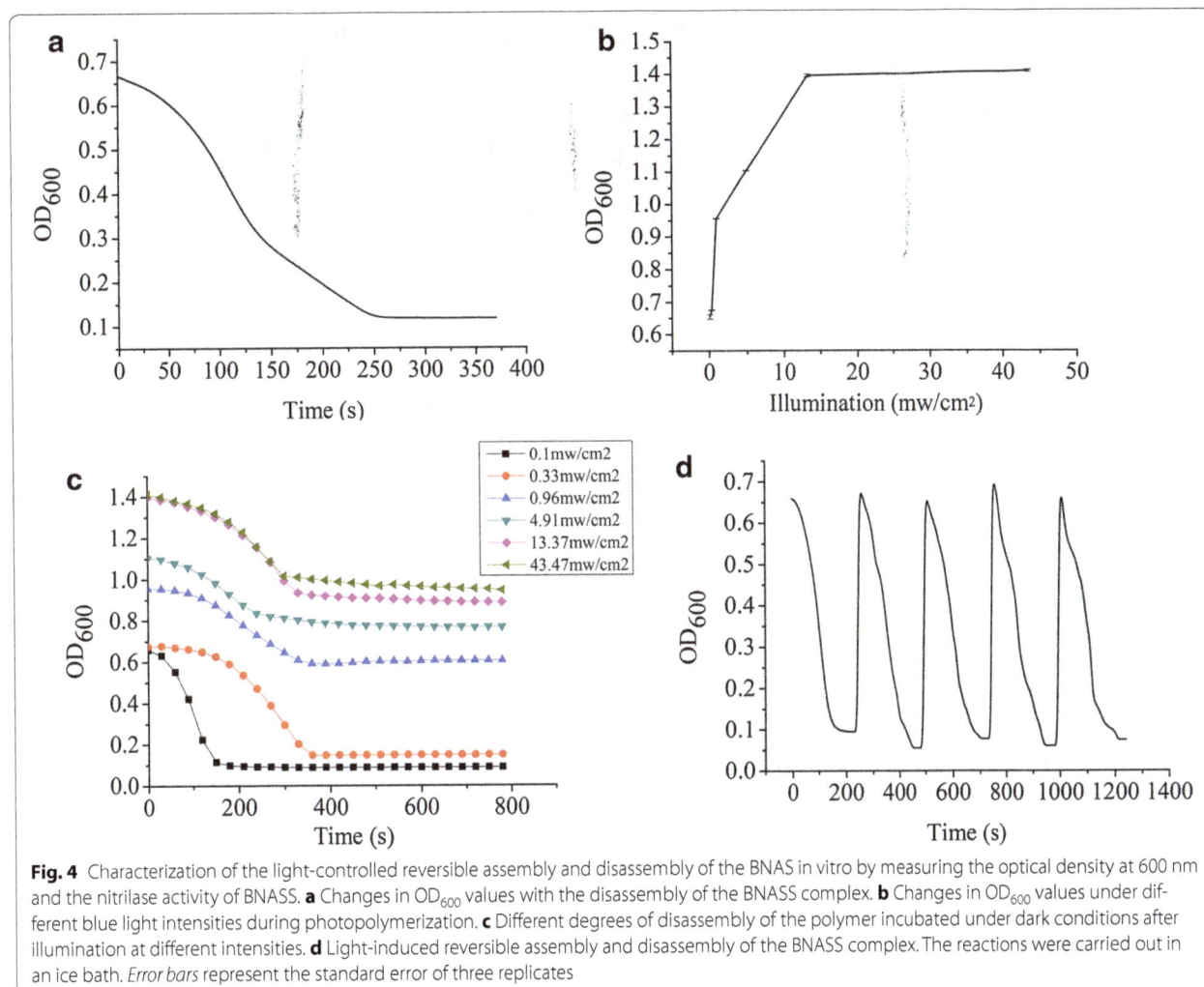

Fig. 4 Characterization of the light-controlled reversible assembly and disassembly of the BNAS in vitro by measuring the optical density at 600 nm and the nitrilase activity of BNASS. **a** Changes in OD_{600} values with the disassembly of the BNASS complex. **b** Changes in OD_{600} values under different blue light intensities during photopolymerization. **c** Different degrees of disassembly of the polymer incubated under dark conditions after illumination at different intensities. **d** Light-induced reversible assembly and disassembly of the BNASS complex. The reactions were carried out in an ice bath. *Error bars* represent the standard error of three replicates

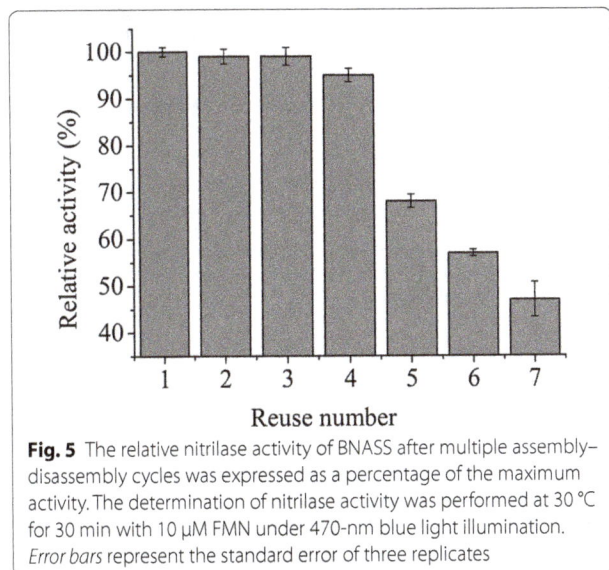

Fig. 5 The relative nitrilase activity of BNASS after multiple assembly–disassembly cycles was expressed as a percentage of the maximum activity. The determination of nitrilase activity was performed at 30 °C for 30 min with 10 μM FMN under 470-nm blue light illumination. *Error bars* represent the standard error of three replicates

in the dark. At the same time, the fluorescence intensity of 5.0 OD_{600} cells collected after induction was measured using a Spectra Max M5 microplate reader (Synergy Mx, Bio-Tek Instruments, Inc., Winooski, VT, USA). As shown in Fig. 7a, the results of fluorescence intensities are exactly the same as those shown in Fig. 6. All results showed that the strong fluorescence signal is due to the presence of AsLOV2-SsrA and blue light. To determine the catalytic activity of BNASS in vivo, BNAS and SspB were coexpressed in *E. coli* BL21. As a control, the same number of cells expressing BNAS and BCNIT were analyzed following the same procedure. Similar to the result obtained in vitro, as shown in Fig. 7b, the activity of BNAS in vivo remained at 93% relative to that of BCNIT. The supramolecular assembly BNASS also maintained approximately 85% activity relative to that of BCNIT regardless of light or dark exposure. The results show that the supramolecular assembly process barely changed the activity of BCNIT.

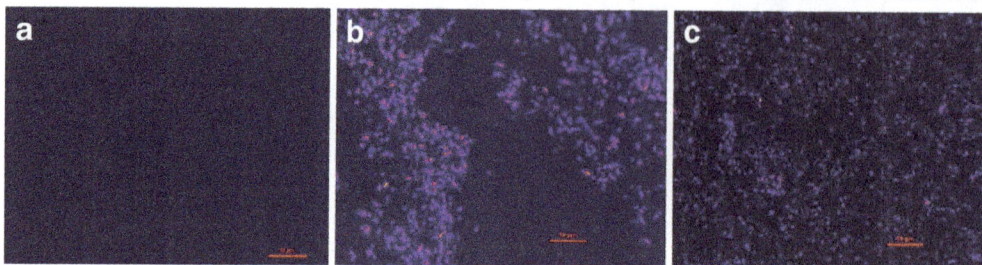

Fig. 6 Fluorescence microscopy visualization of *Escherichia coli* BL21 cells coexpressing **a** mN159, mC160, and SspB under blue light, **b** BNMnAS, BNMcAS, and SspB under blue light. **c** BNMnAS, BNMcAS, and SspB in the dark. *Scale bar* 10 μm

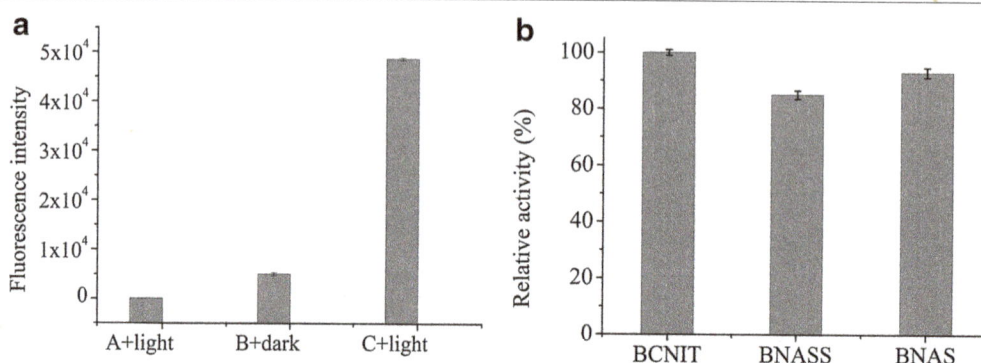

Fig. 7 The fluorescence intensities of fusion protein-coexpressing strains. **a** A + light represents the strains coexpressing mN159, mC160, and SspB under blue light illumination; B + dark represents the strains coexpressing BCNIT-mN159-AsLOV2-SsrA (BNMnAS), BCNIT-mC160-AsLOV2-SsrA (BNMcAS), and SspB in the dark; C + light represents the strains coexpressing BNMnAS, BNMcAS, and SspB under blue light illumination. **b** Comparison of the nitrilase activities of BCNIT, BNAS, and BNASS in *E. coli*. *Error bars* represent the standard error of three replicates. The determination of nitrilase activity was performed at 30 °C for 30 min with 10 μM FMN under 470-nm blue light illumination. Under the experimental conditions, the relative activity was expressed as a percentage of the maximum activity. *Error bars* represent the standard error of three replicates

Conclusions

In conclusion, a strategy for the light-controlled assembly and disassembly of oligomeric enzymes through an optogenetic tool was successfully applied to in vitro and in vivo systems. The light-controlled assembly did not affect the catalytic ability of enzymes, and the assemblies showed good reusability and reversibility. The results indicated that the construction of supramolecular assemblies mediated by AsLOV2 can provide a solid biocatalyst with good reusability and reversibility in cell-free solutions and bacterial cultures, and this method might provide a promising way to precisely and reversibly control protein–protein interactions in mammalian cells. These results indicate that the photocontrolled self-assembly strategy is a powerful tool for achieving the scaffold-free self-assembly and immobilization without carrier of enzymes. Thus, the photocontrolled enzyme assembly strategy may be used in the multienzyme biocatalysis to realize precise and reversible control of the process of metabolic enzyme cascades in vivo and in vitro.

Abbreviations
AsLOV2: LOV2 domain isolated from *Avena sativa*; BCNIT: nitrilase from *Burkholderia cenocepacia* J2315; BNAS: BCNIT-AsLOV2-SsrA domain; BNASS: BNAS-SspB; BNMnAS: BCNIT-mN159-AsLOV2-SsrA; BNMcAS: BCNIT-mC160-AsLOV2-SsrA; CRY: cryptochromes; DLS: dynamic light scattering; FMN: flavin mononucleotide; HPLC: high-performance liquid chromatography; iLID: photoswitch-improved light-inducible dimer; IPTG: isopropyl β-ᴅ-1-thiogalactopyranoside; LB: Luria broth; LOV: light-oxygen-voltage; OD: optical density; PAS: Per-Arnt-Sim; PhyB: phytochrome B; FESEM: field-scanning electron microscopy.

Authors' contributions
QY conducted the experiments and the manuscript writing. YR provided advice in the experiments design and data analysis. All authors read and approved the final manuscript.

Author details
[1] East China University of Science and Technology, Shanghai, China. [2] Key Laboratory of Synthetic Biology, Institute of Plant Physiology and Ecology, Shanghai Institutes for Biological Sciences, Chinese Academy of Sciences, Shanghai, China. [3] Fujian Key Laboratory of Eco-Industrial Green Technology,

Wuyi University, Wuyishan, China. [4] State Key Laboratory of Bioreactor Engineering, New World Institute of Biotechnology, East China University of Science and Technology, Shanghai, China.

Acknowledgements
This work was funded by the National Special Fund for the State Key Laboratory of Bioreactor Engineering (2060204).

Competing interests
The authors declare that they have no competing interests.

Funding
The State Key Laboratory of Bioreactor Engineering (2060204).

References
Brady D, Jordaan J (2009) Advances in enzyme immobilisation. Biotechnol Lett 31(11):1639

Fairman R, Åkerfeldt KS (2005) Peptides as novel smart materials. Curr Opin Struct Biol 15(4):453–463

Gao X, Zhao CC, Yu T, Yang S, Ren YH, Wei DZ (2015) Construction of a reusable multi-enzyme supramolecular device via disulfide bond locking. Chem Commun 51(50):10131–10133

Guntas G, Hallett RA, Zimmerman SP, Williams T, Yumerefendi H, Bear JE, Kuhlman B (2014) Engineering an improved light-induced dimer (iLID) for controlling the localization and activity of signaling proteins. Proc Natl Acad Sci USA 112(1):112–117

Halavaty AS, Moffat K (2007) N- and C-terminal flanking regions modulate light-induced signal transduction in the LOV2 domain of the blue light sensor phototropin 1 from Avena sativa. Biochemistry 46(49):14001–14009

Harper SM, Christie JM, Gardner KH (2004) Disruption of the LOV-J alpha helix interaction activates phototropin kinase activity. Biochemistry 43(51):16184–16192

Holmes TC (2002) Novel peptide-based biomaterial scaffolds for tissue engineering. Trends Biotechnol 20(1):16–21

Jaenicke R (2000) Stability and stabilization of globular proteins in solution. J Biotechnol 79(3):193–203

Kanekura K, Yagi T, Cammack AJ, Mahadevan J, Kuroda M, Harms MB, Miller TM, Urano F (2016) Poly-dipeptides encoded by the C9ORF72 repeats block global protein translation. Hum Mol Genet 25(9):1803–1813

Kennedy MJ, Hughes RM, Peteya LA, Schwartz JW, Ehlers MD, Tucker CL (2010) Rapid blue-light-mediated induction of protein interactions in living cells. Nat Methods 7(12):973–975

King NP, Lai YT (2013) Practical approaches to designing novel protein assemblies. Curr Opin Struct Biol 23(4):632–638

Levskaya A, Weiner OD, Lim WA, Voigt CA (2009) Spatiotemporal control of cell signalling using a light-switchable protein interaction. Nature 461(7266):997–1001

Matsumoto T, Isogawa Y, Minamihata K, Tanaka T, Kondo A (2016) Twigged streptavidin polymer as a scaffold for protein assembly. J Biotechnol 225:61–66

Ni KF, Wang HL, Zhao L, Zhang MJ, Zhang SY, Ren YH (2013) Efficient production of (R)-(−)-mandelic acid in biphasic system by immobilized recombinant E. coli. J Biotechnol 167(4):433–440

Pieters RJ (2009) Maximising multivalency effects in protein–carbohydrate interactions. Org Biomol Chem 7(10):2013–2025

Schoffelen S, van Hest JCM (2012) Multi-enzyme systems: bringing enzymes together in vitro. Soft Matter 8(6):1736–1746

Sengupta P, Jovanovic-Talisman T, Skoko D, Renz M, Veatch SL, Lippincott-Schwartz J (2011) Probing protein heterogeneity in the plasma membrane using PALM and pair correlation analysis. Nat Methods 8(11):969–975

Smith KH, Tejeda-Montes E, Poch M, Mata A (2011) Integrating top-down and self-assembly in the fabrication of peptide and protein-based biomedical materials. Chem Soc Rev 40(9):4563–4577

Strickland D, Yao X, Gawlak G, Rosen MK, Gardner KH, Sosnick TR (2010) Rationally improving LOV domain-based photoswitches. Nat Methods 7(8):623–626

Wang HL, Sun HH, Wei DZ (2013) Discovery and characterization of a highly efficient enantioselective mandelonitrile hydrolase from Burkholderia cenocepacia J2315 by phylogeny-based enzymatic substrate specificity prediction. BMC Biotechnol 13(1):14–24

Yin TF, Wu YI (2013) Guiding lights: recent developments in optogenetic control of biochemical signals. Pfluegers Arch 465(3):397–408

Yu T, Gao X, Ren YH, Wei DZ (2015) Assembly of cellulases with synthetic protein scaffolds in vitro. Bioresour Bioprocess 2(1):16–22

Zimmerman SP, Hallett RA, Bourke AM, Bear JE, Kennedy MJ, Kuhlman B (2016) Tuning the binding affinities and reversion kinetics of a light inducible dimer allows control of transmembrane protein localization. Biochemistry 55(37):5264–5271

16

Electron transport phenomena of electroactive bacteria in microbial fuel cells: a review of *Proteus hauseri*

I-Son Ng[1]*⊕, Chung-Chuan Hsueh[2] and Bor-Yann Chen[2]*

Abstract

This review tended to decipher the expression of electron transfer capability (e.g., biofilm formation, electron shuttles, swarming motility, dye decolorization, bioelectricity generation) to microbial fuel cells (MFCs). As mixed culture were known to perform better than pure microbial cultures for optimal expression of electrochemically stable activities to pollutant degradation and bioenergy recycling, *Proteus hauseri* isolated as a "keystone species" to maintain such ecologically stable potential for power generation in MFCs was characterized. *P. hauseri* expressed outstanding performance of electron transfer (ET)-associated characteristics [e.g., reductive decolorization (RD) and bioelectricity generation (BG)] for electrochemically steered bioremediation even though it is not a nanowire-generating bacterium. This review tended to uncover taxonomic classification, genetic or genomic characteristics, enzymatic functions, and bioelectricity-generating capabilities of *Proteus* spp. with perspectives for electrochemical practicability. As a matter of fact, using MFCs as a tool to evaluate ET capabilities, dye decolorizer(s) could clearly express excellent performance of simultaneous bioelectricity generation and reductive decolorization (SBG and RD) due to feedback catalysis of residual decolorized metabolites (DMs) as electron shuttles (ESs). Moreover, the presence of reduced intermediates of nitroaromatics or DMs as ESs could synergistically augment efficiency of reductive decolorization and power generation. With swarming mobility, *P. hauseri* could own significant biofilm-forming capability to sustain ecologically stable consortia for RD and BG. This mini-review evidently provided lost episodes of great significance about bioenergy-steered applications in myriads of fields (e.g., biodegradation, biorefinery, and electro-fermentation).

Keywords: *Proteus hauseri*, Reductive decolorization, Bioelectricity generation, Bioremediation

Introduction

This review tended to disclose unsolved mysteries of electron transfer phenomena for biomass energy applications with sustainability. Faced with increasingly scarce supply of energy and resources and gradually serious threats of ecological deprivation around the globe, top governing bodies of various nations have pushed forward with efforts for environmental decontamination and materials/energy renewability of the vulnerable ecology. That is,

with strict measures for energy conservation and emission reduction, environmental/ecological protection/prevention and materials/energy recycling/reuses are of great importance to this resource-limited world. Making use of innovative features and environmental benefits of sustainable biomass energy extraction for applications would be inevitable. As a matter of fact, microbial fuel cells (MFCs) are bioelectrochemical devices that use electroactive microbes and mimicked bacterial interactions as biocatalysts to drive current of bioenergy from oxidation of organics for bioelectricity generation and wastewater treatment (Logan et al. 2006; Ren et al. 2012; Pant et al. 2010; Rabaey and Verstraete 2005; Rozendal et al. 2008, 2009). MFCs apparently would be promising platform to evaluate feasibility of bioresources for

*Correspondence: yswu@mail.ncku.edu.tw; bychen@niu.edu.tw;
boryannchen@yahoo.com.tw
[1] Department of Chemical Engineering, National Cheng Kung University,
Tainan 70101, Taiwan
[2] Department of Chemical and Materials Engineering, National I-Lan
University, I-Lan 26047, Taiwan

biorefinery/bioenergy practicability. At ambient temperature, anodic bacteria oxidize organic matter with direct current production for CO_2 production and pollutant decontamination. Meanwhile, protons migrated through separation of an ion exchange membrane (e.g., proton exchange membrane) toward cathode and electrons went through external circuit to the cathode simultaneously. Then, bioelectricity generation would be taken place at (bio)cathode coupled with water formation (i.e., $\frac{1}{2}O_2 + 2H^+ + 2e^- \rightarrow H_2O$). In fact, there were at least several mechanisms in levels of genomics and proteomics for biofilm formation, direct electron transfer, mediator electron transport, and direct interspecies electron transfer dealing with performance of power generation in MFCs (Kumar et al. 2016). Regarding (bio)cathode, selecting appropriate electron acceptor could significantly attenuate the potential loss present on the cathode, effectively augmenting power generation in MFCs (He et al. 2015). Moreover, comparative assessment upon various electron acceptors popularly used in MFCs was also identified. As biocatalysts, exoelectrogenic bacteria are capable to directly transport electrons from extracellular media that enable bacterial cells function in an MFC. Therefore, the electrons produced by oxidation via microbial metabolism are transported through the barriers of biofilm-and-liquid interface, anodic immobilized cells, and solid anode, external circuit to the cathode for effective power generation or further applications (Mohan et al. 2014; ElMekawy et al. 2014). To overcome interphase electron transfer resistance and significantly augment electron flux across the boundary layer on the anodic surface, effective concentrations of biologically secreted or artificially synthesized exogenous mediators (e.g., flavins, pyocyanin, gallic acid and aminophenol, decolorized amines, polyphenolics; Chen et al. 2017a, b; Qin et al. 2016) in extracellular broth media and redox-active complexes (e.g., C type cytochrome) across the outer membrane of microorganisms apparently play crucial roles to attenuate electron transfer resistance across different media interfaces or phases for effective power generation. Exogenous mediators as "electrochemical catalysts" could shuttle electrons to the electrode surface or to electron acceptors or donors if appropriate environmental conditions were provided (e.g., suitable pHs, redox potentials; Chen et al. 2017a). In particular, with assistance of electron shuttles (e.g., o- and p-aminophenol, humic acid, methyl viologen, neutral red), electron transfer from electrochemically inactive bacteria (e.g., Escherichia coli) to the electrode might be effectively facilitated to be mediator MFCs. In contrast, due to self-secreted ESs (e.g., riboflavin, pyocyanin), electrochemically active bacteria—inoculated microbial communities could be successfully acclimatized as

anodophilic consortia to transfer electrons to the cathode in mediator-free environments. That is, appropriate exogenous supplementation or intracellular accumulation of redox mediators all could significantly increase fluxes of electron transfer to simultaneously augment reductive degradation and bioelectricity generation. That is, electron shuttles might somehow still act as catalytic center of mediating electric current in electroactive bacteria-bearing MFCs.

Regarding electrochemically active bacteria, species of genus *Bacillus* (Nimje et al. 2009), *Enterobacter* (Rezaei et al. 2009), *Geobacter* (Richter et al. 2008), *Proteus* (Chen et al. 1999), *Shewanella* (Watson and Logan 2010) have been explored to present practicability for bioelectricity generation. However, compared to pure microbial culture MFCs, mixed culture still exhibited the most promising stable power generation for long-term operation (Hassan et al. 2017; Ishii et al. 2017). In addition, synergistic interactions of mixed microbial communities in MFCs for long-term ecologically stable electrochemical expression could be sustained due to nutrient adaptability, stress resistance, degradation cooperation, in particular under fluctuating pollutant threat to species survival. For example, as Chen et al. (2004) and Chen (2007) indicated, due to the presence of toxic pollutants (e.g., textile dyes), mixed communities would not follow characteristics under competitive exclusion principle (or Gause's law), evolving to be in cooperation for maximal species richness and evenness (Chen 2007; Chen and Chang 2007). This is why in MFCs pure microbial culture of superstar electrochemically active bacteria (e.g., *Shewanella* and *Geobacter*) would very likely not perform better than mixed culture. As a matter of fact, several thorough studies using mixed cultures (Chen et al. 2004; Chen 2007; Chen and Chang 2007) all pointed out cooperative not competitive characteristics to maximize species diversity were the most ecologically stable strategy in face of pollutant threat for "survival to the fittest." In particular, regarding reductive decolorization and bioelectricity generation (Sun et al. 2009, 2011), decolorized metabolites (e.g., aromatic amines; Chen et al. 2011a, 2011b) identified as redox mediators via GC–MS analysis played a crucial role to stimulate operation performance (ca. 40–70% increases in electron transfer capabilities of MFCs; Han et al. 2015). In fact, several reviews have focused on different perspectives of MFCs in terms of basic principles and fundamental practices. However, a comprehensive review to exhibit such lost episodes on electron shuttles in particular used in MFCs is still lacking. Therefore, this review amended such last pieces of the puzzle in MFCs that have been ignored in recent years for clarification.

Considering exoelectrogens for promising adaptation to bioremediation indigenous microbes were first selected to acclimatize from soils and water of contaminated sites. Then, selection pressure would be applied for induced biodegradation if appropriate actions of transformation enzymes could be effectively expressed (e.g., Chen et al. 1995, 2000). For instance, azo dye-bearing effluents from textile dyeing industry could significantly affect water quality in environment due to relatively low biodegradability of textile dyes (ca. $BOD_5/COD \sim 0.2$–0.5). To effectively biodegrade such pollutants, fungi, yeast, bacteria, algae, and plants were mentioned to be used (Saratale et al. 2011). Moreover, with electron-transfer-assisted treatment for green sustainability alternatives to combine with perspectives of dye degradation and energy/material recycling were recently suggested (Solanki et al. 2013). Azo dyes should be first anaerobically reduced to form aromatic amines. Next, aerobic degradation or complete mineralization could be proceeding. Azo linkages ($-N=N-$) usually own sulphonic (SO_3^-) electron-withdrawing groups and thus are electron deficient; thus, they should be effectively reduced in the electron-acceptor-O_2 absent conditions. Therefore, azo dye(s) with sulfonate groups and azo bond(s) strongly resist aerobic biodegradation. Apparently, azo bioreduction was taken place in different very different from oxidative degradation could be indicated (Φ_1 and Φ_2 denoted aromatic compounds) as follows:

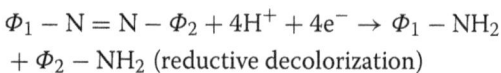

$$\Phi_1 - N = N - \Phi_2 + 4H^+ + 4e^- \rightarrow \Phi_1 - NH_2 + \Phi_2 - NH_2 \text{ (reductive decolorization)}$$

vs.

$$\Phi_1 - N = N - \Phi_2 + 2O_2 + e^- \rightarrow \Phi_1 = O + \Phi_2 + \frac{1}{2}N_2 + NO_3^- \text{ (oxidative degradation)}.$$

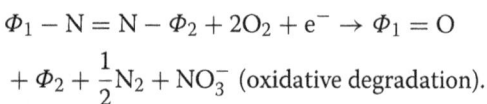

That is, strategies to stimulate effective electron transfer capabilities of pollutant degradation and energy recycling for electrochemically steered decolorization would be top priority for environment-compatible and energy-sustainable treatment.

In addition, considering an energy-efficient bioprocess in common, electrochemically driven fermentation could provide more opportunities to create different redox balancing, directing myriads of metabolic routes for generation of diverse product(s) (Hongo and Iwahara 1979; Rabaey and Rozendal 2010; Rabaey et al. 2011). Regarding electron-transport-assisted bioprocesses, Rabaey and Rozendal (2010) first mentioned the term "electro-fermentation." They used electrical current as alternative source of reducing or oxidizing power to guarantee different redox conditions for production. Evidently, this would be more economically feasible than fermentation

using conventional substrates (e.g., glucose; Rabaey et al. 2011; Schievano et al. 2016). These all showed electron transfer-assisted bioelectrochemical activities would play crucial roles for biorefinery/bioenergy applications. As a matter of fact, many cases of fermentation are redox-imbalanced; electrochemically driving fermentations provided electron transfer capabilities evolved different metabolic pathways to regulate intracellular and/or extracellular electrochemical potential(s) toward generation of value-added product(s). Thus, due to such redox potential-steered bioreactions, more energy-sustainable, substrate-saving, and product diverse-fermentation systems could be electrochemically feasible for maximization of value product formation and toxic pollutant biodegradation. However, to have system optimization for electro-fermentation, complete exploration on characteristics of MFCs was inevitably required for myriads of applications.

Glimpse of microbial fuel cells

In the recent years, He et al. (2017), Kumar et al. (2016), Hernandez-Fernandez et al. (2015), He et al. (2015), and Solanki et al. (2013) provided detailed and prolific reviews with thorough implications of microbial fuel cell (MFC) for wastewater treatment, bioenergy production, azo dye treatment, selection of electron acceptors, and exoelectrogens transfer, respectively. The concept of exoelectrogens of two dissimilatory metal reducing genera as electroactive bacteria (e.g., genomics, proteomics, and metabolomics of nanowire-generating *Shewanella* and *Geobacter* spp.; Table 1 in Kumar et al. 2016) has been well established for bioelectricity-generating applications. However, as aforementioned, sole expression of wild-type or genetically modified "superstar" bioelectricity-generating bacteria (e.g., nanowire-generating *Shewanella* sp.) still could not guarantee maximal electrochemically active bacteria in on-site practice. This was due to synergistic interactions in mixed cultures that would provide optimal expression of bioactivities for simultaneous organic biodegradation and production or bioenergy generation. That was simply due to mixed cultures ecologically acclimatized and electrochemically expressed mixed cultures established for maximal electron transfer-associated performance. The mysteries behind this were very likely due to combined interactions (e.g., feedback compensation of intraspecific/interspecific competition or mutualism and synergistic/antagonistic to achieve maximal species diversity; Chen 2007) of microbial ecology for maximal efficiency of organic oxidation and target product formation under optimal electrochemical activities with appropriate redox balancing. In fact, such competitive interactions of mixed cultures, which generated asymmetric dynamics among species with different levels of electrochemical capability and combined inter-actions, would generally optimize species diversity for long-term coexistence (i.e., die-out of electrochemically promising, but not ecologically favorable species as eco-logically stable outcomes; Briones and Raskin 2003; Chen et al. 2004, 2013a). This was why naturally dominant and ecologically stable, but not the most electrochemically active bacterium-*Proteus hauseri* was selected herein as a model electroactive bacterium to uncover lost episodes for complete understanding on ecologically stable MFCs. Being aware that *P. hauseri* as keystone species could play a crucial role in maintaining the electrochemically active structure of an ecological community in MFCs for stable bioelectricity generation. In fact, in ecosystems extinction of bacteria with outstanding specific functions (e.g., electrochemically active bacteria with extra meta-bolic burden) seemed to be evolutionarily inevitable, and thus bioaugmentation and biostimulation strategies for on-site or in situ bioremediation were regularly imple-mented for practical applications (e.g., bioaugmenta-tion may be needed to prevent extinction of outstanding functioning bacteria; Chen 2007). According to Gause's law—competitive exclusion principle, metabolic bur-den of electrochemically active bacteria might provide condition(s) not favorable for electrochemically active bacteria to compete with other species for long-term sur-vival (Chen and Chang 2007). That was why *P. hauseri* as one of stable electrochemically active bacteria and com-pletely deciphering characteristics of *P. hauseri*-seeded mixed culture MFCs should be disclosed for system sta-bility and operation optimization. As Bajracharya et al. (2016) mentioned, electrochemically anodic biofilm was major biocatalysts of bioelectrochemical systems (BESs) dealing with electron transport phenomena for electric-ity generation. Power-generating capabilities of MFC strongly depended upon whether the capacity of anodic exoelectrogens could significantly reduce mass trans-fer resistance and effectively enhance electron transport phenomena. According to US EPA, complete expression of essential functions of bioactivity (e.g., biodegradation of petroleum hydrocarbon contaminants) would still rely on synergistic assistance of non-functioning bioactivity for pollutant cleanup [e.g., US EPA remedial "technol-ogy"—monitored natural attenuation (MNA)]. To reach this consensus of ecological awareness for cradle-to-cradle treatment, of course selecting ecologically stable and electroactive microorganisms (e.g., *P. hauseri*) would apparently determine whether simultaneous reductive decolorization and energy extraction in MFCs could be stably maintained for long-term operation. In particular, stable maintenance of promising electroactive consor-tia is also projected to increase as electroactive biofilm-forming capability of comrade microbes dominant in

the ecology as "keystone species" (e.g., *P. hauseri*) was significantly better than other species (e.g., dominant nanowire-generating bacteria like *Shewanella* spp.) due to swarming characteristics. Moreover, performance of electron flux in MFCs was strongly controlled by inter-face resistance between cell broth and biofilm and elec-tron shuttling capabilities of redox mediators to pollutant degradation and current generation. This review tended to uncover the redox-mediating and biofilm-forming capabilities of non-nanowire-generating bacteria, for sys-tem design of MFC applications.

History of *Proteus* genus

Regarding *Proteus* genus, *P. hauseri* belonged to the family of *Enterobacteriaceae*, is a gram-negative Proteo-bacterium, but not well characterized like *P. mirabilis*. Recent findings indicated that *P. hauseri* was capable of promising bioelectricity generation and textile dye decol-orization for redox-mediating applications (Chen et al. 2010). The genus of *Proteus* was first classified by Profes-sor Hauser in 1885 to characterize this kind of dimor-phic strains. As indicated, exploration of *Proteus* genus has been implemented for more than one century. Up to now, *Proteus* genus at least contains five species of *P. vul-garis*, *P. mirabilis*, *P. penneri*, *P. hauseri*, and *P. myxofa-ciens* (Pearson et al. 2008, 2011; Poore and Mobley 2003). Regarding classification, Brenner-introduced *P. vulgaris* was a heterogeneous group with at least three biogroups via DHA-hybridizations. The biogroup 1 was separated from *P. vulgaris* and the nomenclature of *P. penneri* was distinguished by negative reactions for indole produc-tion, salicin fermentation, and aesculin hydrolysis. Bio-group 2 was positive for reporter reaction of salicin and aesculin, while biogroup 3 was negative for salicin and aesculin, respectively. According to O'Hara, biogroup 3 could be composed of four distinct DNA groups, further assigned as biogroups 3, 4, 5, and 6. The new biogroup 3 was identified as *P. hauseri* due to l-rhamnose fermen-tation, DNase, lipase production, and Jordan's tartrate utilization (O'Hara et al. 2000). Regard to the other bio-groups, they are still remained open to be further classi-fied via characterization of other biofunctions.

As presented in the milestone history of *Proteus* (Fig. 1), Hauser firstly announced the genus of *Proteus* to the public domain. Brenner and Hickman classified more *Proteus* species into biogroups 1, 2, and 3 includ-ing *P. vulgaris* and *P. penneri*. On the other hand, *P. mirabilis*, a bacterium differentiated from a rod-shaped vegetative cell into an elongated and highly flagellated cell that had swarming motility, was reported as indica-tor in Mobley and Belas (1995). O'Hara finally defined the new biogroups 4, 5, and 6 in 2000. Recently, a new strain, *P. hauseri* ZMd44, was isolated from sewage of sodium hydrogen carbonate hot spring (pH ~ 7.8, [CO_3^{2-}] ~ 297 ppm, [Na^+] ~ 187 ppm) in Chiao-Hsi (or Jiaoxi) on Taiwan's Lanyang Plain. This strain was predominantly selected by high-performance bacterial consortia to simultaneously decolorize monoazo dye direct orange 37 and diazo dye reactive blue 160 (Zhang et al. 2010). Although ZMd44 expressed promising capabilities for reduction of several dyes, acid black 172, acid yellow 42, and direct blue 22 were still not biodegradable. This was perhaps due to their inhibitory potency by a hydroxyl group at *ortho* to azo bond(s) in chemical structures. Due to the above-mentioned selection pressures, ZMd44 expressed promising capabilities in simultaneous bioelectricity generation and reductive decolorization (SBG and RD) through oxidation of secondary organic energy source via co-metabolism (Chen et al. 2010, 2013b, c). This also showed the favorable characteristics of *P. hauseri* for practical applications to contaminant bioremediation.

On the other hand, with clinical significance, *P. vulgaris* and *P. penneri* are opportunistic pathogens responsible for many urinary tract infections to humans. In fact, *Proteus* bacilli are widely distributed in nature (e.g., sewage, manure soil, animal faces). In particular, typical strains of *Proteus* own specific swarming behaviors and tended to be motile. As recent findings suggested, it was suspected that the characteristics of "swarming" mobility were strongly associated with pathogenicity to humans, as active swarming bacteria of *Salmonella typhimurium* exhibited greater tolerance to multiple antibiotics (e.g., polymyxin) (Kim et al. 2003). Thus, it is proposed herein that such swarming mobility very likely could have further practicability for industrial uses. Herein, we would specifically consider on flagellated (swarming, motile)

variant of *Proteus* genus of isolated colonies with surface film (designated as H form of *P. hauseri*) for industrial practice.

Swarming characteristics of *Proteus*

Due to outstanding swarming characteristics, *P. hauseri* could function as "keystone species" to maintain electrochemically active consortia for power generation. Swarming was a kind of specific movement for many microorganisms to migrate and colonize on moist surface. It was powered by rotating of flagella present in *P. hauseri*, *P. mirabilis*, *Bacillus subtilis*, *Pseudomonas aeruginosa*, and many of other kinetic strains (Kearns 2010; Jones et al. 2004). The process of swarming was not driven by a single cell, but the population of bacteria aggregated as multicellular groups (Kearns 2010). Comparing with the motion of swimming depended on individual bacteria and the sliding spread by growth of cell, swarming bacteria move in side-by-side cell groups called raft (Fig. 2a). To trigger the swarming ability, most strains of swarming could simultaneously take place to differentiate from normal vegetable cell into elongated, flagellated, and kinetic swarming cell. In fact, dimorphic phenomenon resulting in the diversity of bacteria lawn pattern emerged in the agar plate. In particular, colonies of *P. hauseri* ZMd44 always emerged in the concentric circle pattern, on agar plate after incubation, (e.g., bull's eye pattern) (Fig. 2b). The initial incubation of ZMd44 in the plate was in vegetable form, as the bacteria were originally grown in broth medium. This type of vegetable cells was short rod-shape with the length about 1.5–2 μm long, and fimbriated around the cell, also (e.g., swimming cell) (Mobley and Belas 1995). After incubation of ZMd44, there would be a swarming

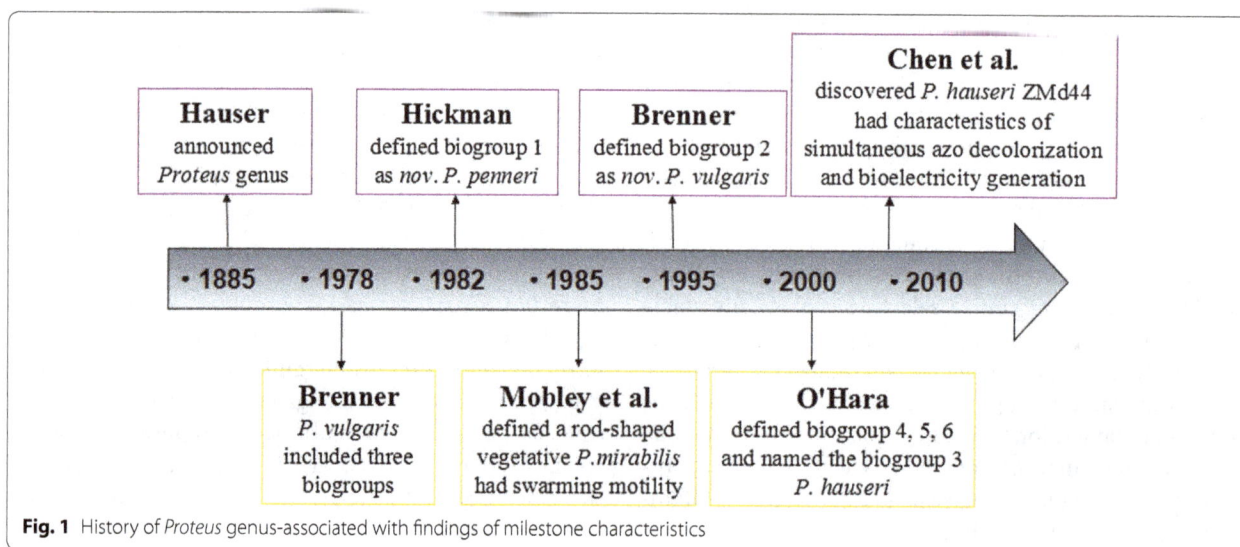

Fig. 1 History of *Proteus* genus-associated with findings of milestone characteristics

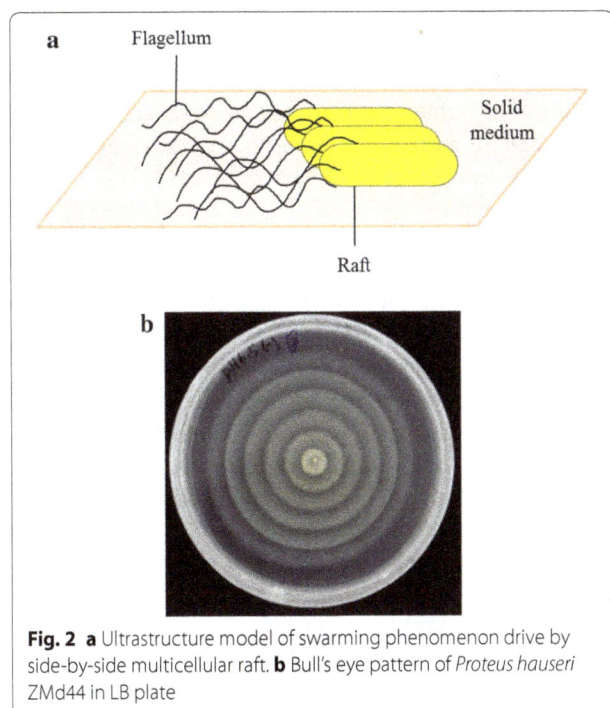

Fig. 2 a Ultrastructure model of swarming phenomenon drive by side-by-side multicellular raft. **b** Bull's eye pattern of *Proteus hauseri* ZMd44 in LB plate

lag without extension of bacteria lawn. Such a lag was still opened to be further explored, but its occurrence suggested that the vegetable cells must regulate its transcription and alter its shape to become the swarming cell. This lag was varied at about 5–8 h which could be shortened by increasing the inoculum density or abolished by particular mutants (Kearns 2010). After the lag, ZMd44 starts to extend and the differentiation has been done. The swimming cell elongated for several times during differentiation, and presented highly flagellated forms called swarming cell. Moreover, the raft of ZMd44 was assembled by this type of swarming cell. The raft would migrate on the surface for certain distance, and then divided back to the swimming cell. The newborn swimming cell would continue the ongoing round of differentiation, but without the lag. In fact, the cyclic lawn was resulted by repeatedly circulation of differentiation during the colonization of ZMd44.

The swarming mobility can be affect by many of environment factors. First, the most swarming strain generally requires an energy-rich, solid medium. Some of the strains like *Salmonella enterica* require the presence of particular supplements. *P. hauseri* ZMd44 could swarm on LB agar plate without additional supplements. However, a lack of energy or carbon source would delay or inhibit such swarming characteristic to be triggered. The reason was the swarming mobility was promoted by high growth rate due to the nutrient energy supports. Several strains of *Proteus* genus have been shown that various

additional amino acids added to the minimal medium (swarming limited) agar would promote the swarming, like alanine, asparagine, and glutamic acid. This phenomenon revealed the specific amino acid could be the energy source and stimulation of swarming (Jones and Park 1967). In our previous study, copper ion significantly inhibited the swarming motility of *P. hauseri*, while Mobley et al. showed that zinc ion restrict the migration of *P. mirabilis*.

Genomic and enzymes studies of *Proteus*

Proteus, as a very special electroactive bacteria genus, genomic studies are important and necessary to understand the cell behaviors. Comparative list of *Proteus* strains with whole genome sequenced has been summarized in Table 1. The known coding genes for proteins or enzymes from *P. hauseri* ZMd44 (Wang et al. 2014) contained three categories of biological process (44.9%), cellular component (23.5%), and molecular function (31.6%). Most of the genes were strongly associated for maintaining bacterial survival (e.g., metabolic process, membrane structure, catalytic activity). As *P. hauseri* ZMd44 owned resistant and tolerant characteristics of copper (Ng et al. 2013, 2014a, b), the related proteins involved the counter-defense metabolic process of copper were top-priority concern for industrial applications. From the perspectives of genomic analysis, there were only four open reading frames (ORFs) associated to copper resistance in the genome of *P. hauseri* ZMd44. These proteins were copper homeostasis protein CutC, copper-resistance protein Crp, multicopper oxidase Mco, and repressor protein FtsI. In particular, Crp was found to be the critical protein-controlled copper resistance (Wang et al. 2014). The genomic mapping of *P. hauseri* ZMd44 was aligned with the control sequence of *P. mirabilis* HI4320

Table 1 Comparative list of *Proteus* strains with whole genome sequenced

Strain	Accession no.	Size (Mb)	GC (%)	Protein	Gene
Proteus hauseri ZMd44	AWXP00000000	3.88	38.1	3147	3499
Proteus penneri ATCC 35198	ABVP00000000	3.75	37.8	4909	4997
Proteus mirabilis HI4320	AM942759	4.06	38.9	3584	3702
Proteus mirabilis ATCC 29906	ACLE00000000	3.98	38.6	3812	3902
Proteus mirabilis BB2000	CP004022	3.85	38.6	3335	3485
Proteus mirabilis WGLW6	AMGT00000000	4.05	39.0	3738	3820
Proteus mirabilis WGLW4	AMGU00000000	3.92	38.8	3508	3598

to specifically locate the position of chromosomal genes. According to BASys prediction, *P. hauseri* seemed to be similar to *P. mirabilis* in genetic characteristics. As a matter of fact, Mobley's group has characterized *P. mirabilis* for more than thirty years (Armbruster et al. 2012, 2013). However, most of the studies only focused on the urinary tract infection caused by *Proteus* bacterial species with medical concerns. According to whole genomic analysis, *P. hauseri* ZMd44 was found to be more proteomically diverse from other *Proteus* genus very possibly due to its outstanding performance in biodecolorization and bioelectricity generation.

As known, the McoA-laccase was also a multicopper oxidase widely distributed among plants, fungi, bacteria (Claus 2003; Mayer and Staples 2002). The laccase was capable to catalyze the oxidation of substituted phenolic and nonphenolic compounds in oxygen-bearing conditions. One crucial property of laccase was oxidative capability toward myriads of nutrient substrates. That is, laccase was not a highly specific functioning and could be used for miscellaneous industrial applications (e.g., dye decolorization, polymers degradation, bioremediation, and organic synthesis) (Mayer and Staples 2002; Santhanam et al. 2011). For example, *P. hauseri* could express McoA-laccase under induction of copper-stimulating condition (Zheng et al. 2013). The further enzymes analysis showed the McoA-laccase from *P. hauseri* ZMd44 more favorable to be induced in thermal and acidic environment possibly as ZMd44 was isolated from sewage of hot spring. The most appropriate condition for maximal expression of McoA-laccase was 60 °C and pH 2.2. Apparently, due to the inhibitory potency of copper ion to bacterial cells, excessive copper ions would repress cell growth and expression of laccase activity of *P. hauseri* ZMd44. Therefore, optimal concentration of copper ion for maximal laccase expression was determined ca. at 3 mM. Electron paramagnetic resonance (EPR) spectra confirmed that this McoA-laccase belonged to typical type 1 (T1) Cu site laccase and revealing that Cu(II) also played crucial roles to McoA-laccase synthesis to provided more binding sites for catalysis (Zheng et al. 2013).

In addition, several synergistic interactions of oxidoreductase-related enzymes such as laccase (EC 1.10.3.2; Dawkar et al. 2009), azoreductase (EC 1.7.1.6; Maier et al. 2004), and NADH dehydrogenase (EC 1.6.5.3; Ng et al. 2013) or cytochrome C oxidase (Wariishi et al. 2002) were involved simultaneously. These enzymes all play crucial roles in dealing with decolorization of azo dyes, suggesting that decolorization seemed not simply to be driven solely by mono enzymatic catalysis or single metabolic route. However, the most significant functioning protein of azo dye decolorization seemed to be azoreductase. In *Proteus* genus, only *P. mirabilis* LAG showed the capability to decolorize the azo dyes via a NADPH-dependent azoreductase (Olukanni et al. 2010). At least two azoreductases and two laccases were systematically reported in public domain, i.e., azoreductase, laccase, NADH-DCIP reductase, riboflavin reductase, lignin peroxidase (Saratale et al. 2011; Han et al. 2012).

Electron-transfer-associated biodecolorization

To elucidate electron transfer characteristics of electroactive bacteria, bacterial capabilities of reductive decolorization and power generation were both evaluated simultaneously. As electron-deficient xenobiotics, azo dyes are capable to be decolorized anaerobically via bacteria through oxygen-sensitive electrophilic oxidoreductase (e.g., laccase and azoreductase). Anaerobic treatment was feasible for treatment of dye-bearing wastewater. That is, to biodegrade such textile dyes, bacterial cells inevitably required capabilities of effective electron transfer (ET) to reduce mass transfer resistance between various extracellular aqueous phases and intracellular compartments. Although accumulation of decolorized intermediates—aromatic amines (AAs)—in cell cultures was inevitable, AAs could act as electron shuttles to feedback stimulate ET phenomena for bioelectricity generation and reductive decolorization. Apparently, both mother dyes and decolorized daughter intermediates owned electrochemical activities for redox mediation at different reduction potentials. Due to interconversion of electrical energy and chemical energy for redox reactions, bioelectrochemical systems that provided appropriate metabolic routes for optimal redox balancing would be the most energy-saving and environmentally friendly means for bioenergy and biorefinery applications. Regarding reductive decolorization of azo dyes, there are several detailed reviews mentioned about microbial degradation from diverse aspects (Pandey et al. 2007; Saratale et al. 2011). However, the systematic review for chemical structure effect on reductive decolorization was still lacking, leading to biodegradability and toxicity potency of azo dyes to be uncertain mysteries for optimal operation. Thus, this section would clarify such content of biodecolorization in term of chemical structure.

Due to diverse structures present in the synthetic dyes, changes in the chemical structures (e.g., isomers or the presence of different substituents) would significantly affect the cellular capability of reductive decolorization and bioelectricity generation. In fact, most of studies have considered color removal associated to dye classes rather than molecular features (Brown and Pagga 1986; Dohanyos et al. 1978; Hitz et al. 1978; Rodman 1971; Shaul 1991). To exhibit detailed figures of chemical structures related to reaction selectivity, it was first noted that the characteristics of substituents and their relative positions to azo linkage all significantly affected the performance of azo dye decolorization. Zimmermann et al. (1982) used purified Orange II azoreductase from a *Pseudomonas* strain KF46 to assess decolorization efficiency of Orange dyes. Evidently, the specificity of Orange II azoreductase toward azo linkage for dye decolorization was strongly dependent upon electron affinity (e.g., electron-withdrawing capability) of substituents near of azo linkage(s). Such characteristics strongly determined whether the dye was susceptible or resistant to biodecolorization. On the other hand, the hydroxyl group on the 2-position of the naphthol ring of the azo dye (e.g., 1-(4′-sulfophenylazo)-2-naphthol) provides a beneficial prerequisite to assist dye decolorization. In contrast, charged groups near azo bond (e.g., 1-(2′-sulfophenylazo)-2-naphthol) significantly hinder the decolorization efficiency. Moreover, Zimmermann et al. (1982) also revealed the correlation between Hammett's substituent constant of various substituents and the decolorization rate of these azo dyes. In addition, electron-withdrawing substituent located to *para* position of azo bond could accelerate reductive decolorization via resonance effect as indicated in Fig. 3 indicated (Hsueh and Chen 2007, 2008; Hsueh et al. 2009).

These results pointed out that the sulfonate group produced the strongest electron-withdrawing effect resulting in the fastest rate of color removal compared to the carboxyl and hydroxyl group. In addition, when the sulfonate or carboxyl group was *ortho* to azo bond, the decolorization rate significantly decreased compared to the *para* substituent to azo bond (e.g., p-MO > o-MO or p-MR_o-MR) likely due to steric hindrance near azo linkage(s) (Hsueh and Chen 2008). The similar phenomena that the more electrophilic electron-deficient azo dyes would be more easily biodecolorized were also observed in *Pseudomonas luteola, Aeromonas hydrophila* NIU01, *Proteus hauseri,* and *Shewanella* sp. WLP72 since they all were capable to reductively biodecolorize azo dyes (Chen et al. 2011c, 2016a, b; Hsueh and Chen 2007, 2008; Hsueh et al. 2009). That is, no matter what bacteria to be used, biodegradability or recalcitrance of azo dyes were possibly strongly associated to chemical structure of target dyes.

In addition, Squella et al. (1999) studied the steric and electronic effects on the electro-reduction of nitro group of β-nitrostyrene derivatives. When electron-donor properties of the substituent at the para position to the ethylenic double bond increased, it would make the reduction potential of nitro group more negative, meaning that the reduction of the nitro group requires more energy to be triggered. Furthermore, β-nitrostyrene derivatives was observed a linear correlation between the Hammett

substituent constant (σ_p) and the half-wave potential was observed. That is, the electrochemical behavior of these derivatives mainly depends on their chemical structures. Furthermore, Pasti-Grigsby et al. (1992) employed *Streptomyces* spp. and *Phanerochaete chrysosporium* to oxidatively decolorize 22 azo dyes to decipher the influence of substituents on azo dye biodegradability. As a matter of fact, the selectivity of *Streptomyces* spp. for removal color of azo dyes was distinct from Orange II azoreductase (Zimmermann et al. 1982). That is, it could decolorize Orange I but not Acid Orange 12 or Orange II owing to hydroxyl group adjacent to azo bond. However, *P. chrysosporium* decolorized Acid Orange 12 and Orange II more effectively than Orange I, and *P. chrysosporium* more extensively decolorized several azo dyes than *Streptomyces* spp. As reduced azo dyes could be reductants (i.e., electron donors), the higher electron density of azo dyes was more liable to be biodecolorized by *Streptomyces* spp and *P. chrysosporium*. Similarly, as Suzuki et al. (2001) studied, the biodegradabilities of 25 sulfonated azo dyes with different substituents correlated to decolorized efficiency with chemical structure of azo dyes were evaluated.

Compared aerobic with anaerobic decolorization of azo dyes, the higher electron density of azo dyes owning to electron-releasing substituents [i.e., hydroxyl group (−OH), alkoxyl group (−OR) or alkyl group (−R)] was more easily to be oxidatively decolorized. In contrast, the lower electron density of azo dyes attached electron-withdrawing substituents [i.e., halogen (−X), nitro group (−NO$_2$), carboxyl group (−COOH) or sulfonate (−SO$_3^-$)] was more easily reductively decolorized. In addition, how and why the electronic effect and steric hindrance effect would affect the efficiency of the decolorization for azo dyes were also presented to compare abiotic photocatalysis (oxidation degradation) with biodecolorization (azo reduction) (Zhang et al. 2012), and such bioreductive phenomena were also shown for reduction of nitroaromatic compound(s) (NACs). This was likely due to azoreductase and nitroreductase both proteomic associated to identical origin in evolution (Rafii and Cerniglia 1993; Rau and Stolz 2003).

Nitroaromatic compound(s) contained strong electron-withdrawing groups (i.e., nitro group, −NO$_2$) as electron deficient as azo dyes. That is, they were susceptible to reduction, and not easy to be oxidatively catabolized in aerobic environments (Ju and Parales 2010; Knackmuss 1996; Kulkarni and Chaudhari 2007). These similar reductive characteristics all supported that nitroreductases and azoreductases may originally derive from similar enzymatic systems (Hsueh et al. 2016, 2017; Misal et al. 2015; Rafii and Cerniglia 1993, 1995; Rau and Stolz 2003).

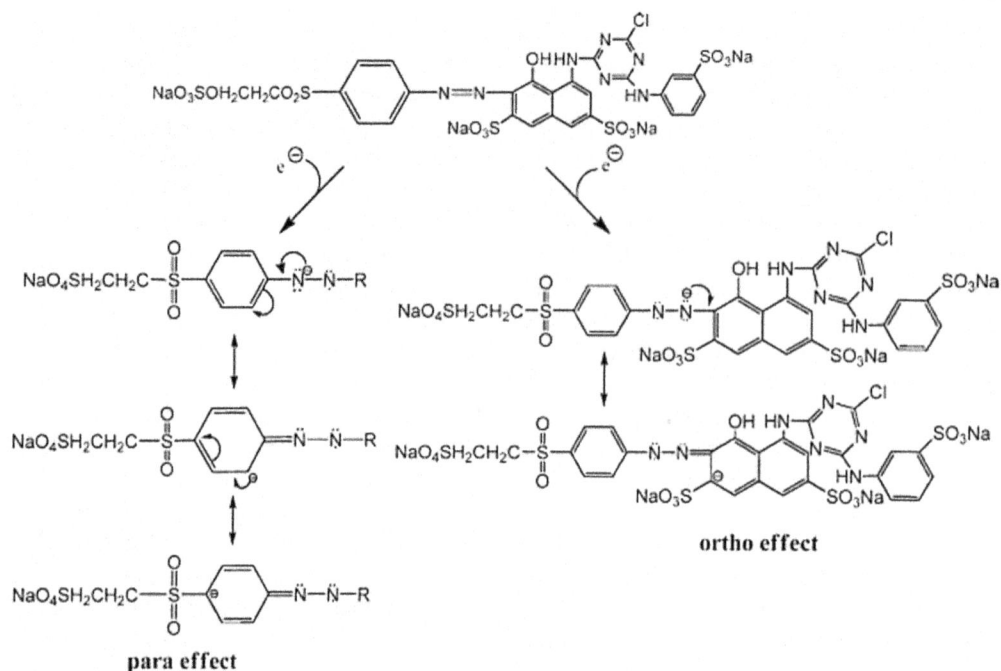

Fig. 3 Example presentation of resonance effect of RR198. The sulfonate at *ortho* and –SO$_2$(CH$_2$)$_2$SO$_4^-$ at *para* to azo bond in RR198 are both high electronegativity substituents. These substituents may stabilize the negative charge present in the reduced intermediates, since azo dyes can be reduced through the inductive effect and resonance effect (Ref. to Fig. 5 in Hsueh et al. 2009)

To explore performance of electron transfer-steered reduction, Hsueh et al. (2017) had used excellent azo dye decolorizers (i.e., *Aeromonas hydrophila* NIU01 and *Shewanella* sp. WLP72) to degrade NACs [i.e., nitrobenzene (NB), 2-nitrophenol (2-NP), 4-nitrophenol (4-NP), 2-nitrotoluene (2-NT), 4-nitrotoluene (4-NT), 2,4-dinitrotoluene (2,4-DNT), 2,6-nitrotoluene (2,6-DNT)]. Evidently, both remarkable decolorizing strains also owned the promising capabilities to reduce nitroaromatic compounds. These all pointed out that bioreduction might not be so proteomically specific. Among all nitroaromatics, 2,4-DNT was the most biodegradable. As 2,4-DNT and 2,6-DNT owned two high electronegativity nitro groups, they very likely possessed higher electrophilicities within these nitroaromatic compounds. Similarly, 2-NP owned higher electrophilicity by inductive effect due to the presence of high electronegativity hydroxyl group in the proximity of nitro group. Thus, they were more biodegradable for reduction. In fact, higher electrophilicities of nitroaromatics were also capable of the more easily bioreductive characteristics. These nitroreduction phenomena were similar to reductive decolorization of azo dyes (Hsueh et al. 2009). In addition, possibly due to less steric hindrance of 2,4-DNT than 2,6-DNT, 2,4-DNT seemed to be more easily to be biodegraded than 2,6-DNT (Fortner et al. 2003).

Therefore, degradation efficiencies of both azo dyes and NACs could be affected due to their chemical structures (i.e., different substituents or positional isomers). Except for toxicity of azo dyes, NACs, or their intermediates of metabolites, the chemical structure would influence the electron density of azo bonds or nitro groups by the electronegativity of the substituents or resonance effect and the steric hindrance to affect the biodegradation efficiency. As aforementioned, the chemical structure was of great significant to affect electrophilic characteristics for oxidation or reduction. According to assessment upon azoreduction and nitroreduction, both reductive reactions were not highly enzymatic specific for redox reactions. It might suggest that electrochemical activities as well as redox potentials directly controlled electron transfer-driven bioelectricity generation. In addition, structures of aromatic diamines or aminophenol like as quinone (hydroquinone) or the positional isomer such as catechol may be suitable as the electron shuttles (Fig. 4; Chen et al. 2013a; Rau et al. 2002; Van der Zee and Cervantes 2009; Xu et al. 2014).

In summary, both nitroreduction and azoreduction were in redox similar characteristics. This could be explained by the fact that intermediate metabolites produced from azo dyes or NACs were used as redox mediators (e.g., electron shuttles) or as the reductants

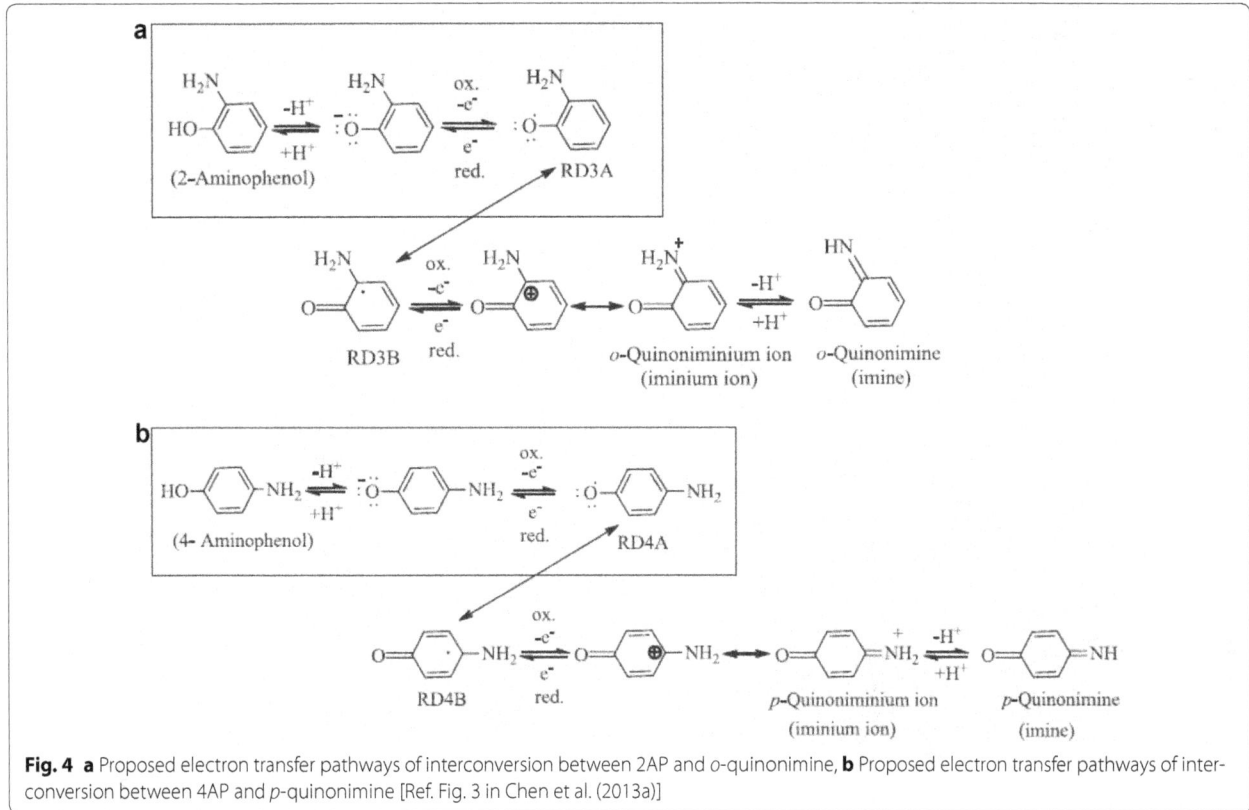

Fig. 4 a Proposed electron transfer pathways of interconversion between 2AP and o-quinonimine, **b** Proposed electron transfer pathways of interconversion between 4AP and p-quinonimine [Ref. Fig. 3 in Chen et al. (2013a)]

(e.g., amines or thiols) (Chang et al. 2001; Chen et al. 2010; Hsueh et al. 2014). These all enhanced the decolorization or microbial production efficiency (Chen et al. 2013a, 2016a, b; Hsueh et al. 2014, 2017; Qin et al. 2016). That was why azo dyes and NACs were electrochemically active chemicals as crucial compositions to drive electron transfer capabilities in MFCs. The comparison of *P. hauseri* with other electroactive bacteria on electron transport efficiency in different MFC conditions is shown in Table 2. Although *P. hauseri* might be not the best electroactive strain, it could still decolorize or degrade myriads of "recalcitrant" dyes for power generation. That was very likely due to competitive interaction between dye decolorization and bioelectricity generation (Chen et al. 2011b).

Electron-transfer-associated applications

When pure microbial cultures or enriched consortia with electrochemically active bacteria were stably acclimatized for maximal electrochemical expression, increases in selectivity of target metabolism redirecting carbon and electron flow away from dye decolorization to bioelectricity generation or microbial propagation toward value product generation could be bioelectrochemically regulated via reorganization of redox balancing. This was all manipulated by electron flux-driven redox potentials

of chemical species that were strongly dependent upon environmental conditions (e.g., ethanol fermentation would be altered to produce acetic acid if oxygen was introduced). That is to say, with supplementation of different species of electron shuttles (ESs), electron transfer characteristics to control metabolic webs of functioning microbes would apparently be altered. With assistance of ESs, rebalancing of various reactants, intermediates, and products of redox potentials in biosystems could thus be established to catalyze myriads of bioprocesses (e.g., fermentation, bioremediation) for system engineering. As aforementioned, ZMd44 isolated from Chiao-Hsi hot spring was capable to express promising SRD and BG capabilities. Regarding reductive decolorization of azo dyes, hydroxyl substituents at *ortho* to azo bond apparently resisted biodegradation of azo dyes likely due to their unfavorable chemical structures for degradation to provoke biotoxicity potency to receptor cells. This failure of reductive decolorization usually led to biosorption or bioaccumulation of such dyes onto microbial surfaces and/or intracellular compartments (Chen et al. 2000; Chen 2002) via metabolism-dependent or -independent mechanisms, respectively. These would impede biodecolorization due to failure in membrane-associated transport, internal compartment of dye(s) into cytoplasm or onto membrane, catalysis of dye-decolorizing

Table 2 Comparisons between *Proteus hauseri* and other electroactive bacteria on electron transport efficiency in different MFC conditions

MFC conditions	v^a(mV)	Total R_{in} (Ω)	Power generation (mW m^{-2})	Refs.
Proteus hauseri ZMd44	82.0–87.4			Chen et al. (2011a)
ZMd44 + 2AP[b]	115.3			Chen et al. (2011a)
ZMd44 + 3AP[b]	104.7			Chen et al. (2011a)
Proteus hauseri ZMd44		94.6		Xu et al. (2014)
ZMd44 + 1A2N[c]		83.2		Xu et al. (2014)
ZMd44 + 4A1N[c]		67.5		Xu et al. (2014)
Aeromonas hydrophila NIU01		659.9		Xu et al. (2014)
NIU01 + 1A2N[c]		462.7		Xu et al. (2014)
NIU01 + 4A1N[c]		383.4		Xu et al. (2014)
Klebsiella pneumoniae ZMd31		463.0		Xu et al. (2014)
ZMd31 + 1A2N[c]		421		Xu et al. (2014)
ZMd31 + 4A1N[c]		381		Xu et al. (2014)
Proteus hauseri ZMd44			21.82	Chen et al. (2014)
ZMd44 + Thionine[d]			38.84	Chen et al. (2014)
ZMd44 + MG[d]			50.04	Chen et al. (2014)
Aeromonas hydrophila NIU01			28.28	Chen et al. (2014)
NIU01 + Thionine[d]			40.87	Chen et al. (2014)
NIU01 + MG[d]			58.86	Chen et al. (2014)
Klebsiella pneumoniae ZMd31			29.63	Chen et al. (2014)
ZMd31 + Thionine[d]			45.97	Chen et al. (2014)
ZMd31 + MG[d]			67.38	Chen et al. (2014)
Proteus hauseri ZMd44		72.17		Chen et al. (2015)
ZMd44 + Thionine[d]		64.39		Chen et al. (2015)
ZMd44 + Azure A[d]		43.96		Chen et al. (2015)
ZMd44 + Azure C[d]		53.61		Chen et al. (2015)
Aeromonas hydrophila NIU01		299.80		Chen et al. (2015)
NIU01 + Thionine[d]		190.69		Chen et al. (2015)
NIU01 + Azure A[d]		56.01		Chen et al. (2015)
NIU01 + Azure C[d]		86.11		Chen et al. (2015)
Klebsiella pneumoniae ZMd31		402.36		Chen et al. (2015)
ZMd31 + Thionine[d]		375.69		Chen et al. (2015)
ZMd31 + Azure A[d]		71.84		Chen et al. (2015)
ZMd31 + Azure C[d]		156.78		Chen et al. (2015)
Shewanella oneidensis DSP10[e]	840			Biffinger et al. (2008)
Shewanella oneidensis MR-1[f]			858.0	Watson and Logan (2010)

[a] Average output voltage. R_{in}: internal resistance

[b] 900 mg L^{-1} of 2AP (2-aminophenol) or 3AP(3-aminophenol) was added

[c] 2 mg L^{-1} of 1A2 N (1-amino-2-naphthol) or 4A1N (4-amino-1-naphthol) was added

[d] 40 mg L^{-1} of Thionine, MG (malachite green), Azure A, or Azure C was added

[e] Glucose as the nutrient in MFC

[f] In cubic MFC

enzymes, or mediation of electron transfer chain for azo reduction. With survey of proteomics, abundant strains nov., identified even as type strains (e.g., *Rahnella* and *Microvirgula* spp.; Han et al. 2012), could be explored; however, detailed bacterial characteristics (e.g., biological or ecological interactions between dye-degrading and/or

electrochemically active capabilities) for diverse applications might be still open to be disclosed. Moreover, to explore new bacterial strains of specific characteristics, critical or hostile selection pressure(s) very different from conventional methods for species isolation should be carried out since some new metabolic functions of identified bacteria might still not be completely explored before. For example, as Han et al. (2012) mentioned, when capabilities of simultaneous color removal of non-azo dyes (e.g., reactive blue 198 and reactive blue 19) and azo dyes (i.e., orange I and reactive blue 160) were specifically selected, new strains of dye decolorizers (i.e., *Rahnella* sp. DX2b and *Microvirgula* sp. SH7b) were thus obtained. Inevitably, due to specific selection pressure applied for bacterial isolation, strains DX2b and SH7b might be simultaneously capable to decolorize dyes with reduction potentials ranging between − 0.8 and − 1.5 V due to different enzymatic catalyses for color removal (e.g., laccase, NADH dehydrogenase, riboflavin reductase with synergistic aid of Fe reductase). Similarly, with different selection pressures implemented for ZMd44, membrane-bound and/or extracellular rather than intracellular enzymatic activities could be expressed for ZMd44 to decolorize textile dyes (Ng et al. 2014a, b). The comparisons between *P. hauseri* and other decolorizers on the efficiencies of decolorizations are shown in Table 3. It was suspected that such characteristics also synergistically reduced mass transfer and electron transport resistance for color removal. Moreover, bacterial decolorization of azo dyes are usually implemented in anaerobic conditions as azo bond(s) should be reduced with ET-aided

color removal to prevent oxygen competing with such electrons for reduction. Due to this, reductive decolorization is usually non-growth-associated, since cellular growth and color removal are competitively excluded. In contrast, for simultaneous reductive decolorization and bioelectricity generation (SRD and BG) in single-chamber MFCs, electrons used for RD and BG were competitive to each other due to nearly fixed number of electrons to be generated. Due to this competitive nature, for SRD and BG, operation mode of reactor directly determines ET favorable toward batch (RD) or CSTR (BG) mode of operation (Hong et al. 2016).

Regarding ET capabilities, there are several bacterial characteristics proteomically functioned through electron transport chain in cells for diverse practical uses. As mentioned previously, azo dye decolorization could be carried out via ET-associated reduction. Considering applications in microbial fuel cells (MFCs), with components of electron transport from intracellular matrix to the solid-state anode, swarming behaviors of *Proteus* spp. could effectively form as anodophiles and/or electrophile-like microcosm coupled with available electron shuttles (ESs) (e.g., decolorized intermediates of orange I and II; Xu et al. 2014). Thus, MFCs could effectively overcome all ET resistances for effective bioremediation (Han et al. 2015; Chen et al. 2015; Chen and Hsueh 2016). This was why bioelectrochemically aided treatment (e.g., MFCs) for pollutant biodegradation could significantly augment the performance of redox-associated biodegradation (e.g., dye decolorization) via SBR and RD. In fact, outer membrane protein-porin expressed by ZMd44 can cross cell membrane, act as a pore, and provide a channel for ET-associated proteins for electron transport with electrochemical effectiveness (Ng et al. 2013). According to LeChatelier's principle, when an MFC at equilibrium is subjected to change(s) in operation condition(s) (e.g., significant increases of ET channels by ESs), then such a system readjusts itself to effectively counteract the applied change(s) (e.g., acceleration of electron "flow" for degradation by increases in reaction rates) for a new equilibrium state. That is, pollutant biodegradation could be effectively stimulated through operation of ES-accumulated MFCs. Although MFC-aided bioremediation could be used to stimulate dye decontamination, the main drawback of aromatic amine(s) generated in azo dye reduction was still inevitable. Therefore, two-stage operation for azo dye mineralization (e.g., combined bioanode and biocathode operation or sequencing batch reactors (SBR) based on temporal separation of anaerobic and aerobic phase; Van der Zee and Villaverde 2005) would be very possibly required. That is, anaerobic or anodic reduction of azo dye(s) should be taken place in first stage and aerobic or

Table 3 Comparisons between *Proteus hauseri* and other decolorizers on the efficiencies of decolorizations. Cited from Hsueh et al. (2017)

Strains	Specific decolorization rate[a]	
	RBk5[b]	RBu160[c]
Acinetobacter johnsonii NIUx72	5.97[d]	5.74[f]
Acinetobacter junii NIUY8	7.86[d]	3.51[i]
Aeromonas hydrophila NIU01	81.6[e]	16.7[f]
NIUx73	1.76[d]	9.46[f]
YT11	65.89[d]	11.76[f]
Aeromonas punctata NIUP9	2.66[d]	4.90[i]
Enterobacter cancerogenus BYm30	1.62[d]	2.51[f]
Exiguobacterium indicum K4	26.4[g]	7.68[g]
Klebsiella pneumoniae ZMd31	2.02[d]	2.71[f]
Klebsiella variicola P11	1.20[g]	1.17[g]
Proteus hauseri ZMd44	3.43[h]	11.16[f]
Shewanella sp. WLP72	48.0[d]	62.0[d]
Shewanella xiamenensis BC01	298.5[j]	169.4[j]
Staphylococcus gallinarum K1	14.11[g]	2.90[g]

[a] Specific decolorization rate (SDR; unit: μM h^{-1} ODU^{-1})

[b] RBk5: reactive black 5

[c] RBu160: reactive blue 160

[d-i] Data adopted from Hsueh et al. (2009, 2017); Zhang et al. (2010); Chen et al. (2011a, b); Han et al. (2011), respectively

[j] Data adopted from Ng et al. (2014c)

cathodic mineralization in the second stage could then be implemented to control appropriate ET capability for complete biodegradation. According to Ng et al. (2014a, b), Cu(II) ions could strongly restrict the swarming motility of *P. hauseri* and possibly significantly decreased number of flagella per cell. With Cu(II) stress applied to change cell morphology of ZMd44 in first stage, aerobic mineralization of aromatic amines at second stage with other aerobic degraders should be efficiently implemented via biodegradation and autoxidation (Campo et al. 2011). However, from electrochemical perspectives, further studies for isolating new dye-mineralizing bacteria should still need to focus on (i) the fate of degraded and autoxidation products, and (ii) the degree of recalcitrant dye residues with environmental friendliness (Van der Zee and Villaverde 2005).

Future directions and prospective

Regarding the model EAB, ZMd44 isolated from Chiao-Hsi hot spring contained several crucial biocharacteristics (e.g., dye decolorization, bioelectricity generation, ecologically stable persistence). These phenotypic properties were associated to metabolism of swarming motility for pathogenicity compared to nanowire-producing *Shewanella* spp. as strains with excellent ET capabilities. With MFC-aided bioremediation, extracellular products or decolorized metabolites (DMs) of *P. hauseri* (e.g., phenyl methadiamine; Chen et al. 2010) as ESs were capable to enhance SBR and RD (Chen et al. 2010). In addition, expression of membrane-bound redox proteins (e.g., cytochromes) of anodic biofilm of MFC was possibly the most feasible means to efficiently mediate ET between biofilm bacteria and solid anode for accelerated dye bioremediation in MFCs (Chen et al. 2011a). These all suggested that *P. hauseri*-inoculated MFC-assisted dye decolorization was possibly the most appropriate due to accumulation of DMs as exogenous ESs. Similarly, with aid of DMs as ESs, MFC-based bioremediation of recalcitrant pollutant(s) (e.g., textile dyes) would be promising due to autocatalysis of such generated intermediate(s) as exogenous mediators. However, toxicity potency of such dyes or derived intermediates to anodic biofilm in MFCs still significantly controlled MFC-aided dye bioremediation (Chen et al. 2010, 2011a). That is, selecting microbial biodegrader(s) highly resistant to toxicity potency of textile dye(s) and amine intermediates would be top-priority concern for effective MFC-assisted dye bioremediation. As a matter of fact, using MFC for dye bioremediation was favorable since DMs (e.g., 2-aminophenol, 1-amino-2-naphthol) as ESs might be accumulated to significantly reduce internal resistance for ET of autocatalysis (Chen et al. 2013a, b, c; Xu et al. 2014). As a matter of fact, due to DMs as ESs, the concept of MFC-assisted dye

biodegradation could be applicable for sustainable bioremediation of recalcitrant chemical(s) (e.g., azo dye(s); Chen et al. 2013c, 2016a, b). Facing the gradually polluted environment, the example of *P. hauseri* showed that even an identified bacterium might still contain crucial functioning capabilities not yet explored for bioremediation and biorefinery. Exploration upon species identified, but not well characterized for myriads of problems bring up recently, is still worthy to be implemented for sustainable development.

In addition, with aid of MFC technology (Fig. 5), fermentation systems could be in redox rebalance to have value product generation via different metabolic routes for energy recycling via augmentation of electron shuttles, reduction of electron transfer resistance, improvement of membrane for ion separation, and modification of electrode materials. Regarding electron transfer-assisted electro-fermentation (Figs. 5, 6), the supply of electric current as additional reducing equivalents would create new target reductive pathways for particular metabolic route(s) under different reduction potentials. H_2/H^+ and redox mediators could act as electron shuttles to stimulate oxidative/reductive fermentation of suspended microbial cultures. As capabilities of ESs strongly affected

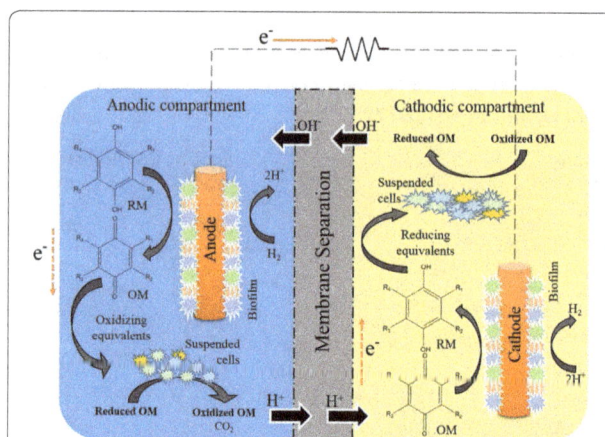

Fig. 5 Possible electron and/or "redox" transfer mechanisms taken place in microbial fuel cells (modified from Schievano et al. 2016). Fermentation systems could be in redox rebalance to have value product generation via different metabolic routes for energy recycling via augmentation of electron shuttles, reduction of electron transfer resistance, improvement of membrane for ion separation, and modification of electrode materials. According to electro-fermentation, the supply of electric current as additional reducing equivalents would create new routes of redox balancing with different reduction potentials. In addition, H_2/H^+ and redox mediators could act as electron shuttles to stimulate oxidative/reductive fermentation of suspended microbial cultures. As capabilities of ESs strongly affected by operation conditions (e.g., pH values, redox balance), appropriate manipulation of ESs and electric current would provide electrochemically favorable environments for cellular growth and target product generation

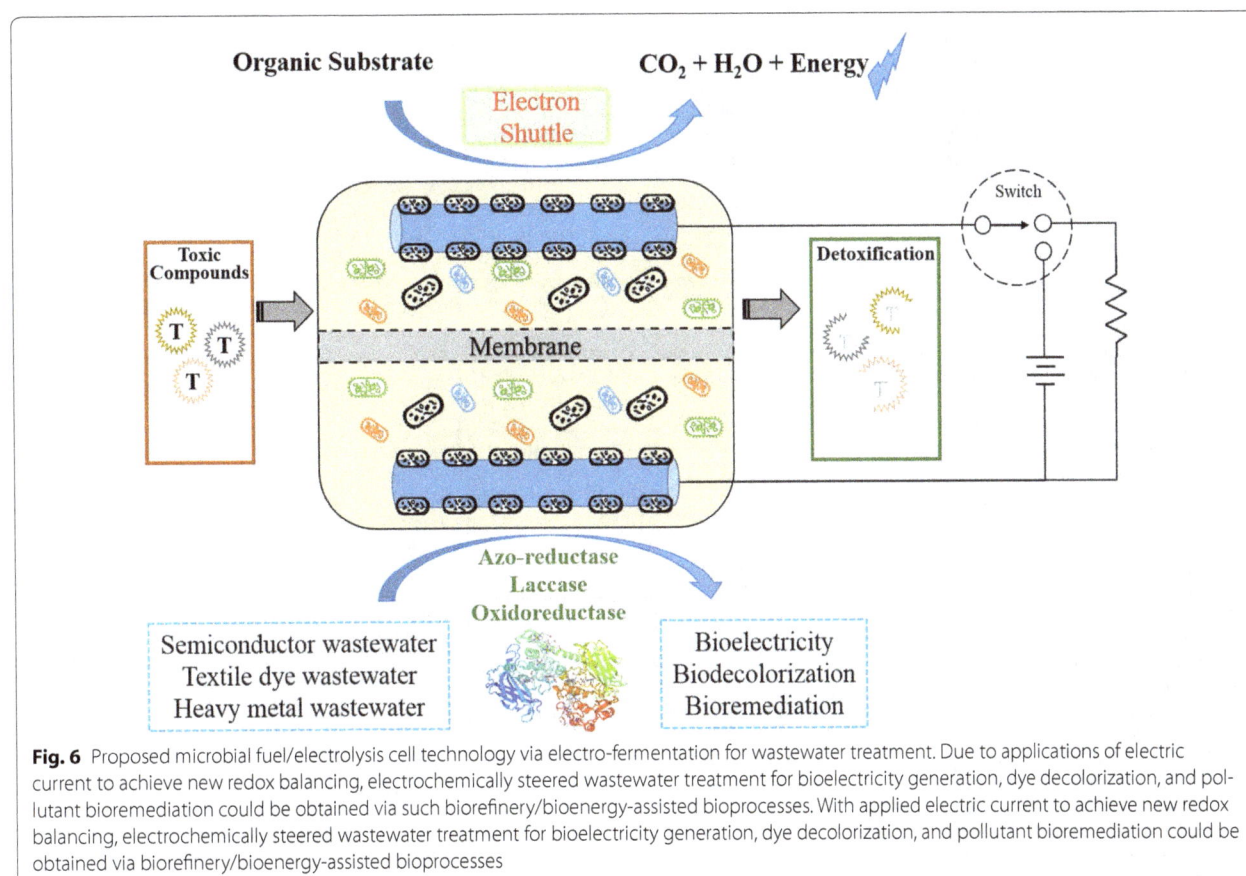

Fig. 6 Proposed microbial fuel/electrolysis cell technology via electro-fermentation for wastewater treatment. Due to applications of electric current to achieve new redox balancing, electrochemically steered wastewater treatment for bioelectricity generation, dye decolorization, and pollutant bioremediation could be obtained via such biorefinery/bioenergy-assisted bioprocesses. With applied electric current to achieve new redox balancing, electrochemically steered wastewater treatment for bioelectricity generation, dye decolorization, and pollutant bioremediation could be obtained via biorefinery/bioenergy-assisted bioprocesses

by operation conditions (e.g., pH values, redox balance; Chen et al. 2017a, b), appropriate manipulation of ESs and electric current would provide electrochemically favorable environments for cellular growth and target product generation. As revealed in Fig. 6, due to applications of electric current to achieve new redox balancing, electrochemically steered wastewater treatment for bioelectricity generation, dye decolorization, and pollutant bioremediation could be obtained via such biorefinery/bioenergy-assisted bioprocesses. That is, MFC could be used as a tool to ripen new era of electrochemical biotechnology via electro-fermentation. Moreover, if this MFC technology moves one step forward, myriads of electrochemical bioprocesses could be achieved to have diverse product generation with assistance of external power (Fig. 6). These should be of great importance to trigger progress of MFC technology with practical value to serve global lives.

Authors' contributions
IS, CC, and BY prepared and wrote the manuscript. All authors read and approved the final manuscript.

Acknowledgements
This study is also dedicated to the memory of Dolloff F. Bishop and Henry Tabak, as most of novel concepts were initiated while the author Bor-Yann Chen worked in National Risk Management Research Laboratory (NRMRL), US Environmental Protection Agency (EPA), Cincinnati, Ohio, U.S.A. under their mentorships.

Competing interests
The authors declare that they have no competing interests.

Funding
Financial supports (MOST106-2621-M-197-001, MOST105-2622-E-197-012-CC3, MOST105-2221-E-197-022, MOST-105-2221-E-006-225-MY3) from the Ministry of Sciences and Technology, Taiwan for the project of Microbial Fuel Cells (MFCs)[sdg] conducted in Biochemical Engineering Laboratory, C&ME NIU and National Cheng Kung University.

References
Armbruster CE, Mobley HL (2012) Merging mythology and morphology: the multifaceted lifestyle of Proteus mirabilis. Nat Rev Microbiol 10(11):743–754
Armbruster CE, Hodges SA, Mobley HL (2013) Initiation of swarming motility by Proteus mirabilis occurs in response to specific cues present in urine and requires excess L-glutamine. J Bacteriol 195(6):1305–1319
Bajracharya S, Sharma M, Mohanakrishna G, Benneton XD, Strik DPBTB, Sarma PM (2016) An overview on emerging bioelectrochemical systems (BESs): technology for sustainable electricity, waste remediation, resource recovery, chemical production and beyond. Renew Energy 98:153–170
Biffinger JC, Byrd JN, Dudley BL, Ringeisen BR (2008) Oxygen exposure promotes fuel diversity for Shewanella oneidensis microbial fuel cells. Biosens Bioelectron 23:820–826

Briones A, Raskin L (2003) Diversity and dynamics of microbial communities in engineered environments and their implications for process stability. Curr Opin Biotechnol 14(3):270–276

Brown D, Pagga U (1986) The degradation of dyestuffs: part II behavior of dyestuffs in aerobic biodegradation test. Chemosphere 15:479–491

Campo P, Platten W III, Suidan MT, Chai Y, Davis JW (2011) Aerobic biodegradation of amines in industrial saline wastewaters. Chemosphere 85(7):1199–1203

Chang JS, Chou C, Lin YC, Lin PJ, Ho JY, Hu TL (2001) Kinetic characteristics of bacterial azo-dye decolorization by Pseudomonas luteola. Water Res 35:2841–2850

Chen BY (2002) Understanding decolorization characteristics of reactive azo dyes by Pseudomonas luteola: toxicity and kinetics. Process Biochem 38(3):437–446

Chen BY (2007) Revealing characteristics of mixed consortia for azo dye decolorization: Lotka-Volterra model and game theory. J Hazard Mater 149(2):508–514

Chen BY, Chang JS (2007) Assessment upon species evolution of mixed consortia for azo dye decolorization. J Chin Inst Chem Eng 38:259–266

Chen BY, Hsueh CC (2016) Deciphering electron shuttles for bioremediation and beyond. Am J Chem Eng 4(5):114–121

Chen BY, Lin CS, Lim HC (1995) Temperature induction of bacteriophage λ; mutants in Escherichia coli. J Biotechnol 40:87–97

Chen KC, Huang WT, Wu JY, Houng JY (1999) Microbial decolorization of azo dyes by Proteus mirabilis. J Ind Microbiol Biotechnol 23(1):686–690

Chen BY, Utgikar VP, Harmon SM, Tabak HH, Bishop DF, Govind R (2000) Studies on biosorption of zinc(II) and copper(II) on Desulfovibrio desulfuricans. Int Biodeterior Biodegrad 46(1):11–18

Chen BY, Chang JS, Chen SY (2004) Bacterial species diversity and dye decolorization of a two-species mixed consortium. Environ Eng Sci 20(4):337–345

Chen BY, Zhang MM, Chang CT, Ding Y, Lin KL, Chiou C-S, Hsueh CC, Xu H (2010) Assessment upon azo dye decolorization and bioelectricity generation by Proteus hauseri. Bioresour Technol 101(12):4737–4741

Chen BY, Wang YM, Ng IS (2011a) Understanding interactive characteristics of bioelectricity generation and reductive decolorization using Proteus hauseri. Bioresour Technol 102(2):1159–1165

Chen BY, Zhang MM, Chamg CT, Ding Y, Chen WM, Hsueh CC (2011b) Deciphering azo dye decolorization characteristics by indigenous Proteus hauseri: chemical structure. J Taiwan Inst Chem Eng 42:327–333

Chen BY, Hsueh CC, Chen WM, Li WD (2011c) Exploring decolorization and halo-tolerance characteristics by indigenous acclimatized bacteria: chemical structure of azo dyes and dose–response assessment. J Taiwan Inst Chem Eng 42:816–825

Chen BY, Hong JM, Ng IS, Wang YM, Ni C (2013a) Deciphering simultaneous bioelectricity generation and reductive decolorization using mixed-culture microbial fuel cells in salty media. J Taiwan Inst Chem Eng 44(3):446–453

Chen BY, Hsueh CC, Liu SQ, Ng IS, Wang YM (2013b) Deciphering mediating characteristics of decolorized intermediates for reductive decolorization and bioelectricity generation. Bioresour Technol 145:321–325

Chen BY, Hsueh CC, Liu SQ, Hung JY, Qiao Y, Yueh PL, Wang YM (2013c) Unveiling characteristics of dye-bearing microbial fuel cells for energy and materials recycling: redox mediators. Int J Hydrog Energy 38(35):15598–15605

Chen BY, Xu B, Qin LJ, Lan JCW, Hsueh CC (2014) Exploring redox-mediating characteristics of textile dye-bearing microbial fuel cells: thionin and malachite green. Bioresour Technol 169:277–283

Chen BY, Xu B, Yueh PL, Han K, Qin LJ, Hsueh CC (2015) Deciphering electron-shuttling characteristics of thionine-based textile dyes in microbial fuel cells. J Taiwan Inst Chem Eng 51:63–70

Chen BY, Ma CM, Han K, Yueh PL, Qin LJ, Hsueh CC (2016a) Influence of textile dye and decolorized metabolites on microbial fuel cell-assisted bioremediation. Bioresour Technol 200:1033–1038

Chen CT, Wu CC, Chen BY, Hsueh CC (2016b) Comparative study on biodecolorization capabilities of indigenous strains to azo dyes. Sci Discov 4:109–115

Chen BY, Ma CM, Liao JH, Hsu AW, Hsueh CC (2017a) Feasibility study on biostimulation of electron transfer characteristics by edible herbs-extracts. J Taiwan Inst Chem Eng 79:125–133

Chen BY, Hsu AW, Wu CC, Hsueh CC (2017b) Feasibility study on biostimulation of dye decolorization and bioelectricity generation by using decolorized metabolites of edible flora-extracts. J Taiwan Inst Chem Eng 79:141–150

Claus H (2003) Laccases and their occurrence in prokaryotes. Arch Microbiol 179(3):145–150

Dawkar VV, Jadhav UU, Ghodake GS, Govindwar SP (2009) Effect of inducers on the decolorization and biodegradation of textile azo dye Navy blue 2GL by Bacillus sp. VUS. Biodegradation 20(6):777–787

Dohanyos M, Madera V, Sedlacek M (1978) Removal of organic dyes by activated sludge. Prog Water Technol 10(5):559–575

ElMekawy A, Hegab HM, Vanbroekhoven K, Pant D (2014) Techno-productive potential of photosynthetic microbial fuel cells through different configurations. Renew Sust Energy Rev 39:617–627

Fortner J, Zhang C, Spain J, Hughes J (2003) Soil column evaluation of factors controlling biodegradation of DNT in the Vadose Zone. Environ Sci Technol 37:3382–3391

Han JL, Liu Y, Chang CT, Chen BY, Chen WM, Xu HZ (2011) Exploring characteristics of bioelectricity generation and dye decolorization of mixed and pure bacterial cultures from wine-bearing wastewater treatment. Biodegradation 22:321–333

Han JL, Ng IS, Wang Y, Zheng X, Chen WM, Hsueh CC, Liu SQ, Chen BY (2012) Exploring new strains of dye-decolorizing bacteria. J Biosci Bioeng 113(4):508–514

Han K, Yueh PL, Qin LJ, Hsueh CC, Chen BY (2015) Deciphering synergistic characteristics of microbial fuel cell-assisted dye decolorization. Bioresour Technol 196:746–751

Hassan H, Jin B, Donner E, Vasileiadis S, Saint C, Dai S (2017) Microbial community and bioelectrochemical activities in MFC for degrading phenol and producing electricity: microbial consortia could make differences. Chem Eng J. In press

He CS, Mu ZX, Yang HY, Wang YZ, Yu HQ (2015) Electron acceptors for energy generation in microbial fuel cells fed with wastewaters: a mini-review. Chemosphere 140:12–17

He L, Du P, Chen Y, Lu H, Cheng X, Chang B, Wang Z (2017) Advances in microbial fuel cells for wastewater treatment. Renew Sust Energ Rev 71:388–403

Hernández-Fernández FJ, de los Ríos A, Salar-García MJ, Ortiz-Martínez VM, Lozano-Blanco LJ, Toms-Alonso F, Quesada-Medina J (2015) Recent progress and perspectives in microbial fuel cells for bioenergy generation and wastewater treatment. Fuel Proc Technol 138:284–297

Hitz HR, Huber W, Reed RH (1978) The adsorption of dyes on activated sludge. J Soc Dyers Colour 94:71–76

Hong JM, Xia YF, Hsueh CC, Chen BY (2016) Unveiling optimal modes of operation for microbial fuel cell-aided dye bioremediation. J Taiwan Inst Chem Eng 67:362–369

Hongo M, Iwahara M (1979) Application of electro-energizing method to L-glutamic acid fermentation. J Agri Biol Chem 43(10):2075–2081

Hsueh CC, Chen BY (2007) Comparative study on reaction selectivity of azo dye decolorization by Pseudomonas luteola. J Hazard Mater 141:842–849

Hsueh CC, Chen BY (2008) Exploring effects of chemical structure on azo dye decolorization characteristics by Pseudomonas luteola. J Hazard Mater 154:703–710

Hsueh CC, Chen BY, Yen CY (2009) Understanding effects of chemical structure on azo dye decolorization characteristics by Aeromonas hydrophila. J Hazard Mater 167:995–1001

Hsueh CC, Wang YM, Chen BY (2014) Metabolite analysis on reductive biodegradation of reactive green 19 in Enterobacter cancerogenus bearing microbial fuel cell (MFC) and non-MFC cultures. J Taiwan Inst Chem Eng 45:436–443

Hsueh CC, You LP, Li JY, Chen CT, Wu CC, Chen BY (2016) Feasibility study of reduction of nitroaromatic compounds using indigenous azo dye-decolorizers. J Taiwan Inst Chem Eng 64:180–188

Hsueh CC, Chen CT, Hsu AW, Wu CC, Chen BY (2017) Comparative assessment of azo dyes and nitroaromatic compounds reduction using indigenous dye-decolorizing bacteria. J Taiwan Inst Chem Eng 79:134–140

Ishii S, Suzuki S, Yamanaka Y, Wu A, Bretschger O (2017) Population dynamics of electrogenic microbial communities in microbial fuel cells started

Jones BV, Young R, Mahenthiralingam E, Stickler DJ (2004) Ultrastructure of *Proteus mirabilis* swarmer cell rafts and role of swarming in catheter-associated urinary tract infection. Infect Immun 72(7):3941–3950

Ju KS, Parales RE (2010) Nitroaromatic compounds, from synthesis to biodegradation. Microbiol Mol Biol Rev 74:250–272

Kearns DB (2010) A field guide to bacterial swarming motility. Nat Rev Microbiol 8(9):634–644

Kim W, Killam T, Sood V, Surette MG (2003) Swarm-cell differentiation in *Salmonella enterica* serovar typhimurium results in elevated resistance to multiple antibiotics. J Bacteriol 185(10):3111–3117

Knackmuss HJ (1996) Basic knowledge and perspectives of bioelimination of xenobiotic compounds. J Biotechnol 51:287–295

Kulkarni M, Chaudhari A (2007) Microbial remediation of nitro-aromatic compounds: an overview. J Environ Manage 85:496–512

Kumar R, Singh L, Zularisam AW (2016) Exoelectrogens: recent advances in molecular drivers involved in extracellular electron transfer and strategies used to improve it for microbial fuel cell applications. Renew Sust Energy Rev 56:1322–1336

Logan BE, Hamelers B, Rozendal R, Schröder U, Keller J, Freguia S, Aelterman P, Verstraete W, Rabaey K (2006) Microbial fuel cells: methodology and technology. Environ Sci Technol 40(17):5181–5192

Maier J, Kandelbauer A, Erlacher A, Cavaco-Paulo A, Gübitz GM (2004) A new alkali-thermostable azoreductase from *Bacillus* sp. strain SF. Appl Environ Microbiol 70(2):837–844

Mayer AM, Staples RC (2002) Laccase: new functions for an old enzyme. Phytochemistry 60(6):551–565

Misal SA, Humne VT, Lokhande PD, Gawai KR (2015) Biotransformation of nitro aromatic compounds by flavin-free NADH azoreductase. J Bioremed Biodeg 6(2):1. https://doi.org/10.4172/2155-6199.1000272

Mobley HL, Belas R (1995) Swarming and pathogenicity of *Proteus mirabilis* in the urinary tract. Trends Microbiol 3(7):280–284

Mohan SV, Velvizhi G, Modestra JA, Srikanth S (2014) Microbial fuel cell: critical factors regulating bio-catalyzed electrochemical process and recent advancements. Renew Sust Energy Rev 40:779–797

Ng IS, Zheng X, Chen BY, Chi X, Lu Y, Chang CS (2013) Proteomics approach to decipher novel genes and enzymes characterization of a bioelectricity-generating and dye-decolorizing bacterium *Proteus hauseri* ZMd44. Biotechnol Bioprocess Eng 18(1):8–17

Ng IS, Xu F, Ye C, Chen BY, Lu Y (2014a) Exploring metal effects and synergistic interactions of ferric stimulation on azo-dye decolorization by new indigenous *Acinetobacter guillouiae* Ax-9 and *Rahnella aquatilis* DX2b. Bioprocess Biosyst Eng 37(2):217–224

Ng IS, Zheng X, Wang N, Chen BY, Zhang X, Lu Y (2014b) Copper response of *Proteus hauseri* based on proteomic and genetic expression and cell morphology analyses. Appl Biochem Biotechnol 173(5):1057–1072

Ng IS, Chen TT, Lin R, Zhang X, Ni C, Sun D (2014c) Decolorization of textile azo dye and congo red by an isolated strain of the dissimilatory manganese-reducing bacterium *Shewanella xiamenensis* BC01. Appl Microbiol Biotechnol 98:2297–2308

Nimje VR, Chen CY, Chen CC, Jean JS, Chen JL (2009) Stable and high energy generation by a strain of *Bacillus subtilis* in a microbial fuel cell. J Power Sources 190(2):258–263

O'Hara CM, Brenner FW, Steigerwalt AG, Hill BC, Holmes B, Grimont P, Hawkey PM, Penner JL, Miller JM, Brenner DJ (2000) Classification of *Proteus vulgaris* biogroup 3 with recognition of *Proteus hauseri* sp. nov., nom. rev. and unnamed Proteus genomospecies 4, 5 and 6. Int J Syst Evol Microbiol 50(5):1869–1875

Olukanni O, Osuntoki A, Kalyani D, Gbenle G, Govindwar S (2010) Decolorization and biodegradation of reactive blue 13 by *Proteus mirabilis* LAG. J Hazard Mater 184(1):290–298

Pandey A, Singh P, Iyengar L (2007) Bacterial decolorization and degradation of azo dyes. Int Biodeterior Biodegrad 59(2):73–84

Pant D Van, Bogaert G, Diels L, Vanbroekhoven K (2010) A review of the substrates used in microbial fuel cells (MFCs) for sustainable energy production. Bioresour Technol 101(6):1533–1543

Pasti-Grigsby MB, Paszczynski A, Goszczynski S, Crawford DL, Crawford RL (1992) Influence of aromatic substitution patterns on azo dye degradability by *Streptomyces* spp. and *Phanerochaete chrysosporium*. Appl Environ Microbiol 58:3605–3613

Pearson MM, Sebaihia M, Churcher C, Quail MA, Seshasayee AS, Luscombe NM, Abdellah Z, Arrosmith C, Atkin B, Chillingworth T, Hauser H, Jagels K, Moule S, Mungall K, Norbertczak H, Rabbinowitsch E, Walker D, Whithead S, Thomson NR, Rather PN, Parkhill J, Mobley HLT (2008) Complete genome sequence of uropathogenic *Proteus mirabilis*, a master of both adherence and motility. J Bacteriol 190(11):4027–4037

Pearson MM, Yep A, Smith SN, Mobley HL (2011) Transcriptome of *Proteus mirabilis* in the murine urinary tract: virulence and nitrogen assimilation gene expression. Infect Immun 79(7):2619–2631

Poore CA, Mobley HL (2003) Differential regulation of the *Proteus mirabilis* urease gene cluster by UreR and H-NS. Microbiology 149(12):3383–3394

Qin LJ, Han K, Yueh PL, Hsueh CC, Chen BY (2016) Interactive influences of decolorized metabolites on electron-transfer characteristics of microbial fuel cells. Biochem Eng J 109:297–304

Rabaey K, Rozendal RA (2010) Microbial electrosynthesis-revisiting the electrical route for microbial production. Nat Rev Microbiol 8:706–716

Rabaey K, Verstraete W (2005) Microbial fuel cells: novel biotechnology for energy generation. Trends Biotechnol 23(6):291–298

Rabaey K, Girguis P, Nielsen LK (2011) Metabolic and practical considerations on microbial electrosynthesis. Curr Opin Biotechnol 22(3):371–377

Rafii F, Cerniglia CE (1993) Comparison of the azoreductase and nitroreductase from *Clostridium perfringens*. Appl Environ Microbiol 59:1731–1734

Rafii F, Cerniglia CE (1995) Reduction of azo dyes and nitroaromatic compounds by bacterial enzymes from the human intestinal tract. Environ Health Perspect 103(5):17–19

Rau J, Stolz A (2003) Oxygen-insensitive nitroreductases NfsA and NfsB of Escherichia coli function under anaerobic conditions as lawsone-dependent azo reductases. Appl Environ Microbiol 69:3448–3455

Rau J, Knackmuss HJ, Stolz A (2002) Effects of different quinoid redox mediators on the anaerobic reduction of azo dyes by bacteria. Environ Sci Technol 36:1497–1504

Ren H, Lee HS, Chae J (2012) Miniaturizing microbial fuel cells for potential portable power sources: promises and challenges. Microfluid Nanofluid 13(3):353–381

Rezaei F, Xing D, Wagner R, Regan JM, Richard TL, Logan BE (2009) Simultaneous cellulose degradation and electricity production by *Enterobacter cloacae* in a microbial fuel cell. Appl Environ Microbiol 75(11):3673–3678

Richter H, McCarthy K, Nevin KP, Johnson JP, Rotello VM, Lovley DR (2008) Electricity generation by *Geobacter sulfurreducens* attached to gold electrodes. Langmuir 24(8):4376–4379

Rodman CA (1971) Removal of colour from textile dye wastes. Text Chem Colorist 3:239

Rozendal RA, Hamelers HVM, Rabaey K, Keller J, Buisman CJN (2008) Towards practical implementation of bioelectrochemical wastewater treatment. Trends Biotechnol 26(8):450–459

Rozendal RA, Leone E, Keller J, Rabaey K (2009) Efficient hydrogen peroxide generation from organic matter in a bioelectrochemical system. Electrochem Commun 11(9):1752–1755

Santhanam N, Vivanco JM, Decker SR, Reardon KF (2011) Expression of industrially relevant laccases: prokaryotic style. Trends Biotechnol 29(10):480–489

Saratale RG, Saratale GD, Chang JS, Govindwar SP (2011) Bacterial decoloriza-
    tion and degradation of azo dyes: a review. J Taiwan Inst Chem Eng
    42(1):138–157

Schievano A, Sciarria TP, Vanbroekhoven K, Wever HD, Puig S, Andersen SJ,
    Rabaey K, Pant D (2016) Electro-fermentation-merging electrochem-
    istry with fermentation in industrial applications. Trends Biotechnol.
    34(11):866–878

Shaul G (1991) Fate of water soluble azo dyes in activated sludge process.
    Chemosphere 22:107–119

Solanki K, Subramanian S, Basu S (2013) Microbial fuel cells for azo dye
    treatment with electricity generation: a review. Bioresour Technol
    131:564–571

Squella JA, Sturm JC, Weiss-Lopez B, Bonta M, Nuñez-Vergara LJ (1999) Elec-
    trochemical study of b-nitrostyrene derivatives: steric and electronic
    effects on their electroreduction. J Electroanal Chem 466:90–98

Sun J, Hu YY, Bi Z, Cao YQ (2009) Simultaneous decolorization of azo dye
    and bioelectricity generation using a microfiltration membrane
    air-cathode single-chamber microbial fuel cell. Bioresour Technol
    100(13):3185–3192

Sun J, Hu YY, Hou B (2011) Electrochemical characteriztion of the bioanode
    during simultaneous azo dye decolorization and bioelectricity genera-
    tion in an air-cathode single chambered microbial fuel cell. Electrochim
    Acta 56(19):6874–6879

Suzuki T, Timofei S, Kurunczi L, Dietze U, Schuurmann G (2001) Correlation
    of aerobic biodegradability of sulfonated azo dyes with the chemical
    structure. Chemosphere 45:1–9

Van der Zee FP, Cervantes FJ (2009) Impact and application of electron shuttles
    on the redox (bio)transformation of contaminants: a review. Biotechnol
    Adv 27:256–277

Van der Zee FP, Villaverde S (2005) Combined anaerobic–aerobic treat-
    ment of azo dyes—a short review of bioreactor studies. Water Res
    39(8):1425–1440

Wang N, Ng IS, Chen PT, Li Y, Chen YC, Chen B-Y, Lu Y (2014) Draft genome
    sequence of the bioelectricity-generating and dye-decolorizing bacte-
    rium Proteus hauseri strain ZMd44. Genome Announc 2(1):e00992-13

Wariishi H, Kabuto M, Mikuni J, Oyadomari M, Tanaka H (2002) Degradation of
    water-insoluble dyes by microperoxidase-11, an effective and stable perox-
    idative catalyst in hydrophilic organic media. Biotechnol Prog 18(1):36–42

Watson VJ, Logan BE (2010) Power production in MFCs inoculated with
    Shewanella oneidensis MR-1 or mixed cultures. Biotechnol Bioeng
    105(3):489–498

Xu B, Chen BY, Hsueh CC, Qin LJ, Chang CT (2014) Deciphering characteristics
    of bicyclic aromatics—mediators for reductive decolorization and bio-
    electricity generation. Bioresour Technol 163:280–286

Zhang MM, Chen WM, Chen BY, Chang CT, Hsueh CC, Ding Y, Lin K-L, Xu H
    (2010) Comparative study on characteristics of azo dye decolorization
    by indigenous decolorizers. Bioresour Technol 101(8):2651–2656

Zhang Q, Jing YH, Shiue A, Chang CT, Chen BY (2012) Deciphering effects of
    chemical structure on azo dye decolorization/degradation charac-
    teristics: bacterial vs. photocatalytic method. J Taiwan Inst Chem Eng
    43:760–766

Zheng X, Ng IS, Ye C, Chen BY, Lu Y (2013) Copper ion-stimulated McoA-lac-
    case production and enzyme characterization in Proteus hauseri ZMd44.
    J Biosci Bioeng 115(4):388–393

Zimmermann T, Kulla HG, Leisinger T (1982) Properties of purified Orange
    II azoreductase, the enzyme initiating azo dye degradation by Pseu-
    domonas KF46. Eur J Biochem 129:197–203

17

Magnetic ZIF-8/cellulose/Fe$_3$O$_4$ nanocomposite: preparation, characterization, and enzyme immobilization

Shi-Lin Cao[1,2†], Hong Xu[2,3,4†], Lin-Hao Lai[1], Wei-Ming Gu[1], Pei Xu[2,4], Jun Xiong[2], Hang Yin[2], Xue-Hui Li[3], Yong-Zheng Ma[5], Jian Zhou[3], Min-Hua Zong[2,3] and Wen-Yong Lou[2,4*] (iD)

Abstract

Background: The ZIF-8-coated magnetic regenerated cellulose-coated nanoparticles (ZIF-8@cellu@Fe$_3$O$_4$) were successfully prepared and characterized. The result showed that ZIF-8 was successfully composited on to the surface of the cellulose-coated Fe$_3$O$_4$ nanoparticles by co-precipitation method. Moreover, the glucose oxidase (GOx, from *Aspergillus niger*) was efficiently immobilized by the ZIF-8@Cellu@Fe$_3$O$_4$ nanocarriers with enhanced catalytic activities. The enzyme loading was 94.26 mg/g and the enzyme activity recovery was more than 124.2%. This efficiently immobilized enzyme exhibits promising applications in biotechnology, diagnosis, biosensing, and biomedical devices.

Conclusions: A new core–shell magnetic ZIF-8/cellulose nanocomposite (ZIF-8@Cellu@Fe$_3$O$_4$) was fabricated and structurally characterized. Glucose oxidase (GOx) was successfully immobilized by the biocompatible ZIF-8@Cellu@Fe$_3$O$_4$ with high protein loading (94.26 mg/g) and enhanced relative activity recovery (124.2%).

Keywords: Metal–organic frameworks, Glucose oxidase, Zeolitic imidazolate framework

Background

Metal–organic frameworks (MOFs) show attractive applications in various fields including gas adsorption (Li et al. 2009) and chemical separation (Maes et al. 2010), and catalysis (Lee et al. 2009). Zeolitic imidazolate framework (ZIF) materials belong to an important class of MOF (Phan et al. 2010), which exhibit the tunable pore size, chemical functionality of classical MOFs, exceptional chemical stability, and structural diversity of zeolites (Wu et al. 2007). Because of these features, ZIFs show great promise for enzyme immobilization (Hou et al. 2017; Wu et al. 2017). Hou et al. (2015) reported the construction of mimetic multi-enzyme systems by embedding GOx in ZIF-8 and application of this system

as biosensors for glucose detection, exhibiting extraordinary electro-detection performance, and lower detection limit. Lyu et al. (2014) reported one-step immobilization process for protein-embedded metal–organic frameworks with enhanced activities, in which the Cyt c immobilized by ZIF-8 carrier exhibited an enhancement of enzyme activity compared with free Cyt c.

In order to separate the MOF-based materials easily, previous studies have reported on MOF-functionalized magnetic nanoparticles which can be recycled under magnetic field and have excellent physical and chemical characters of the MOF shell (Ke et al. 2012). Further investigations on magnetic MOFs with core–shell structure are still needed because the controllable growth of the MOF crystals on the magnetic nanoparticles remains a great challenge (Nong et al. 2015). For instance, before the MOF crystal growth process, the magnetic nanoparticles need to be surface-modified by styrene sulfonate (Zhang et al. 2013), polyacrylic acid (Jin et al. 2014), chitosan (Xia et al. 2017), and SiO$_2$ (Wehner et al. 2016). It is worthy to note that cellulose may be acted as a promising

*Correspondence: wylou@scut.edu.cn

†Hong Xu and Shi-Lin Cao are both co-first author and contribute equally to this work.

[4] Guangdong Province Key Laboratory for Green Processing of Natural Products and Product Safety, South China University of Technology, No. 381 Wushan Road, Guangzhou 510640, China

Full list of author information is available at the end of the article

surface material, because of its abundance of hydroxyl groups. These hydroxyl groups may promote the adsorption of metal ions for the formation of MOF crystal (Liu et al. 2012). Cellulose is the most abundant renewable polysaccharide on earth, which is sustainable, biocompatible, biodegradable, and non-toxic (Lavoine et al. 2012). Previous studies showed that cellulose can dissolve in NaOH/urea aqueous media under − 12 °C, and surface modified the Fe_3O_4 magnetic nanoparticles (Cai and Zhang 2005). Thus, investigating the growth of the MOF onto the cellulose-modified Fe_3O_4 is of interest.

The glucose oxidase (GOx) is an aerobic dehydrogenation enzyme, which has played an important role on deoxidization, glucose removal, and gluconic acid synthesis. It is widely used in forage, medicine, and other fields (Wong et al. 2008). In recent years, in order to overcome the disadvantages of the free GOx such as poor mechanical stability, difficult separation, and non-recyclability (Cao et al. 2016), several nanoparticles, such as titanium dioxide nanotubes (Ravariu et al. 2011), Fe_3O_4/APTES (França 2014), Ag@Zn-TSA (Dong et al. 2016), and ZIF-8 (Wu et al. 2015) were attempted to be used as enzyme carriers for the immobilization of GOx. Among these nanoparticles, carriers containing metal–organic frameworks (MOFs) have received more and more concern because of their excellent physical and chemical properties mentioned above.

In this study, a new core–shell magnetic ZIF-8-coated magnetic regenerated cellulose-coated nanoparticle (ZIF-8@Cellu@Fe_3O_4) was fabricated. The as-prepared ZIF-8@Cellu@Fe_3O_4 was structurally characterized in detail. The glucose oxidase (GOx) was embedded in the pores of the ZIF-8@Cellu@Fe_3O_4 with a high relative activity recovery and protein loading.

Methods

Preparation of magnetic regenerated cellulose-coated nanoparticle (Cellu@Fe_3O_4)

The Fe_3O_4 nanoparticles were fabricated by co-precipitation method according to our previous literatures (Deng et al. 2016; Cao et al. 2017): 2.43 g $FeCl_3 \cdot 6H_2O$ and 0.9 g $FeCl_2 \cdot 4H_2O$ were dissolved in 200 mL deionized water at room temperature. The mixture was added dropwise into a 25% ammonia solution with stirring, N_2 purge, and the pH at 10. The temperature was raised to 60 °C and kept for 1 h; the magnetite precipitate was collected with an external magnet and washed three times with deionized water.

150 mg of Fe_3O_4 was dispersed in 30 mL aqueous solution containing 7 wt% of NaOH and 12 wt% of urea and pre-cooled to − 12 °C for more than 1 h. Then, 100 mg of microcrystalline cellulose was added into the above suspension. After 1 h of freezing, the microcrystalline cellulose was dissolved completely. Then the deionized water was mixed with the above mixture and the

cellulose-coated Fe_3O_4 (Cellu@Fe_3O_4) was formed. The Cellu@Fe_3O_4 was collected with an external magnet and washed three times with deionized water.

Preparation of magnetic ZIF-8 nanoparticles (ZIF-8@Cellu@Fe_3O_4)

Zinc nitrate hexahydrate was dissolved in deionized water (40 mM, 2 mL) mixed with 10 mg Cellu@Fe_3O_4 under stirring for 20 min. Then 2-methylimidazole (160 mM, 2 mL) was added into the mixture and stirred for 3 h (Liang et al. 2015).

Preparation of GOx-loaded ZIF-8@Cellu@Fe_3O_4 nanocomposite

The synthesis processes of GOx-loaded ZIF-8@Cellu@Fe_3O_4 nanocomposites are illustrated in Scheme 1. Zinc nitrate hexahydrate was dissolved in deionized water (40 mM, 2 mL) mixture and stirring with 10 mg of Cellu@Fe_3O_4 for 20 min. Then 2-methylimidazole (160 mM, 2 mL) was added into the mixture and stirred for 10 min (Lyu et al. 2014; Liang et al. 2015; Du et al. 2017). Free GOx was dissolved in buffer (200 mM, pH 4.0–8.0); 0.2 mL free enzyme (100 U/mL) was added into solutions. The reaction lasted for (0.5–3 h) and immobilized at (10–50 °C, 200 rpm). The immobilized GOx was separated through an external magnetic field. Then, the un-immobilized GOx was removed by continuous washing until no protein was detected. The washing solutions were collected to detect the amount of un-immobilized GOx. The amount of immobilized GOx loaded on the ZIF-8@Cellu@Fe_3O_4 was calculated as the difference between the initial and the un-immobilized GOx. The GOx-loaded ZIF-8@Cellu@Fe_3O_4 was named as GOx-ZIF-8@Cellu@Fe_3O_4.

Enzyme activity assay and protein concentration

Protein concentration was determined according to the Bradford method using bovine serum albumin as standard (Lowry et al. 1951).

The activities of free and immobilized glucose oxidase were determined by indigo carmine method (Zhou et al. 2008). Glucose oxidase was dissolved in 1 mL phosphate buffer (200 mM, pH 7.0) and then 4 mL 0.2 mol/L glucose solution was added. The solution was mixed at 37 °C for 10 min. 3 mL acetic acid–sodium acetate (0.1 M acetic acid 500 mL and 0.1 M sodium acetate 30 mL, pH 3.5) was added as buffer solution, and 1.3 mL indigo carmine (0.1 mM) was used as redox indicator. After treated at 100 °C (boiling water) for 13 min, the absorbance of solution at wavelength of 615 nm was measured.

Activity recovery (%) was calculated as follows:

$$= 100 \times \frac{\text{Activity of immobilized enzyme (U)}}{\text{Activity of free enzyme used for immobilization (U)}}.$$

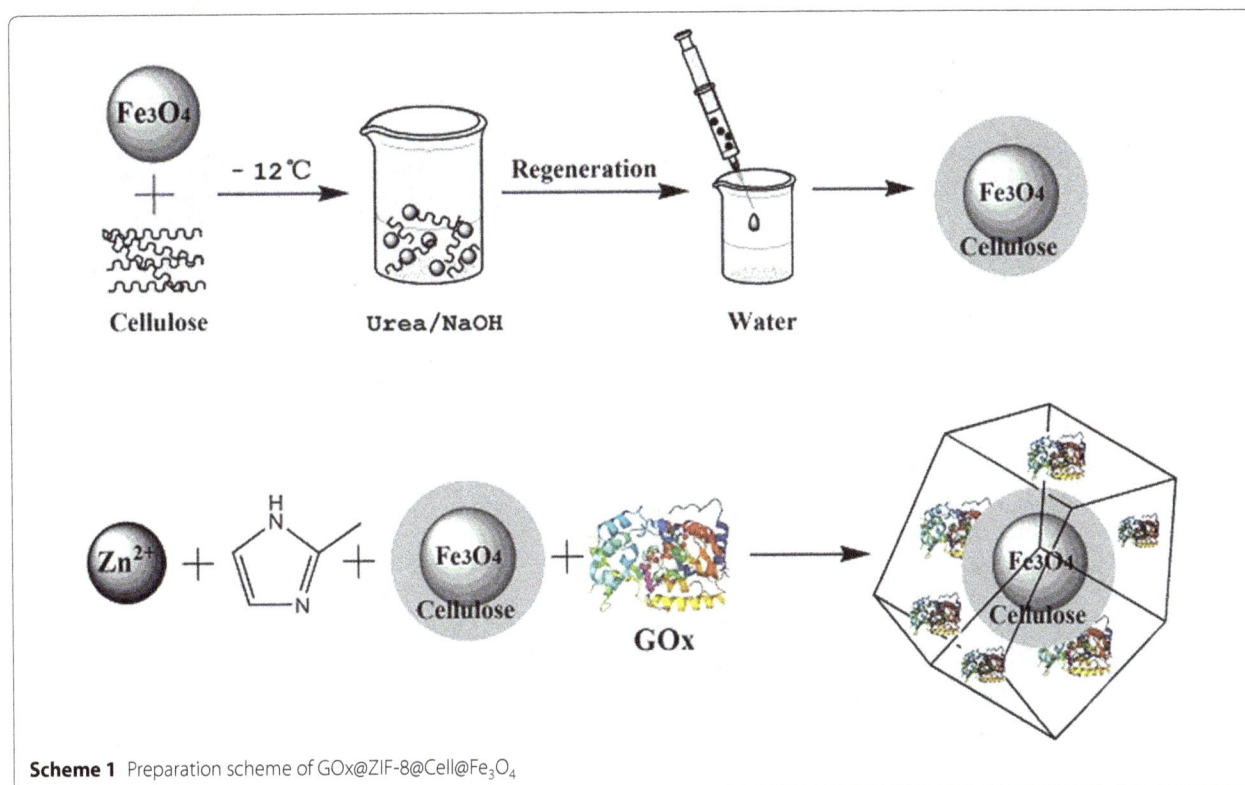

Scheme 1 Preparation scheme of GOx@ZIF-8@Cell@Fe$_3$O$_4$

Enzyme loading (%) was calculated as follows:

$$= 100 \times \frac{\text{Enzyme content of immobilized enzyme (mg)}}{\text{Content of enzyme used for immobilization (mg)}}.$$

Results and discussion

Characterization of Cellu@Fe$_3$O$_4$ and ZIF-8@Cellu@Fe$_3$O$_4$

The X-ray diffraction patterns of the ZIF-8@Cellu@Fe$_3$O$_4$, Cellu@Fe$_3$O$_4$, microcrystalline cellulose, and naked Fe$_3$O$_4$ are shown in Fig. 1. However, the

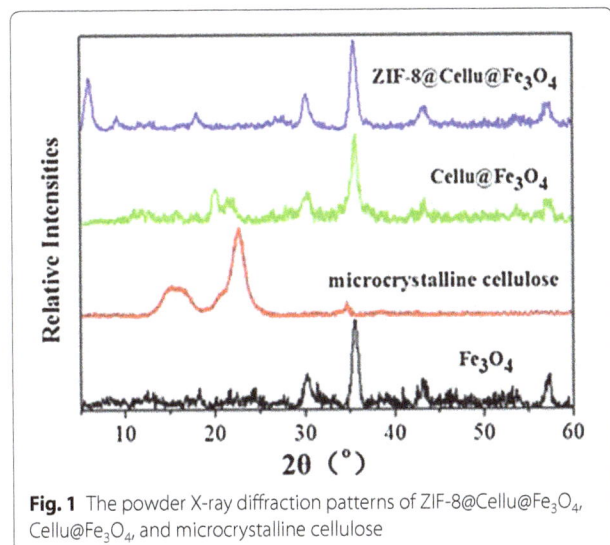

Fig. 1 The powder X-ray diffraction patterns of ZIF-8@Cellu@Fe$_3$O$_4$, Cellu@Fe$_3$O$_4$, and microcrystalline cellulose

microcrystalline cellulose showed three peaks at $2\theta = 14.8°$, $16.5°$, and $22.7°$ assigned to the (110), (110), and (200) planes which were characteristic peaks for the cellulose crystalline (I) (Edwards et al. 2012). By comparison, the Cellu@Fe$_3$O$_4$ displayed three diffraction peaks at $2\theta = 12.4°$, $20.2°$, and $22.2°$ assigned to the (110), (110), and (200) planes of cellulose crystalline (II) (Togawa and Kondo 1999). This illustrated that after dissolving-regeneration process, the cellulose crystalline form changed from cellulose (I) to cellulose (II) (Carrillo et al. 2004). Moreover, the Cellu@Fe$_3$O$_4$ showed four distinct peaks at $2\theta = 30.24°$, $35.60°$, $43.24°$, and $57.16°$, ascribing to the crystal plane diffraction peaks of the (220), (400), (422), and (511) diffraction peaks for Fe$_3$O$_4$ (JCPDS Card No. 19–0629) (Cao et al. 2014). The ZIF-8@Cellu@Fe$_3$O$_4$ XRD pattern is also shown in Fig. 1. The result showed that visible diffraction peaks at about $2\theta = 7.3°$, $10.5°$, and $18.0°$ were assigned to the characteristic diffraction peak of ZIF-8 (Pan et al. 2011). These results indicated the formation of the ZIF-8@Cellu@Fe$_3$O$_4$.

Figure 2 shows the FTIR spectra of ZIF-8@Cellu@Fe$_3$O$_4$, Cellu@Fe$_3$O$_4$, and microcrystalline cellulose. For the microcrystalline cellulose (Fig. 2a), the band at 1431 cm^{-1} was attributed to C–O–H stretching vibration. However, for Cellu@Fe$_3$O$_4$ (Fig. 2b), the peak at 1431 cm^{-1} disappeared and peak at 1421 cm^{-1} was observed (Colom and Carrillo 2002). This also illustrated that after dissolving-regeneration process, the cellulose crystalline form of Cellu@Fe$_3$O$_4$ changed from cellulose (I) to cellulose (II) (Colom and Carrillo 2002). Moreover, there was a strong absorption peak of Cellu@Fe$_3$O$_4$ at around 594 cm^{-1} assigned to the characteristic peak of Fe$_3$O$_4$ (Cornell et al. 1999). Also, the bands of Cellu@Fe$_3$O$_4$ at 3440 cm^{-1}, attributed to hydrogen bonding of cellulose, became broader and weaker, illustrating the strong interaction between Fe$_3$O$_4$ and cellulose layer which were observed (Kondo et al. 1994; Kondo and Sawatari 1996; Zhang et al. 2001). The FTIR spectrum shown in Fig. 2c displays the chemical composition of the ZIF-8@Cellu@Fe$_3$O$_4$. A strong peak at 421 cm^{-1} is ascribed to the Zn–N stretch mode (Zhang et al. 2013). The broad bands around 500–1350 and 1350–1500 cm^{-1} were assigned as the plane bending and stretching of imidazole ring, respectively (Lu et al. 2012). These results showed that the ZIF-8 was successfully composited on to the surface of the Cellu@Fe$_3$O$_4$ by co-precipitation method.

As shown in the scanning electron microscope (SEM) graphy (Fig. 3), the Cellu@Fe$_3$O$_4$ has an average diameter of around 29.7 nm and displays uniform structure and morphology. The size of ZIF-8@Cellu@Fe$_3$O$_4$ was approximated to 170 nm.

The vibrating specimen magnetometer (VSM) magnetization curves of the Cellu@Fe$_3$O$_4$ and ZIF-8@Cellu@Fe$_3$O$_4$ are shown in Fig. 4. Saturation magnetization (M_S) was used to measure the magnetization of samples defined as the maximum magnetic response of a material in an external magnetic field (Xiao et al. 2014). It is observed that the M_S of Fe$_3$O$_4$ is 21.37 emu/g and of ZIF-8@Cellu@Fe$_3$O$_4$ (4.9 emu/g) is lower than that of Cellu@Fe$_3$O$_4$ nanoparticles (12.8 emu/g).

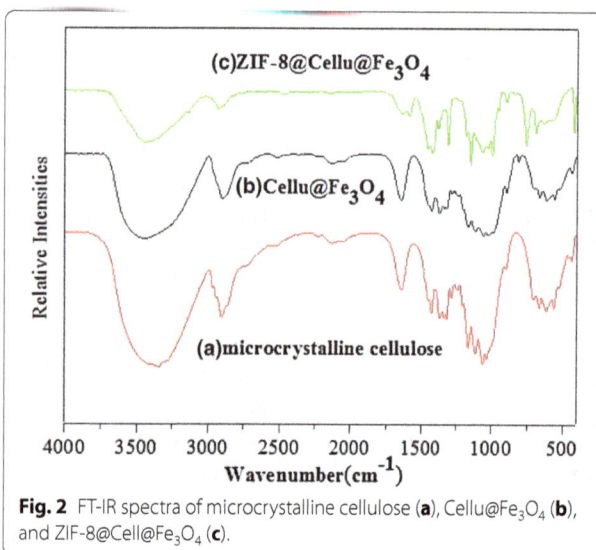

Fig. 3 SEM graphy of Cellu@Fe$_3$O$_4$ (**A**), ZIF-8@Cellu@Fe$_3$O$_4$ (**B**)

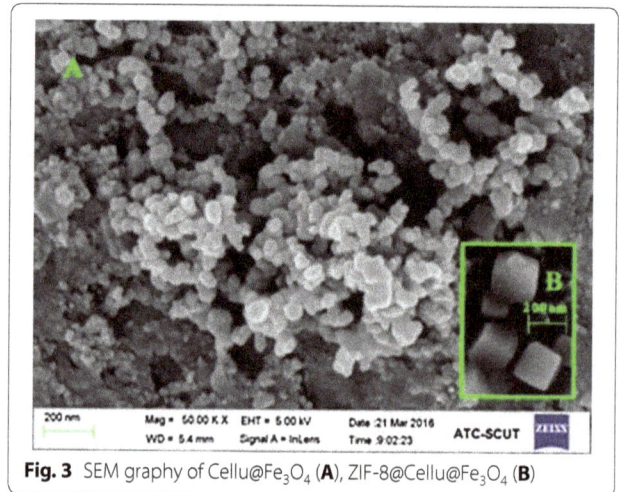

Fig. 4 Hysteresis loops of Fe$_3$O$_4$, Cellu@Fe$_3$O$_4$, ZIF-8@Cellu@Fe$_3$O$_4$, and GOx-ZIF-8@Cellu@Fe$_3$O$_4$

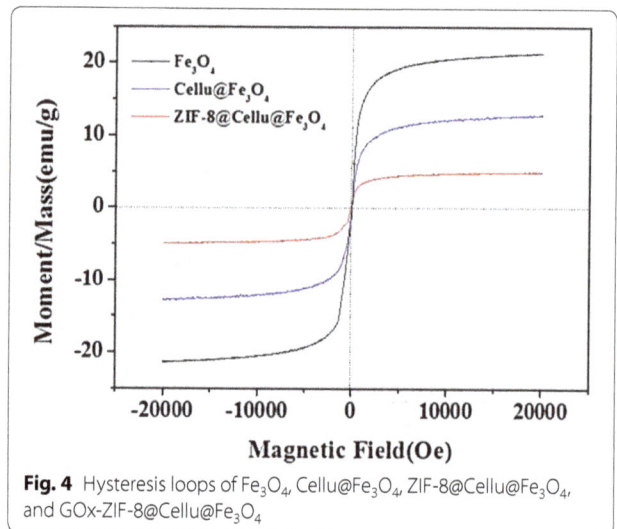

Fig. 2 FT-IR spectra of microcrystalline cellulose (**a**), Cellu@Fe$_3$O$_4$ (**b**), and ZIF-8@Cell@Fe$_3$O$_4$ (**c**).

Immobilization of GOx by ZIF-8@Cellu@Fe$_3$O$_4$ nanocomposite

Figure 5 shows the effect of buffer pH on enzyme activity recovery and enzyme loading. The results show that the highest activity recovery of immobilized glucose oxidase (GOx-ZIF-8@Cellu@Fe$_3$O$_4$) displayed at pH 7.0 (123.7%), with the protein loading 91.3 mg/g. The GOx-ZIF-8@Cellu@Fe$_3$O$_4$ exhibited relative high activity at faintly acid and neutral conditions (pH 6.5–7.0), and became deactivated in the both the acid and alkaline conditions (Cao et al. 2008). Also, ZIF-8 is very stable in neutral and basic conditions (Jian et al. 2015) and exhibit best enzyme encapsulation capacity at pH 7.5–8.0. Thus, the protein loading capacity changed depending on different pH values and the optimal pH value for GOx immobilization was 7.0.

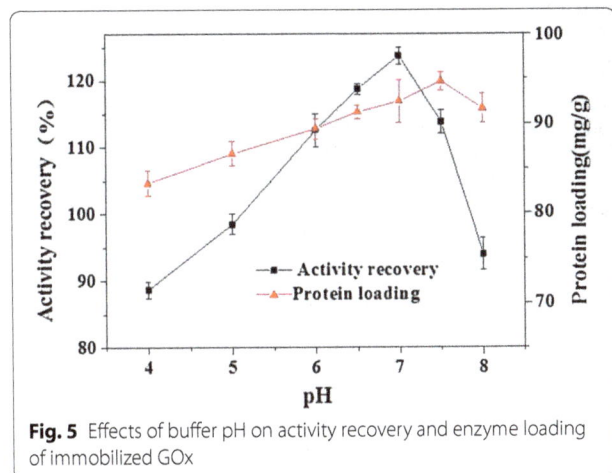

Fig. 5 Effects of buffer pH on activity recovery and enzyme loading of immobilized GOx

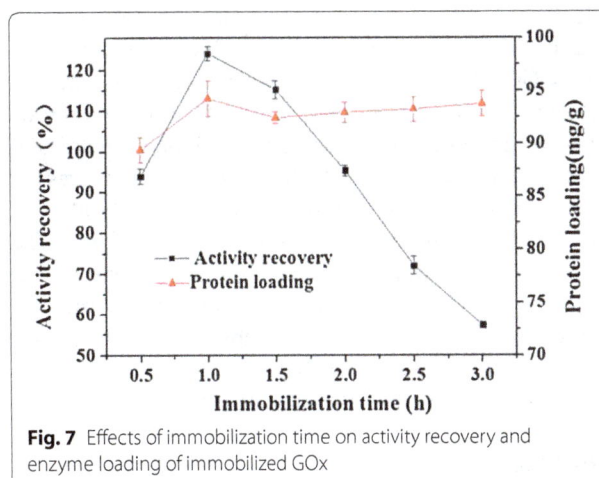

Fig. 7 Effects of immobilization time on activity recovery and enzyme loading of immobilized GOx

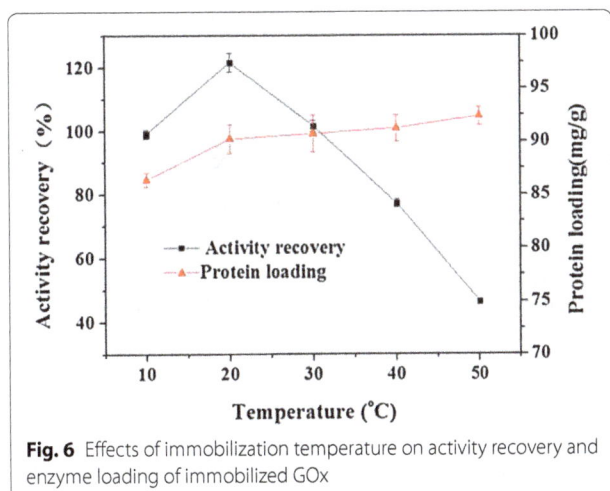

Fig. 6 Effects of immobilization temperature on activity recovery and enzyme loading of immobilized GOx

Figure 6 shows the effect of immobilization temperature on activity recovery and enzyme loading. The results showed that the GOx-ZIF-8@Cellu@Fe$_3$O$_4$ with highest activity recovery was obtained at 20 °C. When the temperature was higher than 30 °C, enzyme activity recovery decreased significantly. Thus, 20 °C was selected as the immobilization temperature in the following experiment.

Figure 7 shows the effect of immobilization time on enzyme activity recovery and enzyme loading. The result shows that when the immobilization time was 1 h, the highest relative activity was obtained. After 1 h, the relative activity decreased gradually, which was possibly due to the partial inactivation of the enzyme with immobilized time prolonged. In conclusion, the optimum immobilization time was 1 h at which the activity recovery of GOx attained 124.2% and the enzyme loading was 94.26 mg/g.

Generally speaking, the optimal condition of the immobilization process was that buffer pH 7, temperature 20 °C, and immobilization time 1 h. At this condition, the activity recovery was 124.2%, and the enzyme loading was 94.26 mg/g. The specific enzyme activity values of immobilized and free GOx were 12.42 and 10.00 U/mg, respectively. The thermal stability of free and immobilized GOx at 65 °C was performed. The activities of both free and immobilized GOx decreased gradually with increasing of the incubation time. The immobilized GOx exhibited more than 40% of its initial activity after 4 h of incubation, while that of free GOx was about 13%. This result showed that the thermal stability of immobilized GOx was enhanced after immobilization (Zhou et al. 2012a, b). As a comparison, GOx immobilized onto a 3-(aminopropyl)triethoxysilane (APTES)-coated Fe$_3$O$_4$ nanocarrier which retained less than 50% of the native GOx activity (Park et al. 2011). Thus the immobilization process in the present study is promising for enzyme immobilization.

A comparative study shown in Fig. 8 explores the mechanism of the enhanced enzyme relatively activity after immobilization. The presence of Cellu@Fe$_3$O$_4$,

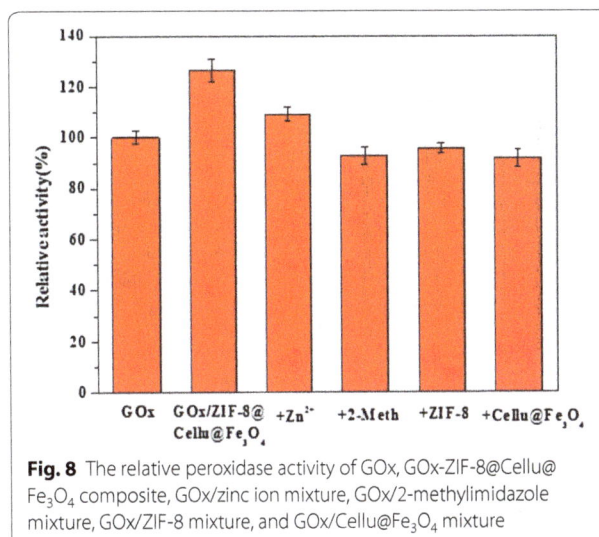

Fig. 8 The relative peroxidase activity of GOx, GOx-ZIF-8@Cellu@Fe$_3$O$_4$ composite, GOx/zinc ion mixture, GOx/2-methylimidazole mixture, GOx/ZIF-8 mixture, and GOx/Cellu@Fe$_3$O$_4$ mixture

ZIF-8 and 2-methylimidazole could not enhance the activity of GOx, while that of Zn^{2+} increased the activity of GOx by about 9.3%, which was similar with the previous literature (Lyu et al. 2014). Comparison shows that the immobilization of GOx in ZIF-8@Cellu@Fe_3O_4 enhanced the activity of GOx by 24.2%. These results showed that the GOx-ZIF-8@Cellu@Fe_3O_4 exhibited an increased relative activity compared to the free GOx. This may be attributed to the following reason: the enzyme immobilization process changed the enzyme conformation and increased the substrate affinity toward the glucose; the interaction between the GOx and Zn^{2+} in ZIF-8 enhanced the catalytic activity (Lyu et al. 2014).

Conclusions

In conclusion, Glucose oxidase (GOx) was successfully immobilized onto the biocompatible ZIF-8@Cellu@Fe_3O_4 via co-precipitation process. Its morphology, structure, and magnetic properties were determined. The GOx immobilized in ZIF-8@Cellu@Fe_3O_4 had high protein loading (94.26 mg/g) and enhanced relative activity recovery (124.2%). The results of the present work provide an efficient enzyme immobilization process and promote the application and development of immobilized enzyme catalysis.

Abbreviations
GOx: glucose oxidase; MOFs: metal–organic frameworks; ZIF: zeolitic imidazolate framework; Cellu@Fe_3O_4: magnetic regenerated cellulose-coated nanoparticle; ZIF-8@Cellu@Fe_3O_4: ZIF-8-coated magnetic regenerated cellulose nanoparticle; GOx-ZIF-8@Cellu@Fe_3O_4: GOx-loaded ZIF-8@Cellu@Fe_3O_4 nanocomposite.

Authors' contributions
Conceived and designed the experiments: SLC, XH, WYL, and MHZ. Performed the experiments: SLC, XH, WMG, and LHL. Analyzed the data: XH, SLC, XP, JX, HY, and YZM. Contributed reagents/materials/analysis tools: MHZ, WYL, and JZ. Wrote the paper: XH, SLC, YZM, and WYL. All authors read and approved the final manuscript.

Author details
[1] Department of Food Science, Foshan University (Northern Campus), Nanhai, Foshan 528231, China. [2] Laboratory of Applied Biocatalysis, School of Food Science and Engineering, South China University of Technology, No.381 Wushan Road, Guangzhou 510640, China. [3] School of Chemistry and Chemical Engineering, South China University of Technology, No. 381 Wushan Road, Guangzhou 510640, China. [4] Guangdong Province Key Laboratory for Green Processing of Natural Products and Product Safety, South China University of Technology, No. 381 Wushan Road, Guangzhou 510640, China. [5] School of Marine Science and Technology, Tianjin University, 92 Weijin Road, Tianjin 300072, China.

Acknowledgements
Not applicable.

Competing interests
The authors declare that they have no competing interests.

Funding
The National Natural Science Foundation of China (21336002; 21676104; 21376096), the Fundamental Research Funds for the Chinese Universities (2015PT002; 2015ZP009), the Program of State Key Laboratory of Pulp and Paper Engineering (2017ZD05), the Open Funding Project of the State Key Laboratory of Bioreactor Engineering, and High-Level Talent Start-Up Research Project of Foshan University (GG07016).

References
Cai J, Zhang L (2005) Rapid dissolution of cellulose in LiOH/urea and NaOH/urea aqueous solutions. Macromol Biosci 5:539
Cao L, Ye J, Tong L et al (2008) A new route to the considerable enhancement of glucose oxidase (GOx) activity: the simple assembly of a complex from CdTe quantum dots and GOx, and its glucose sensing. Chemistry 14:9633–9640
Cao S-L, Li X-H, Lou W-Y et al (2014) Preparation of a novel magnetic cellulose nanocrystal and its efficient use for enzyme immobilization. J Mater Chem B 2:5522–5530
Cao S, Xu P, Ma Y et al (2016) Recent advances in immobilized enzymes on nanocarriers. Chin J Catal 37:1814–1823
Cao S-L, Deng X, Xu P et al (2017) Highly efficient enzymatic acylation of dihydromyricetin by the immobilized lipase with deep eutectic solvents as co-solvent. J Agric Food Chem 65:2084–2088
Carrillo F, Colom X, Suñol JJ et al (2004) Structural FTIR analysis and thermal characterisation of lyocell and viscose-type fibres. Eur Polym J 40:2229–2234
Colom X, Carrillo F (2002) Crystallinity changes in lyocell and viscose-type fibres by caustic treatment. Eur Polym J 38:2225–2230
Cornell RM, Schwertmann U, Cornell R (1999) The iron oxides: structure, properties, reactions, occurrences and uses. Clay Miner 34:209–210
Deng X, Cao S, Li N et al (2016) A magnetic biocatalyst based on mussel-inspired polydopamine and its acylation of dihydromyricetin. Chin J Catal 37:584–595
Dong S, Zhang D, Suo G et al (2016) Exploiting multi-function metal–organic framework nanocomposite Ag@Zn-TSA as highly efficient immobilization matrixes for sensitive electrochemical biosensing. Anal Chim Acta 934:203–211
Du Y, Gao J, Zhou L et al (2017) Enzyme nanocapsules armored by metal–organic frameworks: a novel approach for preparing nanobiocatalyst. Chem Eng J 327:1192–1197
Edwards JV, Prevost NT, Condon B et al (2012) Immobilization of lysozyme–cellulose amide-linked conjugates on cellulose I and II cotton nanocrystalline preparations. Cellulose 19:495–506
França TCP (2014) Preparation and characterization of hybrid Fe3O4/APTES for immobilization of GOX. Mater Sci Forum 798–799:460–465
Hou C, Wang Y, Ding Q et al (2015) Facile synthesis of enzyme-embedded magnetic metal–organic frameworks as a reusable mimic multi-enzyme system: mimetic peroxidase properties and colorimetric sensor. Nanoscale 7:18770
Hou M, Zhao H, Feng Y et al (2017) Synthesis of patterned enzyme–metal organic framework composites by ink-jet printing. Bioresour Bioprocess 4:40
Jian M, Liu B, Zhang G et al (2015) Adsorptive removal of arsenic from aqueous solution by zeolitic imidazolate framework-8 (ZIF-8) nanoparticles. Colloids Surf A Physicochem Eng Aspects 465:67–76
Jin T, Yang Q, Meng C et al (2014) Promoting desulfurization capacity and separation efficiency simultaneously by the novel magnetic Fe3O4@PAA@MOF-199. RSC Adv 4:41902–41909
Ke F, Qiu LG, Yuan YP et al (2012) Fe3O4@MOF core–shell magnetic micro spheres with a designable metal–organic framework shell. J Mater Chem 22:9497–9500
Kondo T, Sawatari C (1996) A Fourier transform infra-red spectroscopic analysis of the character of hydrogen bonds in amorphous cellulose. Polymer 37:393–399
Kondo T, Sawatari C, Manley RSJ et al (1994) Characterization of hydrogen bonding in cellulose-synthetic polymer blend systems with regioselectively substituted methylcellulose. Macromolecules 27:210–215
Lavoine N, Desloges I, Dufresne A et al (2012) Microfibrillated cellulose—its barrier properties and applications in cellulosic materials: a review. Carbohydr Polym 90:735
Lee J, Farha OK, Roberts J et al (2009) Metal–organic framework materials as catalysts. Chem Soc Rev 38:1450–1459

Li J-R, Kuppler RJ, Zhou H-C (2009) Selective gas adsorption and separation in metal–organic frameworks. Chem Soc Rev 38:1477–1504

Liang K, Ricco R, Doherty CM et al (2015) Biomimetic mineralization of metal–organic frameworks as protective coatings for biomacromolecules. Nat Commun 6:7240

Liu Z, Wang H, Liu C et al (2012) Magnetic cellulose–chitosan hydrogels prepared from ionic liquids as reusable adsorbent for removal of heavy metal ions. Chem Commun 48:7350

Lowry OHNG, Rosebrough NJJ, Farr AL et al (1951) Protein measurement with folin phenol reagent. J Biol Chem 193:265–275

Lu G, Li S, Guo Z et al (2012) Imparting functionality to a metal–organic framework material by controlled nanoparticle encapsulation. Nat Chem 4:310

Lyu F, Zhang Y, Zare RN et al (2014) One-pot synthesis of protein-embedded metal–organic frameworks with enhanced biological activities. Nano Lett 14:5761

Maes M, Alaerts L, Vermoortele F et al (2010) Separation of C(5)-hydrocarbons on microporous materials: complementary performance of MOFs and zeolites. J Am Chem Soc 132:2284–2292

Nong J, Zhao W, Qin X et al (2015) Recent progress in the study of core–shell-structured materials with metal organic frameworks (MOFs) as shell. Chem Ind Eng Prog 34:774–783

Pan Y, Liu Y, Zeng G et al (2011) Rapid synthesis of zeolitic imidazolate framework-8 (ZIF-8) nanocrystals in an aqueous system. Chem Commun 47:2071–2073

Park HJ, McConnell JT, Boddohi S et al (2011) Synthesis and characterization of enzyme–magnetic nanoparticle complexes: effect of size on activity and recovery. Colloids Surf B Biointerfaces 83:198–203

Phan A, Doonan CJ, Uriberomo FJ et al (2010) Synthesis, structure, and carbon dioxide capture properties of zeolitic imidazolate frameworks. Acc Chem Res 43:58–67

Ravariu, Manea, Parvulescu et al (2011) Titanium dioxide nanotubes on silicon wafer designated for GOX enzymes immobilization. Dig J Nanomater Biostruct 6:703–707

Togawa E, Kondo T (1999) Change of morphological properties in drawing water-swollen cellulose films prepared from organic solutions. A view of molecular orientation in the drawing process. J Polym Sci, Part B: Polym Phys 37:451–459

Wehner T, Mandel K, Schneider M et al (2016) Superparamagnetic luminescent MOF@Fe3O4/SiO2 composite particles for signal augmentation by magnetic harvesting as potential water detectors. ACS Appl Mater Interfaces 8:5445

Wong CM, Wong KH, Chen XD (2008) Glucose oxidase: natural occurrence, function, properties and industrial applications. Appl Microbiol Biotechnol 78:927–938

Wu H, Zhou W, Yildirim T (2007) Hydrogen storage in a prototypical zeolitic imidazolate framework-8. J Am Chem Soc 129:5314

Wu X, Ge J, Yang C et al (2015) Facile synthesis of multiple enzyme-containing metal–organic frameworks in a biomolecule-friendly environment. Chem Commun 51:13408

Wu X, Yang C, Ge J (2017) Green synthesis of enzyme/metal–organic framework composites with high stability in protein denaturing solvents. Bioresour Bioprocess 4:24

Xia GH, Cao SL, Xu P et al (2017) Preparation of a nanobiocatalyst by efficiently immobilizing Aspergillus niger lipase onto magnetic metal–biomolecule frameworks (BioMOF). ChemCatChem 9:1794–1800

Xiao F, Feng C, Jin C et al (2014) Magnetic and electromagnetic properties of Fe3O4/C self-assemblies. Mater Lett 122:103–105

Zhang L, Ruan D, Zhou J (2001) Structure and properties of regenerated cellulose films prepared from cotton linters in NaOH/urea aqueous solution. Ind Eng Chem Res 40:5923–5928

Zhang T, Zhang X, Yan X et al (2013) Synthesis of Fe3O4@ZIF-8 magnetic core–shell microspheres and their potential application in a capillary microreactor. Chem Eng J 228:398–404

Zhou JQ, Chen SH, Wang JW (2008) A simple and convenient method to determine the activity of glucose oxidase. Exp Technol Manag 12:15

Zhou L, Jiang Y, Gao J et al (2012a) Oriented immobilization of glucose oxidase on graphene oxide. Biochem Eng J 69:28–31

Zhou L, Jiang Y, Gao J et al (2012b) Graphene oxide as a matrix for the immobilization of glucose oxidase. Appl Biochem Biotechnol 168:1635–1642

18

Comparative evaluation of wastewater-treatment microbial fuel cells in terms of organics removal, waste-sludge production, and electricity generation

Yusuke Asai[1], Morio Miyahara[1,2], Atsushi Kouzuma[1] and Kazuya Watanabe[1*]

Abstract

Microbial fuel cells (MFCs) are devices that exploit living microbes for electricity generation coupled to organics degradation. MFCs are expected to be applied to energy-saving wastewater treatment (WWT) as alternatives to activated-sludge reactors (ASRs). Although extensive laboratory studies have been performed to develop technologies for WWT-MFCs, limited information is available for comparative evaluation of MFCs and ASRs in terms of organics removal and waste-sludge production. In the present study, laboratory WWT experiments were performed using cassette-electrode MFCs and ASRs that were continuously supplied either with artificial domestic wastewater (ADW) containing starch and peptone or with artificial industrial wastewater (AIW) containing methanol as the major organic matter. We found that these two types of WWT reactors achieved similar organics-removal efficiencies, namely, over 93% based on chemical oxygen demands for the ADW treatment and over 97% for the AIW treatment. Sludge was routinely removed from these reactors and quantified, showing that amounts of waste sludge produced in MFCs were approximately one-third or less compared to those in ASRs. During WWT, MFCs continuously generated electricity with Coulombic efficiencies of 20% or more. In reference to ASRs, MFCs are demonstrated to be attractive WWT facilities in terms of stable organics removal and low waste-sludge production. Along with the unnecessity of electric power for aeration and the generation of power during WWT, the results obtained in the present study suggest that MFCs enable substantial energy saving during WWT.

Keywords: Wastewater treatment, Microbial fuel cells, Activated sludge, Exoelectrogens, Power generation, Waste sludge

Background

Activated-sludge reactors (ASRs) are widely used for the treatment of domestic and industrial wastewater (Eckenfelder and O'Conner 1961). Although ASRs have been successfully used for wastewater treatment (WWT), intrinsic limitations associated with the use of ASRs include the consumption of large amounts of electric energy (Rosso et al. 2008) and the production of substantial amounts of waste sludge (Hall 1995). In Japan, annual

*Correspondence: kazuyaw@toyaku.ac.jp
[1] School of Life Science, Tokyo University of Pharmacy and Life Sciences, Tokyo 192-0392, Japan
Full list of author information is available at the end of the article

electric power consumption in municipal WWT plants exceeds 80 billion KWh that accounts for approximately 0.7% of the total electric power consumption in this country (Mizuta and Shimada 2010). In addition, waste sludge annually produced in municipal WWT plants in Japan exceeds 70 million tons that accounts for over 20% of the total industrial waste (Imai et al. 2010).

Microbial fuel cells (MFCs) are devices that exploit living microbes for the conversion of organic matter into electricity (Logan et al. 2006). Using naturally occurring microbiomes, MFCs are able to generate electricity from organic wastes and wastewater (Watanabe 2008). In particular, MFCs are expected to be applied to energy-saving WWT (Li et al. 2014), and extensive work has been performed to develop MFC technologies applicable to WWT (Miyahara et al. 2013; Zhang et al. 2013). It has been demonstrated that ASRs can be converted to MFCs by removing aeration apparatuses from aeration tanks and inserting cassette-type electrodes instead (Yoshizawa et al. 2014). Merits expected in using MFCs for WWT include no need of energy for aeration, power generation from pollutants, and possible reduction in waste-sludge production. Although previous studies have evaluated organics-removal efficiencies and power generation in WWT-MFCs (Miyahara et al. 2013; Yoshizawa et al. 2014), limited experimental data are available for the amounts of waste sludge produced in MFCs (Zhang et al. 2013).

To date, there have been few studies that comparatively evaluated WWT performances of MFCs and ASRs operated under same conditions; in particular, it has not been experimentally demonstrated whether or not waste sludge produced in MFCs is actually less than that in ASRs. In the present study, we operated MFCs and ASRs in parallel by continuously supplying either with artificial domestic wastewater (ADW) containing starch, yeast extract, peptone, and urea as the major organic components or with artificial industrial wastewater (AIW) containing methanol as the major organic component, and their WWT performances (organics removal and waste-sludge production) were compared. Methanol was selected as a substrate in AIW, since it is widely used in industrial processes and known to be a major pollutant in industrial wastewater (Yamamuro et al. 2014). Results obtained are considered to serve as fundamental datasets for the practical development of MFC technologies for WWT.

Methods

Reactors used in WWT experiments

Photos of laboratory ASR and MFC used in the present study are presented in Fig. 1. An ASR comprised an aeration tank (approximately 1.5 L) and settling tank (approximately 0.5 L) that were separated by a partition board (Fig. 1a). Bottom parts of the aeration and settling tanks were connected, and sludge settled in the setting tank was returned to the aeration tank by gravity.

MFC used in the present study was a cassette-electrode (CE) reactor (approximately 1.5 L in water content; panels b and c in Fig. 1) that were equipped with 6 CEs prepared as described elsewhere (Shimoyama et al. 2008; Miyahara et al. 2013). As indicated with arrows in Fig. 1c, water flowed up and down between CEs and partition boards in CE-MFC. A CE had two anode/separator/cathode sets on both sides, between which air was filled (5 mm in thickness). An anode ($126\ cm^2$ in area) was made of a graphite-felt sheet (3 mm in thickness; Sohgoh Carbon, Yokohama, Japan), while a cathode ($126\ cm^2$) was an air cathode produced as described previously (Cheng et al. 2006). A separator (punctured polypropylene sheets, 1 mm in thickness) was used to separate between an anode and cathode. The water surface in CE-MFC was covered with floating boards to reduce the exposure to oxygen (Miyahara et al. 2015).

Operation of WWT reactors

Sludge used as inocula for ASR and MFC was obtained from a return-sludge line in a municipal WWT plant (Asakawa Water Reclamation Center in Tokyo, Japan). Sludge was diluted with a mineral medium containing (per liter) 50 mg BBL yeast extract, 175 mg NH_4Cl,

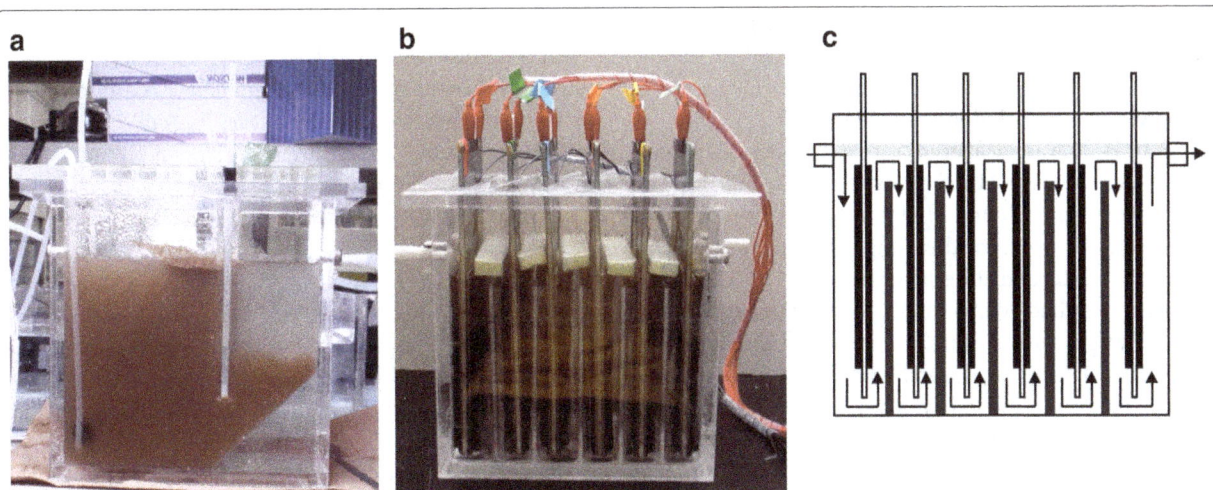

Fig. 1 Reactors used in the WWT experiments. **a** A side view of ASR. **b** A side view of CE-MFC. **c** A schematic diagram of CE-MFC. Water flows are indicated with *arrows*. *Light gray bars* above the water surface are floating boards that prevent water from the contamination with oxygen, while *dark gray bars* between CEs are partition boards that facilitate the up and down flows of water

5.26 mg KH_2PO_4, 22.05 mg $CaCl_2 \cdot 2H_2O$, 0.43 mg $MgSO_4 \cdot 7H_2O$, 21.3 mg KCl, 8.76 mg $NaHCO_3$, and 1 mL of trace-element solution (DSMZ 663; Deutsche Sammlung von Mikroorganismen und Zellkulturen GmbH) (pH 7.0), and the reactors were filled with the sludge suspension. An initial mixed-liquor suspended solid (MLSS) concentration was approximately 2000 mg L^{-1}. The reactors were continuously suppled with either ADW containing starch, yeast extract, peptone, and urea as the major organic components (Miyahara et al. 2013) or AIW containing methanol as the major organic matter in the mineral medium. Chemical oxygen demand (COD, mg L^{-1}) concentrations in ADW and AIW were approximately 500 and 1500 mg L^{-1}, respectively. ADW with a COD value of 500 mg L^{-1} was used for comparing results obtained in the present study with those of previous studies (Miyahara et al. 2013; Yoshizawa et al. 2014). A hydraulic retention time (HRT, D) was 24 h in all experiments. ASR was supplied with air at a rate of 3 L min^{-1}.

When commencing the operation of MFC, all anodes and cathodes were connected in parallel via an external resister (R_{ext}, Ω), and a voltage across the resister (E, mV) was monitored using a data logger (HA-1510, Graphtec, Yokohama, Japan).

Evaluation of organics removal

Effluents from the reactors were sampled at the effluent ports, and sludge was removed by centrifugation at 8000×g for 5 min. A COD concentration (mg L^{-1}) in an effluent was measured using a COD reactor and a COD 0–1500 ppm range kit (Hach, Loveland, CO, USA). A COD-removal efficiency (CRE, %) was calculated from the influent COD (COD_{in}, mg L^{-1}) and effluent COD (COD_{ef}, mg L^{-1}) as CRE = [$COD_{in} - COD_{ef}$]/COD_{in}. Methanol in effluents was measured using gas chromatography (Yamamuro et al. 2014).

Evaluation of waste-sludge production

MLSS in ASR was measured at certain intervals (generally 1 week). Before MLSS measurements, the partition board was removed, and sludge in ASR was uniformly suspended. A portion of the sludge suspension was sampled in triplicate, and, after being dried at 105 °C for 24 h, they were weighed. MLSS (mg L^{-1}) was determined from the dry weights and a sample volume. In order to keep MLSS at 2000 mg L^{-1}, an appropriate amount of sludge suspension was removed (removed sludge suspension; RSS, L) from ASR, and the fresh mineral medium was infused into ASR for compensating for the volume loss. Thereafter, the partition board was inserted again, and the operation of ASR was continued. During the operation, MLSS in effluent ($MLSS_{eff}$; sludge was precipitated by the

centrifugation and dried) was also measured by weighing dried suspended solids. Daily amounts of waste-sludge production (WSR, mg L^{-1} day^{-1}) were calculated from MLSS and $MLSS_{eff}$ as follows: WSR = {(MLSS × RSS)/Interval of measurement (D) + $MLSS_{eff}$ × HRT}/Reactor volume.

To measure MLSS in MFC, the operation was temporarily halted at certain intervals. CEs were removed from the MFC reservoir, and loosely attached biofilms onto CEs were washed away by water flush and were mixed with the sludge suspension in the MFC reservoir. A portion of the mixed sludge suspension was sampled in triplicate, and, after being dried at 105 °C for 24 h, they were weighed. MLSS was calculated from the dry weight and a sample volume. The mixed sludge suspension was subsequently discarded, and, after MFC was equipped with the spent CEs and filled with the fresh mineral medium, the operation was re-started by supplying with ADW or AIW. During the operation, sludge in an effluent (precipitated by the centrifugation) was also measured at certain intervals as described above. Daily amounts of waste-sludge production were calculated from MLSS and $MLSS_{eff}$ as described above.

Evaluation of electricity generation in MFCs

Current (I, mA) was calculated from E and R_{ext} using equations $I = E/R_{ext}$. A current density (J, mA m^{-2}) was estimated by dividing I by the total projected area of the anodes (1512 cm^2). Power (P, mW) was estimated according to an equation $P = IE$, while a power density (PD, mW m^{-2}) was estimated by dividing P by the total projected area of the anodes. A Coulombic efficiency (ε_c, %) was calculated based on a COD removal ($COD_{in} - COD_{ef}$) and a measured current as described previously (Miyahara et al. 2013). Polarization and power density curves were drawn using a potentiostat (HZ-5000, Hokuto Denko, Tokyo, Japan) as described previously (Miyahara et al. 2013), and the maximum power density (the peak in a power curve; P_{max}, mW m^{-2}; based on the projected anode area) and open-circuit voltage (OCV, mV) were determined as described elsewhere (Logan et al. 2006).

Results

COD removal

ASR and MFC were continuously supplied with either ADW or AIW, and COD_{ef} was routinely measured (panels a and b in Fig. 2). During the operation of MFC, R_{ext} was changed as described below. Figure 2 shows that ~20 days were needed for ASR and MFC to sufficiently treat ADW and AIW, and the initial 20 days were therefore considered to be the acclimatization periods. These acclimatization periods were relatively long with unknown reasons. After the initial acclimatization periods, COD_{ef} values for ADW-treating ASR and MFC were mostly below 50 mg L^{-1} (Fig. 2a), and those for AIW-treating reactors were also below 50 mg L^{-1} (Fig. 2b). During these periods, methanol was not detected in effluents from the AIW-treating reactors (data not shown). Accordingly, CRE values for the ADW treatment were mostly over 90%, while those for AIW treatment was over 95%. During the stable operation (from days 80 to 100), mean CRE values were estimated (Table 1), showing that CRE values for ASR and MFC were not significantly different from each other.

Waste-sludge production

At certain intervals, sludge suspensions were removed from ASR and MFC as described in the "Methods" section, and amounts of waste sludge removed from these reactors (waste sludge) were quantified. Figure 3 shows

normalized amounts of waste sludge produced during the stable operation of ASR and MFC (from day 80 to day 100) treating either ADW or AIW. It was found that the amounts of waste sludge produced in MFC were significantly lower than those in ASR. The amount of waste sludge produced in ADW-treating MFC (ADW-MFC) was approximately one-third of that in ADW-treating ASR, while that in AIW-treating MFC (AIW-MFC) was approximately one-fifth of that in AIW-treating ASR (Table 1).

Electricity generation in MFC

Changes in E for ADW-MFC and AIW-MFC were monitored using data loggers (Fig. 4), and I was calculated from E and R_{ext} (Fig. 4). As presented in this figure, R_{ext} was changed to maintain E at around 400 mV; previous studies have shown that efficient WWT and electricity generation were achieved by maintaining E at around

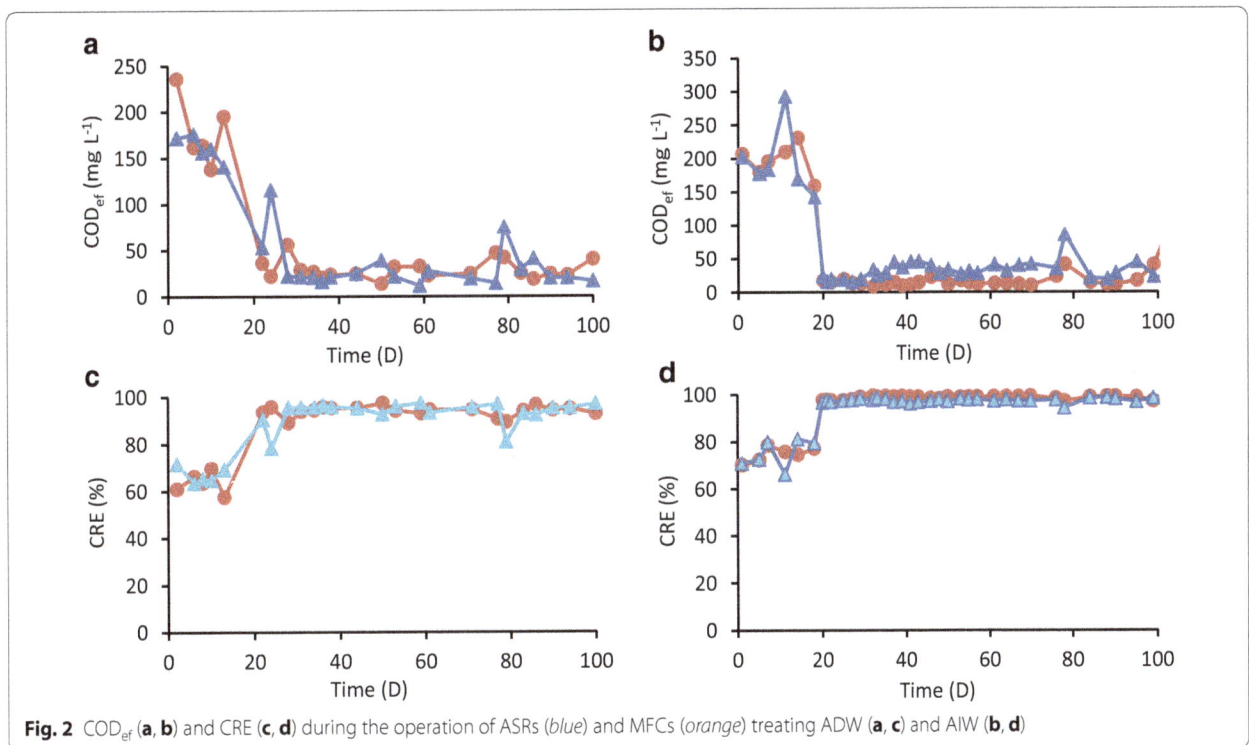

Fig. 2 COD_{ef} (**a, b**) and CRE (**c, d**) during the operation of ASRs (*blue*) and MFCs (*orange*) treating ADW (**a, c**) and AIW (**b, d**)

Table 1 Summary of performance data obtained during stable operation (days 80 to 100)

Reactor	Wastewater	CRE (%)	Waste sludge (mg L^{-1} D^{-1})	ε_c (%)	P_{max} (mW m^{-2})	OCV (mV)
ASR	ADW	94 ± 5	44 ± 15	–	–	
	AIW	97 ± 4	88 ± 14	–	–	
MFC	ADW	93 ± 4	10 ± 2	26 ± 5	124 ± 11	770 ± 23
	AIW	98 ± 2	30 ± 4	20 ± 3	160 ± 10	780 ± 26

Values are means ± SDs ($n > 3$)

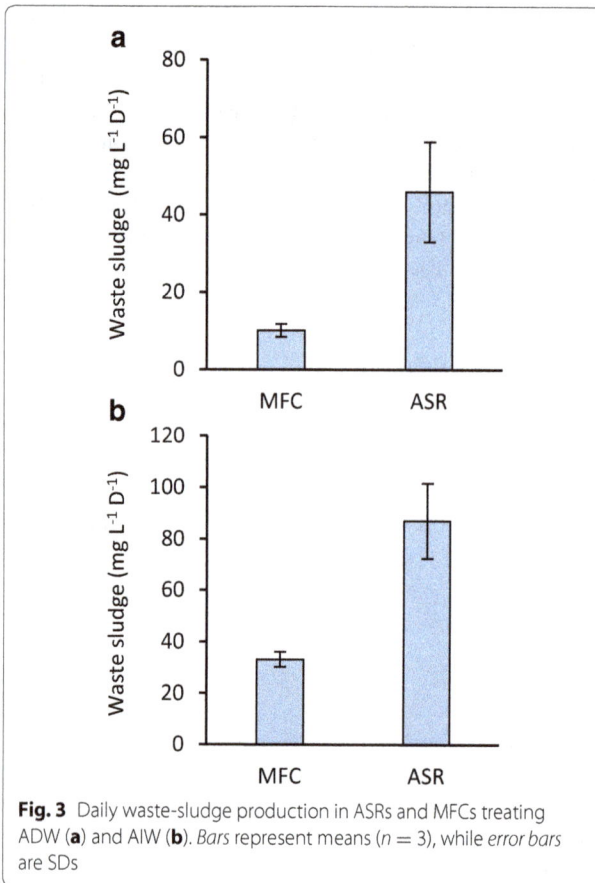

Fig. 3 Daily waste-sludge production in ASRs and MFCs treating ADW (**a**) and AIW (**b**). *Bars* represent means ($n = 3$), while *error bars* are SDs

such values (Miyahara et al. 2013; Yoshizawa et al. 2014). As a result, R_{ext} for ADW-MFC was finally kept at 20 Ω, while that for AIW-MFC was 10 Ω. This figure shows that AIW-MFC generated more current than ADW-MFC in the last 20 days. Based on measured I values and COD removal, ε_c was determined as an index for electron recovery (Table 1). The mean ε_c value for ADW-MFC was higher than that for AIW-MFC, despite that more current was generated in AIW-MFC than that in ADW-MFC. This was due to the high COD load to AIW-MFC compared to ADW-MFC.

Polarization analyses were routinely conducted (generally once in every 5 days) for monitoring changes in electrochemical performances of ADW-MFC and AIW-MFC. Representative polarization and power curves are presented in Fig. 5, and changes in P_{max} for these MFCs are illustrated in Fig. 6. Figure 5 shows that ADW-MFC and AIW-MFC were properly functioned as fuel cells that produced catalytic currents. We therefore employed P_{max} as an index for monitoring electrochemical performances of these MFCs. As indicated in Fig. 6, P_{max} relatively rapidly increased in ADW-MFC compared to that in AIW-MFC, whereas higher P_{max} values were finally

observed for AIW-MFC than those for ADW-MFC. This trend would have been related to the minor occurrence of anaerobic methanol-dissimilatory microbial populations in the activated sludge used as the inoculum for MFCs. The competition between denitrifiers and exoelectrogens also needs to be considered. It is also likely that the final high power density observed for AIW-MFC was due to the high COD in AIW compared to that in ADW. A possible reason for inconsistency in trends between the COD removal and P_{max} would be that microbial terminal electron-accepting reactions other than current generation also contributed to the COD removal, as indicated with ε_c.

Discussion

ADW used in the present study was the same as that used in our previous laboratory experiments for assessing CE-MFCs for WWT (Miyahara et al. 2013; Yoshizawa et al. 2014); in these studies, however, despite that HRT and COD_{in} were also the same as those in the present study, CRE values were reported to be around 80%. On the other hand, as presented in Fig. 1 and Table 1, CRE values observed in the present work were higher than 90% (93% in average). We consider that the up and down flow of water in MFC was effective for improving the quality of effluent water; this flow system was newly employed in the present study. We had employed right and left flows of water in CE-MFC (named the slalom flow) in previous studies (Miyahara et al. 2013; Yoshizawa et al. 2014), while we found that this system was associated with a problem of undercurrent that was not effectively contacted with electrode surfaces. The present study therefore proposes that the up and down flow of wastewater is effective for gaining high WWT efficiencies in CE-MFCs. In order to further improve water flow in CE-MFCs, computational fluid dynamics analyses may be necessary.

It has long been predicted that waste sludge produced in MFC during WWT may be much less than that in ASR (Oh et al. 2010; Li et al. 2014), while limited information is available for amounts of sludge produced in MFCs (Zhang et al. 2013). Zhang et al. (2013) monitored suspended solids in MFCs treating municipal wastewater, while amounts of waste sludge produced in MFC were not precisely compared with that in ASR operated under same conditions. To our knowledge, the present study was the first to empirically evaluate waste-sludge production in MFC and ASR operated under same WWT conditions, demonstrating that substantial reduction in waste-sludge production is possible in MFC compared to that in ASR.

Previous studies have demonstrated that excess biofilms, particularly those loosely adhere to reservoir and electrode surfaces, should be removed for maintaining

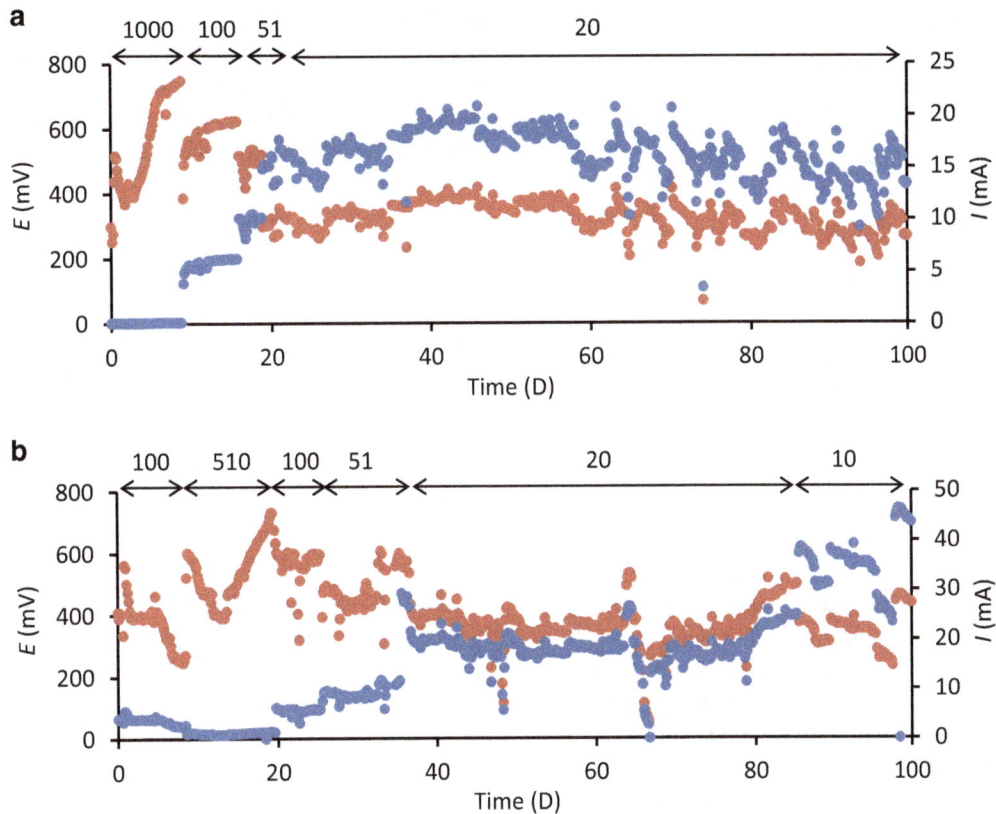

Fig. 4 *E* (*orange*) and *I* (*blue*) during the operation of ADW-MFC (**a**) and AIW-MFC (**b**). *Arrows* and *numbers* above the graphs indicate R_{ext} for these MFCs

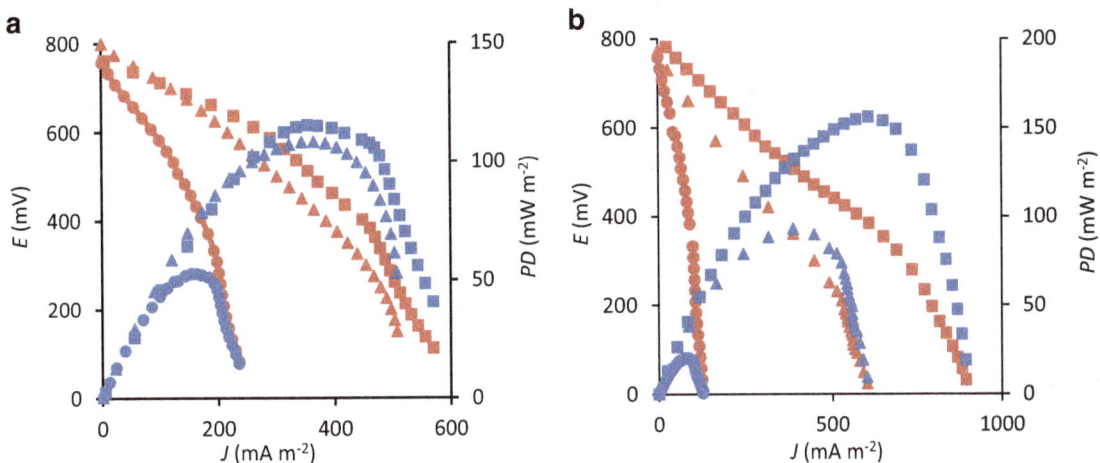

Fig. 5 Representative polarization (*orange*) and power curves (*blue*) for ADW-MFC (**a**) and AIW-MFC (**b**). *Circles* day 20, *triangles* day 60, *squares* day 100

the electricity-generation capacity of continuous-flow MFCs (Miyahara et al. 2013). These biofilms are the major components in waste sludge in MFCs, and it is therefore important to understand how much biofilm is produced in MFCs during WWT, and also how often these biofilms should be removed from MFC inner surfaces. Despite that the present study routinely removed biofilms and discarded them as waste sludge, electricity

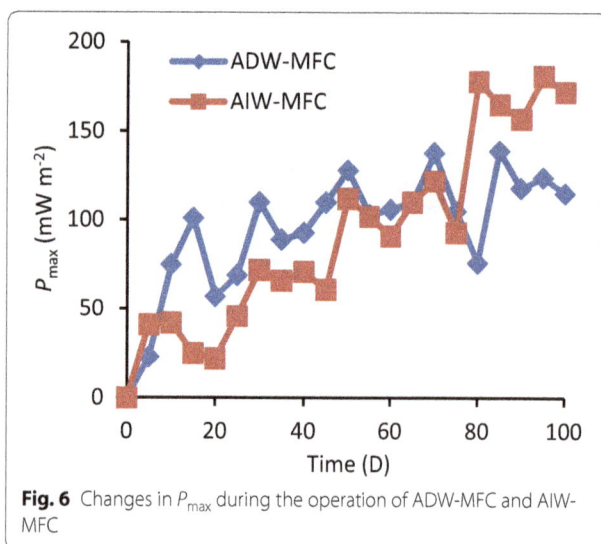

Fig. 6 Changes in P_{max} during the operation of ADW-MFC and AIW-MFC

generation was kept relatively constant in ADW-MFC and AIW-MFC. These results support the idea that excess biofilms (mostly not involved in electricity generation) should be routinely removed for maintaining electric outputs from WWT-MFCs, and the present study provides with a benchmark for sludge-removal practices for WWT-MFCs. Further studies will be necessary to develop protocols to determine an appropriate frequency of sludge removal from WWT-MFCs depending on wastewater qualities.

Concerning electric outputs from MFCs, values for ε_c and P_{max} reported in the present study were similar to those reported in our previous studies despite the different water-flow systems (Miyahara et al. 2013; Yoshizawa et al. 2014). Although the mean ε_c values of 20 and 26% reported in the present study (Table 1) are relatively high compared to those for other WWT-MFC experiments, e.g., approximately 10% for MFC treating municipal wastewater (Zhang et al. 2013), further increases in ε_c are desirable for developing more useful MFCs. Since a previous study has shown that large portions of electrons released from organics by exoelectrogens are consumed by aerobic respiration using oxygen that entered through air cathodes in MFCs (Shimoyama et al. 2008), improvement of air cathodes (e.g., optimization of the oxygen-transfer rate) will be necessary for further improving ε_c in WWT-MFCs.

Conclusions

The present study comparatively evaluated MFCs and ASRs using two different types of wastewater, suggesting that MFCs are attractive WWT facilities in terms of stable organics removal and low waste-sludge production. Along with the unnecessity of electric power for aeration and the generation of power during WWT, the results obtained in the present study further suggest that MFCs

are able to save substantial amounts of energy needed for WWT. Although the possibility for energy saving in wastewater-treatment MFCs has already been suggested (Li et al. 2014), the results presented in this study (in particular, those related to waste-sludge reduction), for the first time, facilitate reasonable estimation of energy saving in wastewater-treatment MFCs. In municipal WWT plants, electric energy is typically used for aeration (approx. 40%), sludge treatment (approx. 30%), and others (30%) (Shi 2011). On the other hand, electric energy generated in MFC during the treatment of typical municipal wastewater (e.g., COD 200 mg L^{-1}) is estimated to be equivalent to 20% of the energy needed for WWT in municipal plants. It is therefore suggested that the use of MFC for WWT can save approximately 80% in total (a sum of 40% from no aeration, 20% from 2/3 sludge reduction, and 20% from electricity generation) of energy consumed in ASR-type WWT plants. For practical application of MFCs to WWT, further studies are necessary for developing technologies for the production of large and cheap electrodes. Experiments using real wastewater and those with short HRT should also be done.

Authors' contributions

YA and MM performed experiments, AK contributed general advice, and KW edited the manuscript. All authors read and approved the final manuscript.

Author details

[1] School of Life Science, Tokyo University of Pharmacy and Life Sciences, Tokyo 192-0392, Japan. [2] Meidensha Corporation, Shinagawa, Tokyo 141-8616, Japan.

Acknowledgements

We thank Nanako Amano for technical assistance.

Competing interests

The authors declare that they have no competing interests.

Funding

This work was supported by the New Energy and Industrial Technology Development Organization (NEDO) of Japan.

References

Cheng S, Liu H, Logan BE (2006) Increased performance of single-chamber microbial fuel cells using an improved cathode structure. Electrochem Commun 8(7):489–494

Eckenfelder WW, O'Conner DJ (1961) Biological waste treatment. Pergamon Press, Oxford

Hall JE (1995) Sewage sludge production, treatment and disposal in the European Union. Water Environ J 9(4):335–343

Imai T, Liu Y, Ukita M, Hung YT (2010) Solubilization of sewage sludge to improve anaerobic digestion. In: Wang LK, Tay JH, Tay STL, Hung YT (eds) Environmental bioengineering. Humana Press, New York, pp 75–122

Li WW, Yu HQ, He Z (2014) Towards sustainable wastewater treatment by using microbial fuel cells-centered technologies. Energy Environ Sci 7(3):911–924

Logan BE, Hamelers B, Rozendal R, Schröder U, Keller J, Freguia S, Aelterman P, Verstraete W, Rabaey K (2006) Microbial fuel cells: methodology and technology. Environ Sci Technol 40(17):5181–5192

Miyahara M, Hashimoto K, Watanabe K (2013) Use of cassette-electrode microbial fuel cell for wastewater treatment. J Biosci Bioeng 115(2):176–181

Miyahara M, Yoshizawa T, Kouzuma A, Watanabe K (2015) Floating boards improve electricity generation from wastewater in cassette-electrode microbial fuel cells. J Water Environ Technol 13(3):221–230

Mizuta K, Shimada M (2010) Benchmarking energy consumption in municipal wastewater treatment plants in Japan. Water Sci Technol 62(10):2256–2262

Oh ST, Kim JR, Premier GC, Lee TH, Kim C, Sloan WT (2010) Sustainable wastewater treatment: how might microbial fuel cells contribute. Biotechnol Adv 28(6):871–881

Rosso D, Larson LE, Stenstrom MK (2008) Aeration of large-scale municipal wastewater treatment plants: state of the art. Water Sci Technol 57(7):973–978

Shi CY (2011) Mass flow and energy efficiency of municipal wastewater treatment plants. IWA Publishing, London

Shimoyama T, Komukai S, Yamazawa A, Ueno Y, Logan BE, Watanabe K (2008) Electricity generation from model organic wastewater in a cassette-electrode microbial fuel cell. Appl Microbiol Biotechnol 80(2):325–330

Watanabe K (2008) Recent developments in microbial fuel cell technologies for sustainable bioenergy. J Biosci Bioeng 106(6):528–536

Yamamuro A, Kouzuma A, Abe T, Watanabe K (2014) Metagenomic analyses reveal the involvement of syntrophic consortia in methanol/electricity conversion in microbial fuel cells. PLoS ONE 9(5):e98425

Yoshizawa T, Miyahara M, Kouzuma A, Watanabe K (2014) Conversion of activated-sludge reactors to microbial fuel cells for wastewater treatment coupled to electricity generation. J Biosci Bioeng 118(5):533–539

Zhang F, Ge Z, Grimaud J, Hurst J, He Z (2013) Long-term performance of liter-scale microbial fuel cells treating primary effluent installed in a municipal wastewater treatment facility. Environ Sci Technol 47:4941–4948

19

Mycoremediation potential of *Pleurotus* species for heavy metals

Meena Kapahi[1,2]* ⓘ and Sarita Sachdeva[1]

Abstract

Mycoremediation is one of the biotechniques that recruits fungi to remove toxic pollutants from environment in an efficient and economical manner. Mushrooms, macro-fungi, are among the nature's most important mycoremediators. *Pleurotus* species (also called oyster mushrooms) are considered to be the most popular and widely cultivated varieties worldwide and this might be attributed to their low production cost and higher yields. Apart from their nutritive and therapeutic properties, *Pleurotus* species have high biosorption potential due to their extensive biomass, i.e. mycelial production. The genus has been reported to accumulate high levels of heavy metals. The current state-of-the art review mainly summarises previous investigations carried out by researchers on different roles and mechanisms played by *Pleurotus* species on heavy metals mycoremediation.

Keywords: *Pleurotus* species, Heavy metals, Biosorption, Mycoremediation, Laccase, Manganese peroxidase

Introduction

Indiscriminate use of chemicals has led to severe contamination of environmental segments by heavy metals. Heavy metals are non-biodegradable and tend to be biomagnified in the food chain (Singh et al. 2008). They pose a risk to human health when transferred via food chain and can further lead to toxic effects in organisms even in trace amounts. These metals can hinder different cellular processes. Their effects are generally concentration dependent and also differ with respect to individual toxicity. Hence, it becomes crucial to remove them prior to final discharge of effluents in environmental segments.

Conventional methods like chemical precipitation, adsorption, ion exchange, reverse osmosis and electrodialysis, to get rid of heavy metal burden of the environment, have their own shortcomings. These methods offer limitations like slow metal precipitation and incomplete removal (Aziz et al. 2015), generation of contaminated sludge requiring careful disposal (Gunatilake 2015; Ayangbenro 2017), high cost involved in the processes (Firdousi 2017), high energy and reagent requirements and clogging of membranes (Ahalya et al. 2003).

In this scenario, it is important to opt for an economically feasible and effective treatment method which is free from these limitations and is able to translate the need of removal of heavy metals in terms of eco-friendly approach. Bioremediation is a way of cleaning up heavy metals using biomass (or microorganisms) through the processes of biodegradation, biosorption, bioaccumulation and bioconversion operating in different ways (Kulshreshtha et al. 2014; Mosa et al. 2016). The microorganisms bind to heavy metals and concentrate them (Joutey et al. 2015). Biosorption is a passive process and heavy metals get adsorbed on the surface of the biosorbent (Velásquez and Dussan 2009) exhibiting the tolerance of biosorbent towards heavy metals. The mechanisms like extracellular (chelation and cell wall binding) and intracellular (binding to compounds like proteins) sequestration of heavy metals have been proposed as mechanisms for heavy metals tolerance in fungi (Fawzy et al. 2017). Biosorbent from mushrooms can be prepared from mycelium or fruit body (live or dead) and spent mushroom substrate (SMS). The factors like the presence of microbial population, the availability of contaminants to these organisms, metal ion concentration and environmental factors like temperature, pH and the presence of nutrients affect the biosorption process in totality (Prakash 2017). The process includes precipitation, ion

*Correspondence: meenakapahi@mru.edu.in
[1] Department of Biotechnology, Manav Rachna International University, Sector 43, Faridabad 121004, India
Full list of author information is available at the end of the article

exchange, electrostatic interaction, the redox process, etc. (Yang et al. 2015).

The biological process of remediation display features like economic viability (Ayangbenro 2017) and repeated use of biomass, selective metal binding, effective desorption and recycling of desorbents. Different microorganisms like algae, bacteria, fungi, yeast have been employed to carry out biosorption. The potential of fungal biomass as biosorbent has been accepted for the removal of heavy metals and radionuclides from polluted waters because of their excellent metal binding properties and tolerance towards metals and adverse environment like diverse pH and temperature conditions (Qazilbash 2004; Anand et al. 2006; Yazdani et al. 2010; Salman et al. 2014). Fungi have been reported to exhibit the ability to chemically modify or affect their bioavailability (Prakash 2017). Fungi have chitin in their walls which can tolerate high concentrations of metals and are capable of growing on medium at low pH and temperature exhibiting excellent mycoremediation potential.

Mushrooms, macro-fungi, have fruiting bodies that grow out of a mass of mycelium. They are a favourite delicacy in many parts of the world. The consumption of edible mushrooms is increasing due to a good content of proteins and trace minerals. Mushrooms have also been reported as nutraceuticals having anti-oxidant, anti-cancer, immunostimulatory, anti-inflammatory and anti-diabetic therapeutic properties (Barros et al. 2007; Kim et al. 2007; Sarikurkcu et al. 2008; Synytsya et al. 2009). These functional characteristics are mainly due to their chemical composition.

Apart from this, mushrooms can be employed for decontamination of the polluted environment. Mushrooms can build up heavy metals in high concentrations in their bodies above maximum permissible concentrations (Kalac and Svoboda 2000) and can act as an effective biosorption tool (Das 2005). High accumulation potential and shorter life span are some of the advantages of using mushrooms as biosorbents. Mushrooms belonging to the genera including *Agaricus, Boletus, Armillaria, Polyporus, Russula, Pleurotus, Termitomyces* have been investigated by some researchers for the uptake of heavy metals (Raj et al. 2011).

Pleurotus species

The genus *Pleurotus*, commonly called Oyster mushroom, is a type of gilled mushrooms which grows normally on wood. It encompasses many species, for example *P. ostreatus, P. pulmonary, P. sajor-caju, P. cornucopiae, P. sapidus, P. platypus* and *P. ostreatoroseus*. It is found all around the world, mainly in forest environments. The genus has enzymes like laccase (LAC) and Mn-peroxidase (MnP), which degrade the lignocellulosic residues into food and enable them to grow on a variety of agricultural wastes with broad adaptability to varied agro-climatic conditions (Agrahar-Murugkar and Subbuakshmi 2005). A number of substrates like wheat straw, corn and sawdust can be used for its cultivation. They are popular and are widely cultivated throughout the world for food owing to simple production technology and their higher biological efficiency (Manzi et al. 2001). The genus is considered to be rich in proteins, fibres, carbohydrates, vitamins and minerals and owns a very pleasant taste. It is rich in immense therapeutic properties (Kalac and Svoboda 2000). There has been a rise in research activities related to the genus because of its multiple uses including biosorption.

Pleurotus species—sequestering heavy metals

Pleurotus species have been found to demonstrate a very effective biosorption potential for a wide range of environmental contaminants including heavy metals (Table 1). The accumulation of heavy metals in the fruit bodies tends to increase with an increase of the metals in the substrate (Ogbo and Okhuoya 2011). Heavy metals have become concentrated in certain areas, such as traffic congested highways, emission areas and cement- and battery-waste polluted sites. *Pleurotus* species growing near these polluted sites have the ability to accumulate heavy metals in high concentrations in their bodies. Mushrooms growing in heavily polluted areas like vicinity of the smelters have been reported to accumulate as much as 1540 times more than the background level of nickel (Barcan et al. 1998). The bioaccumulation potential of *P. ostreatus* from metal scrap sites has also been evaluated for Cu, Fe, Zn and Mn (Boamponsem et al. 2013). However, the accumulation potential of the species varies with the metallic species. Differences in accumulation potential for different heavy metals may be ascribed to the various types of growth substrates found in ecosystems. In a study conducted by Brunnert and Zadražil (1983), more Hg than Cd has been found to be accumulated in fruiting bodies in *P. ostreatus*, while more Cd has been found in *P. flabellatus*. Purkayastha et al. (1994) reported highest uptake of Cu and Cd as compared to Co and acid Hg ions by *P. sajor-caju*. However, the uptake of Cd was reduced in the presence of Cu in *P. sajor-caju* owing to the chemical interference.

Biosorption from the substrates

They have the ability to enhance the nutritional content of the soil found in these areas (Adenipekun 2008) and bioremediate (Radulescu et al. 2010). A considerable decrease in Cu, Mn and Ni in cement-contaminated soil and a slow decrease in lead content of battery-polluted soils in case of *P. pulmonarius* have been observed (Adenipekun et al. 2011). The bioaccumulation potential of Hg by *P. ostreatus* grown on artificial compost has also been studied by Bressa et al. (1988). Uptake and bioaccumulation studies have been done on *Pleurotus* species grown on metal-enriched substrates (Jain et al. 1988, 1989). *Pleurotus sajor-caju*, grown on metal-enriched substrate duckweed, has been found to accumulate Cd content above permissible limits (Jain et al. 1988).

Heavy metals distribution after biosorption

Subsequent to uptake, the metals are distributed unevenly within the fruiting bodies of mushrooms. The highest concentrations have been observed in the spore-forming part followed by rest of the cap and stipe (Gabriel et al. 1996). Cd has been found to be present in higher concentrations in caps (22–56 mg/kg dry wt) than in stipe (13–36 mg/kg dry wt) (Favero et al. 1990a) in *P. ostreatus*. The fruit body production has been found to be unaffected when exposed to a concentration up to 285 mg Cd/kg of dried substrate. Cadmium has been found to be accumulated to a higher concentration of 20 mg/g dry weight in *P. ostreatus* when grown in liquid cultures of malt broth (Favero et al. 1991). *Pleurotus* species have been found to show resistance to high Cd concentrations (Gabriel et al. 1996). Their capacity to accumulate the heavy metals can lead to their immobilisation but ingestion by other organisms can result in transfer along food chain (Osman and Bandyopadhyay 1999). The amounts of Pb, As, Fe, Cd and Hg in *P. ostreatus* available in the market (Accra, Ghana) have been found to be unsafe (Quarcoo and Adotey 2013).

Factors affecting the biosorption process

It has generated interest in the researchers to use the species for biosorption of heavy metals from wastewater. Influence of a range of operational parameters like pH, temperature, biomass and initial metal ion concentrations and contact time have been considered while assessing their biosorption potential. The biosorption by the target species varies with the type of metal, its concentration and composition of substrate (Javaid and Bajwa 2008; Ogbo and Okhuoya 2011). In the biosorption study conducted by Adhikari et al. (2004), *P. florida* has been found to sorb heavy metals in the order of Cd > Cr and to accumulate 1.2–2.5% more Cd than *Fusarium oxysporum*, *Penicillium* species and *Aspergillus awamorii*. The dead biomass can bind metals at levels higher, equivalent to or lesser than live biomass depending on the method used to kill the biomass (Zhu et al. 2010). Boamponsem

et al. (2013) reported that the age of the fungal fruiting body or its size is of less importance in the accumulation of heavy metals by mushrooms. The interval between the fructifications affects the same. *P. florida*, *P. ostreatus* and *P. djamour* recorded the highest maximum accumulation (1.63–2.58 ppm) in the third flush of fructification (Dulay et al. 2015).

Pleurotus florida, *P. ostreatus*, *P. sajor-caju*, *P. djamor*, *P. salmoneo-stramineus* have been reported to be affected by Pb. The concentration of 100 ppm resulted in the lowest mycelial growth (Dulay et al. 2015). *P. ostreatus* have carboxylic, amino, thiol, phosphate and hydroxide groups on the cell wall helping in the biosorption of heavy metals (Banerjee and Nayak 2007; Javaid et al. 2011). IR analysis of lyophilised cells *P. eryngii* revealed the presence of carboxylic, amino, hydroxyl and methyl groups (Joo et al. 2011). *P. ostreatus* and *P. sapidus* have been reported to show affinity towards Cu and Zn as compared to Cd and Pb (Ita et al. 2006, 2008). This is in consensus with reports by Zhu et al. (2010). However, fruiting bodies of *P. ostreatus* immobilised in calcium alginate were shown to be effective in removing Pb and Co from solution (Xiangliang et al. 2005, 2009). *P. ostreatus* displayed tremendous removal potential in the order of Ni > Cu > Cr > Zn ions from effluents of electroplating units (Javaid and Bajwa, 2008). *P. floridianus* and *P. sajor-caju* have been reported to exhibit affinity (biosorption efficiency) in the order of Cd > Zn > Ni > Pb > Cu > Fe (Lamrood and Ralegankar 2013). Uptake of heavy metals by *Azolla* species and its further translocation in *P. sajor-caju* have been studied by Jain et al. (1989). Javaid et al. (2011) conducted a study to assess the biosorption potential of *P. ostreatus* in single and multi-metal ion systems for Cr, Cu, Ni and Zn. Similarly, the SMS biosorbent of the species has been reported to exhibit higher selectivity for Ni than Cu in a bi-metal biosorption study conducted by Tay et al. (2016). *P. sajor-caju* has been demonstrated to remove metals like Cu, Fe, Mg, Mn, Zn (in the pro-degradant additive) on modified polyethylene films (Klein et al. 2012).

Both chemisorption and ion-exchange have been reported to be the involved mechanisms in metals biosorption. Lyophilised cells of *P. eryngii* showed higher bioconcentration values for Pb and Cd (Joo et al. 2011). Studies were conducted on removal of Pb, Zn, Cu and Mn from artificially contaminated soil using *P. tuberregium*. More than 90% of the metals were removed. There was a significant increase in the level of heavy metals in the pileus of the mushroom after biosorption process (Oyetayo et al. 2012). It has been reported to show preference towards Fe, Al, Zn and Mn followed by Pb and Hg (Nnorom et al. 2013). It has further been reported that *P. tuber-regium* has more bioaccumulative properties when grown from spawn rather than from sclerotia (Oghenekaro et al. 2008). In the packed bed column study on Cd employing *P. platypus* using industrial wastewater, the effect of parameters like bed depth and flow rate has been assessed (Vimala et al. 2011a). Biosorption of Cd by *P. mutilus* in packed bed column has also been done by Khitous et al. (2015). The packed biosorbent can be used for three regeneration cycles. *Pleurotus* SMS has been employed in a fixed bed study to remove Mn(II) ions from aqueous solutions. Flow rate of 1 ml/min, bed height of 30 cm, and metal ion concentration of 10 mg/l have been found to be suitable for biosorption (Kamarudzaman et al. 2015). *Pleurotus* species have also been assessed for the removal of different heavy metals from chemical laboratory waste in the form of live mycelia (Arbanah et al. 2012, 2013). The highest biosorption efficiency for Fe and Cu has been found to be 80.52 and 45.20%, respectively (Arbanah et al. 2012). In a similar study conducted by Akyüz and Kirbağ (2010), *P. eryngii* grown on various agro-wastes has been reported to show maximum uptake of K and the lowest of Cu contents.

The pH values of a solution should be considered as an important factor impacting the biosorption process. The pH influences the toxicity and solution chemistry of the heavy metals (Frutos et al. 2016), hydrolysis and complexation properties by bringing changes in ionic form (Deng et al. 2009). Hence, the ionic charge of the functional groups and the metal speciation at varied pH values affect biosorption process. Under acidic environment, positively charged metal ions get attached to the negatively charged biomass. Under high pH, metal ions precipitate as metal hydroxides (Hlihor et al. 2014). The optimum pH for live and heat-inactivated *P. sapidus* encapsulated in calcium alginate beads has been found to be 6 (Yalçinkaya et al. 2002). In a study assessing the potential of *P. ostreatus* as a biosorbent in removing Pb(II) from electroplating industrial wastewater, the maximum Pb(II) biosorption of 92% in aqueous solution has been achieved at an unadjusted pH of 5.2 (Tay et al. 2009). Similarly, pH range of 2.5–6 for the biosorption of Ni, Zn, Cr, Cu, Fe and Pb has been reported for *P. ostreatus* (Arbanah et al. 2012; Osman and Bandyopadhyay 1999; Tay et al. 2010). Tay et al. (2010) also carried out a study regarding the removal of Pb and Cu ions from aqueous solution. Cu(II) removal sharply increased from 38.21% at pH 2.0 to 81% at pH 5.0 in *P. cornucopiae* as reported by Danış (2010). The maximum biosorption of Pb(II) by *P. ferulae* with pH up to 3, temperature 30°, and initial metal concentration 100 mg/l has been reported by Adebayo (2013). Optimum biosorption of divalence

cations [Ni(II) and Cu(II)] by *Pleurotus* mushroom SMS has also been reported to be between pH of 5 and 6 (Tay et al. 2012). Pre-concentration and determination of Cd(II) and Co(II) in vegetables, using *P. eryngii* immobilised on Amberlite XAD-16 as a solid-phase biosorbent, have also been reported by Özdemir et al. (2012). The optimum extraction conditions were determined at a pH of 6.0 for Cd(II) and 5.0 for Co(II). In a similar study, pH range of 4–5 has been optimised for *P. ostreatus* immobilised on Amberlite XAD-4 for the biosorption of Cr, Cd and Cu (Kocaoba and Arısoy 2011). In the research on hybrid of *P. sajor-caju* and sunflower waste biomass immobilised on sodium alginate, the maximum equilibrium uptake for lead was found to be at pH 4.5 (Majeed et al. 2012). *P. cornucopiae* has been used to remove Cr from aqueous solution with bubbling fluidised bed (Xu et al. 2016).

Pre-treating the biomass with heat, alkalies or acids has a significant effect on the biosorption process depending upon the type of metal and fungal species. Pre-treatment of living biomass by physical and chemical methods resulted in an improvement in cadmium biosorption in comparison with living biomass of *P. florida* (Das et al. 2007). Methods like freeze drying (FD), oven drying (OD) and sun drying (SD) have been used for *P. ostreatus* for analysing the contents of different heavy and trace elements. Among the detected elements, K ranked the highest by 2.59, 1.31 and 2.30% in FD, OD and SD samples, respectively. OD biomass of *P. ostreatus* showed an increase in removal rate on increasing metal ion concentration (Javaid and Bajwa 2007). The other conditions affecting the biosorption as reported are ionic strength, other ions and complexing agents. The presence of high ionic strength and appreciable quantities of a complexing agent like EDTA significantly reduce the Pb(II) removal (Osman and Bandyopadhyay 1999).

Heavy metals vis-a-vis effects

The uptake of heavy metals has its consequent deleterious effects on the growth, productivity and cellular proteins. Gabriel et al. (1996) reported fructification of *Pleurotus* species in Cd-contaminated environment. Baldrian et al. (2000) demonstrated inhibition of mycelial penetration into soil by Cd and Hg. Effect of Hg on the highest cadmium uptake (between 88.9 and 91.8%) was observed with aerobic fungal biomass from the exponential growth phase in *P. sajor*-caju (Cihangir and Saglam 1999). Cadmium up to 150 µg/ml slowly inhibited mycelia development in case of *P. ostreatus* but never blocked it completely (Favero et al. 1991). Effect of Hg on the growth of wood-rotting basidiomycetes including *P.*

ostreatus was studied by Mandal et al. (1998). The growth of the mushroom was significantly inhibited. Purkayastha et al. (1994) reported more than 85% reduction of growth in *P. sajor-caju* at 15 and 6 µg/ml of Pb(II) and Hg(II), respectively. Pb reduced mycelial protein significantly (36%), but Hg caused maximum reduction (30%) of proteins in sporocarps. Pb reduced biological efficiency of sporocarp production. Mercury has been reported to prevent growth and fruit body production in *P. tuberregium*, while stipe length, stipe diameter and cap diameter were affected by lead followed by cadmium (Akpaja et al. 2012). Mineral (Fe, Zn, Li) enrichment reduced anti-oxidant activity in *P. ostreatus* owing to polyphenol complexation with these elements leading to decreased free radical availability (Fontes et al. 2013).

Heavy metals and enzyme regulation

The saprotrophic basidiomycetes utilise a variety of extracellular enzymes including ligninolytic enzymes for the utilisation of complex nutrients (Kapoor and Viraraghavan 1998). Factors controlling enzyme production among white rot fungi have also been widely studied. The main factors that influence the enzyme production are the nutrients, inhibitory compounds, temperature and interrelationships with other fungi (Baldrian and Gabriel 2002). Extracellular ligninolytic and cellulolytic enzymes are regulated by heavy metals on transcription level and during the course of their action. The effect of the heavy metals on enzymatic activities influences the energy flux in the ecosystem. In a study, a positive regulation of laccase and isoenzymes on copper application has been reported in the case of *P. ostreatus* (Baldrian and Gabriel 2002; Palmieri et al. 2000). The Mn-peroxidase activity decreased with increasing Cd concentration, whereas activities of endo-1,4-L-glucanase, 1,4-L-glucosidase and laccase highly increased in the presence of metal (Baldrian and Gabriel 2003). It has been reported that the *P. sajor-caju* laccase isozyme genes (phenol oxidase A1b (POXA1b), POXA2 and POXC) are differentially regulated at the transcriptional level in response to copper and manganese (Collins and Dobson 1997; Soden and Dobson 2001). The addition of Hg has been found to decrease the activity of laccase immediately and reduce the stability of the enzyme (Collins and Dobson 1997; Baldrian and Gabriel 2002). Interestingly, Cu and also Hg increased MnP activity slightly. However, when incubated in the presence of all three metals, the activity of MnP decreased even at low concentrations of Cd, Cu and Hg (Baldrian and Gabriel 2002) showing the synergetic effect of the heavy metals. Manganese has also been found to affect MnP gene transcription and enzyme activity in a positive way in some fungi like *Pleurotus*

spp. (Ruiz-Duenas et al. 1999). A study was conducted by Drzewiecka et al. (2010) to assess the effect morphology and physiology of *Pleurotus eryngii* after incubating the spawn in the Zn-, Cu-, Co-, Cd- and Ni-enriched substrate. Laccase activity was stimulated by Ni and Cu even at low concentrations during incubation stage; but inhibited during fruiting stage. The inhibition effect was more pronounced when exposed to multi-metal solution.

To consider a fungal species as a biosorbent, desorption of the adsorbed metal ions and subsequent reuse and efficiency of the biomass in biosorption need to be taken into account. Acidic solution desorption has been reported to be more effective than alkaline solution desorption (Prasad et al. 2013). Under acidic conditions, protons compete for the sites releasing metal ions in the medium. Ninety-seven percent desorption of the adsorbed Hg from immobilised and heat-treated *P. sajor-caju* resulted when eluted with HCl (Arica et al. 2003). Ninety-nine percent of lead could be desorbed from *P. ostreatus* using HCl for a contact period of 1 h. The used biomass of *P. florida* could be regenerated and reused for biosorption of lead for six times (Prasad et al. 2013). A regeneration rate of 59% of Cu has been reported for *P. mutilus* (Henini et al. 2011). However, they can be improved by coupling the chemical desorption method with a copper recovery; the regenerated biomass for a content 10 g/l has a maximum adsorption capacity smaller but still significant 59.75 mg/g.

Conclusion

Different methods are being adopted to remove heavy metals from wastewater. Keeping in mind the financial aspects, it is necessary to produce low-cost, effective and recyclable adsorbents for their widespread use. There are some limitations of using mushrooms for biosorption. Biosorption potential of different species is also being assessed in a comparative way. Looking at the amount of work done on *Pleurotus* spp., the species holds a promise to be used as a biosorbent for heavy metals. The degree of tolerance is different for the species for different heavy metals. For performance assessment studies in the future, multi-component sorption studies should be stressed upon as the industrial wastewater is a cocktail of metal ions in solution and that plays an important role in the sorption efficiency of the species. The biosorption potential of the species is yet to be tapped and used commercially. Mushrooms being a food crop and looking at the potential of mushroom mycelia, the SMS produced after harvesting the mushroom can be used for the mycoremediation of the degraded sites. The aged mycelia, SMS, are otherwise generated in huge amounts by the mushroom farms and pose a disposal problem.

Table 1 Previous contributions of heavy metals biosorption using different forms of *Pleurotus* species

Biosorbent type	*Pleurotus* species	Heavy metals	References
Oven- and freeze-dried, autoclaved mycelia	*P. florida*	Cd	Das et al. (2007)
Oven-dried mycelia	*P. ostreatus*	Cr	Javaid and Bajwa (2007), Puentes-Cárdenas et al. (2012)
		Pb	Tay et al. (2009), Liew et al. (2010)
		Cd	Tay et al. (2011)
		Cu, Cr, Ni, Zn	Javaid et al. (2011)
	P. florida	Pb	Prasad et al. (2013)
Live mycelia	*P. ostreatus*	Cd	Favero et al. (1990a, b)
		Hg	Mandal et al. (1998)
		Cu, Cr, Ni, Zn	Javaid and Bajwa (2008)
		Cu, Cr, Fe, Zn	Arbanah et al. (2012)
		Cr	Arbanah et al. (2013)
	P. ostreatus, P. florida, P. djamour, P. salmoneo-stramineus, P. cystidiosus	Pb	Dulay et al. (2015)
	P. eryngii	Mn	Wu et al. (2016)
	P. floridianus, P. sajor-caju	Cu, Cd, Fe, Ni, Pb, Zn	Lamrood and Ralegankar (2013)
Biomass immobilised on calcium alginate	*P. ostreatus*	Pb	Xiangliang et al. (2005)
		Co	Xiangliang et al. (2009)
		Cu, Pb	de Almeida and Burgess (2013)
	P. sapidus	Cd, Hg	Yayçinkaya et al. (2002)
	P. sajor-caju and sunflower waste biomass hybrid	Pb	Majeed et al. (2012, 2014)
Biomass immobilised on XAD-4	*P. ostreatus*	Cu, Cr, Cd	Kocaoba and Arisoy (2011)
	P. eryngii	Cd, Co	Özdemir et al. (2012)
SMS	*P. ostreatus*	Cu	Tay et al. (2010)
		Cr	Carol et al. (2012)
		Cu, Ni	Tay et al. (2012, 2016)
		Cd, Pb, Cu	Frutos (2016)
Fruit body accumulation	*P. ostreatus*	Cd	Favero et al. (1990a, b)
		Hg	Bressa et al. (1988)
	P. cornucopiae	Cu	Danis (2010)
		Cr	Xu et al. (2016)
	P. platypus	Cd	Vimala and Das (2011b)
	P. ostreatus, P. tuber-regium.	Hg	Nnorom et al. (2012)
	P. ferulae	Pb	Adebayo (2013)
	P. ostreatus, P. florida, P. djamour, P. salmoneo-stramineus, P. cystidiosus	Pb	Dulay et al. (2015)
	P. eryngii	Pb	Jiang et al. (2016)
	P. ostreatus	Pb	Jiang et al. (2017)
Sun-dried fruit	*P. ostreatus*	Pb	Osman and Bandyopadhyay (1999)
Oven-dried fruit	*P. ostreatus*	Cu	Huo et al. (2011)
		Cu, Pb, Zn, Mn	Oyetayo et al. (2012)
	P. platypus	Cd	Vimala and Das (2011b)
	P. eous	Cr, Ni, Pb	Suseem and Mary Saral (2014)
Freeze-dried fruit	*P. eryngii*	Cd, Pb	Joo et al. (2011)

Authors' contributions
The authors (MK and SS) have made substantial contributions to conception and design, or acquisition of data, or analysis and interpretation of data, and they have been involved in drafting the manuscript or revising it critically for important intellectual content. The authors have given final approval of the version to be published. Each author should have participated sufficiently in the work to take public responsibility for appropriate portions of the content, and agreed to be accountable for all aspects of the work in ensuring that questions related to the accuracy or integrity of any part of the work are appropriately investigated and resolved. Both authors read and approved the final manuscript.

Author details
[1] Department of Biotechnology, Manav Rachna International University, Sector 43, Faridabad 121004, India. [2] Manav Rachna University, Sector 43, Faridabad 121004, India.

Acknowledgements
NA.

Competing interests
The authors declare that they have no competing interests. The Editor may ask for further information relating to competing interests.

References

Adebayo AO (2013) Investigation on *Pleurotus ferulae* potential for the sorption of Pb(II) from aqueous solution. Bull Chem Soc Ethiop 27:25–34

Adenipekun CO (2008) Bioremediation of engine-oil polluted soil by *Pleurotus tuber-regium* Singer, a Nigerian white-rot fungus. Afr J Biotechnol 7:55–58

Adenipekun CO, Ogunjobi AA, Ogunseye AO (2011) Management of polluted soils by a white-rot fungus: *Pleurotus pulmonarius*. Assumption Univ J Technol 15:57–61

Adhikari T, Manna MC, Singh MV, Wanjari RH (2004) Bioremediation measure to minimize heavy metals accumulation in soils and crops irrigated with city effluent. J Food Agric Environ 2(1):266–270

Agrahar-Murugkar D, Subbuakshmi G (2005) Nutritional value of edible wild mushrooms collected from the Khasi hills of Meghalaya. Food Chem 89:599–603

Ahalya N, Ramachandra TV, Kanamadi RD (2003) Biosorption of heavy metals. Res J Chem Environ 7(4):71–79

Akpaja EO, Nwogu NA, Odibo EA (2012) Effect of some heavy metals on the growth and development of *Pleurotus tuber-regium*. Mycosphere 3:57–60

Akyüz M, Kirbað S (2010) Element contents of *Pleurotus eryngii* (DC. ex Fr.) Quel. var. eryngii grown on some various agro-wastes. Ekoloji 19(74):10–14

Anand P, Isar J, Saran S, Saxena RK (2006) Bioaccumulation of copper by *Trichodermaviride*. Bioresour Technol 97:1018–1025

Arbanah M, Miradatul Najwa MR, Ku Halim KH (2012) Biosorption of Cr(III), Fe(II), Cu(II), Zn(II) ions from liquid laboratory chemical waste by *Pleurotus ostreatus*. Int J Biotechnol Wellness Ind 1:152–162

Arbanah M, Miradatul Najwa MR, Ku Halim KH (2013) Utilization of *Pleurotus ostreatus* in the removal of Cr(VI) from chemical laboratory waste. Int Refreed J Eng Sci 2(4):29–39

Arica MY, Arpa C, Kaya B (2003) Comparative biosorption of mercuric ions from aquatic systems by immobilized live and heat-inactivated *Trametes versicolor* and *Pleurotus sajur-caju*. Bioresour Technol 89:145–154

Ayangbenro Babalola (2017) A new strategy for heavy metal polluted environmental a review of microbial biosorbents. Int J Environ Res Public Health 14:94

Aziz HA, Adlan MN, Ariffin KS (2015) Heavy metals (Cd, Pb, Zn, Ni, Cu and Cr(III)) removal from water in Malaysia: post treatment by high quality limestone. Bioresour Technol 99(6):1578–1583

Baldrian P, Gabriel J (2002) Copper and cadmium increase laccase activity in *Pleurotus ostreatus*. FEMS Microbiol Lett 206:69–74

Baldrian P, Gabriel J (2003) Lignocellulose degradation by *Pleurotus ostreatus* in the presence of cadmium. FEMS Microbiol Lett 220:235–240

Baldrian P, In Der Wiesche C, Gabriel J, Nerud F, Zadraz̆il F (2000) Influence of cadmium and mercury on activities of ligninolytic enzymes and degradation of polycyclic aromatic hydrocarbons by *Pleurotus ostreatus* in soil. Appl Environ Microbiol 66:2471–2478

Banerjee A, Nayak D (2007) Biosorption of no-carrier-added radio-nuclides by calcium alginate beads using 'tracer packet' technique. Bioresour Technol 98:2771–2774

Barcan VS, Kovnatsky EF, Smetannikova MS (1998) Absorption of heavy metals in wild berries and edible mushrooms in an area affected by smelter emissions. Water Air Soil Pollut 103:173–195

Barros L, Baptista P, Estevinho LM, Ferreira ICFR (2007) Bioactive properties of the medicinal mushroom *Leucopaxillus giganteus* mycelium obtained in the presence of different nitrogen sources. Food Chem 105:179–186. doi:10.1016/j.foodchem.2007.03.063

Boamponsem GA, Obeng AK, Osei-Kwateng M, Badu AO (2013) Accumulation of heavy metals by *Pleurotus ostreatus* from soils of metal scrap sites. Int J Curr Res Rev 5(4):01–09

Bressa G, Coma L, Costa P (1988) Bioaccumulation of Hg in the mushroom *Pleurotus ostreatus*. Ecotoxicol Environ Safe 16:85–89

Brunnert H, Zadraz̆il F (1983) The translocation of mercury and cadmium into the fruiting bodies of six higher fungi. A comparative study on species specificity in five lignocellulolytic fungi and the cultivated mushroom *Agaricus bosporus*. Eur J Appl Micorbiol Biotechnol 17:358–364

Carol D, Kingsley SJ, Vincent S (2012) Hexavalent chromium removal from aqueous solutions by *Pleurotus ostreatus* spent biomass. Int J Eng Sci Technol 4(1):7–22

Cihangir N, Saglam N (1999) Removal of cadmium by *Pleurotus sajor-caju* basidiomycetes. Acta Biotechnol 19:171–177

Collins PJ, Dobson A (1997) Regulation of laccase gene transcription in *Trametes versicolor*. Appl Environ Microbiol 63:3444–3450

Danış Ü (2010) Biosorption of copper(II) from aqueous solutions by *Pleurotus cornucopiae*. BALWOIS 2010, Ohrid, Republic of Macedonia, 25–29 May 2010

Das N (2005) Heavy metals biosorption by mushrooms. Indian J Natl Prod Resour 4:454–459

Das N, Charumathi D, Vimala R (2007) Effect of pretreatment on Cd^{2+} biosorption by mycelia biomass of *Pleurotus florida*. Afr J Biotechnol 6:2555–2558

de Almeida LK, Burgess JE (2013) Biosorption and bioaccumulation of copper and lead by *Phanerochaete* and *Pleurotus ostreatus*. http://www.ewisa.co.za/literature/files/182_133%20Burgess.pdf. Accessed 20 June 2016

Deng L, Zhang Y, Qin J, Wang X, Zhu X (2009) Biosorption of Cr(VI) from aqueous solutions by nonliving green algae *Cladophora albida*. Miner Eng 22:372–377

Drzewiecka K, Siwulski M, Mleczek M, Golinski P (2010) The influence of elevated heavy metals content in substrate on morphology and physiology of King Oyster mushroom (*Pleurotus eryngii*) effects on human health. In: 15th International conference on heavy metals in the environment. http://www.chem.pg.gda.pl/ichmet/

Dulay RMR, De Castro MAEG, Coloma NB, Bernardo AP, Cruz AGD, Tiniola RC, Kalaw SP, Reyes RG (2015) Effects and myco-remediation of lead (Pb) in five *Pleurotus* mushrooms. Int J Biol Pharm Allied Sci 4(3):1664–1677

Favero N, Bressa G, Costa P (1990a) Response of *Pleurotus ostreatus* to cadmium exposure. Ecotoxicol Environ Safe 20(1):1–6

Favero N, Costa P, Paolo Rocco G (1990b) Role of copper in cadmium metabolism in the basidiomycetes *Pleurotus ostreatus*. Comp Biochem Physiol Part C Comp Pharmacol 97(2):297–303

Favero N, Costa P, Massimino ML (1991) In vitro cadmium uptake by basidiomycetes *Pleurotus ostreatus*. Biotechnol Lett 13:701–704

Fawzy EM, Abdel-Motaal FF, EL-zayat SA (2017) Biosorption of heavy metals onto different eco-friendly substrates. J Toxicol Environ Health Sci 9(5):35–44

Firdousi SA (2017) Bioaccumulation and bio-absorptions of heavy metals by the mushroom from the soil. J Med Chem Drug Discov 2(3):25–33

Fontes Vieira PA, Gontijo DC, Vieira BC, Fontes EAF, Soares de Assunção L, Leite JPV, Oliveira MGdA, Kasuya MCM (2013) Antioxidant activities, total phenolics and metal contents in *Pleurotus ostreatus* mushrooms enriched with iron, zinc or lithium. LWT Food Sci Technol 54(2):421–425

Frutos I, García-Delgado C, Gárate A, Eymar E (2016) Biosorption of heavy metals by organic carbon from spent mushroom substrates and their raw materials. Int J Environ Sci Technol 13(11):2713–2720

Gabriel J, Capelari M, Rychlovský P, Krenželok M, Zadražil F (1996) Influence of cadmium on the growth of *Agrocybe perfecta* and two *Pleurotus* spp. and translocation from polluted substrate and soil to fruit bodies. Toxicol Environ Chem 56:141–146

Gunatilake SK (2015) Methods of removing heavy metals from industrial wastewater. J Multidiscip Eng Sci Stud 1(1):12–18

Henini G, Laidani Y, Fatiha Souahi F (2011) Study of adsorption of copper on biomass *Pleurotus mutilus* and the possibility of its regeneration by desorption. Energy Proced 6:441–448

Hlihor RM, Bulgariu L, Sobariu DL, Diaconu M, Tavares T, Gavrilescu M (2014) Recent advances in biosorption of heavy metals: support tools for biosorption equilibrium, kinetics and mechanism. Rev Roum Chim 59:527–538

Huo C-L, Shang Y-Y, Zheng J-J, He R-X, He XS (2011) The adsorption effect of three mushroom powder on Cu^{2+} of low concentration.

Ita BN, Essien JP, Ebong GA (2006) Heavy metal levels in fruiting bodies of edible and non-edible mushrooms from the Niger delta region of Nigeria. J Agric Soc Sci 2:84–87

Ita BN, Ebong GA, Essien JP, Eduok SI (2008) Bioaccumulation potential of heavy metals in edible fungal sporocarps from the Niger delta region of Nigeria. Pak J Nutr 7:93–97

Jain SK, Gujral GS, Jha NK, Vasudevan P (1988) Heavy metal uptake by Pleurotus sajor-caju from metal-enriched duckweed substrate. Biol Wastes 24:275–282

Jain SK, Gujral GS, Vasudevan P, Jha NK (1989) Uptake of heavy metals by Azolla pinnata and their translocation onto the fruit bodies of Pleurotus sajor-caju. J Ferment Bioeng 68(1):64–67

Javaid A, Bajwa R (2007) Biosorption of Cr(III) ions from tannery wastewater by Pleurotus ostreatus. Mycopathologia 5:71–79

Javaid A, Bajwa R (2008) Biosorption of electroplating heavy metals by some basiodiomycetes. Mycopathologia 6:1–6

Javaid A, Bajwa R, Shafique U, Anwar J (2011) Removal of heavy metals by adsorption on Pleurotus ostreatus. Biomass Bioenergy 35:1675–1682

Jiang Y, Hao R, Yang S (2016) Equilibrium and kinetic studies on biosorption of Pb(II) by common edible macrofungi: a comparative study. Can J Microbiol 62(4):329–337

Jiang Y, Has R, Yang S (2017) Natural bioaccumulation of heavy metals onto common edible macrofungi and equilibrium and kinetic studies on biosorption of Pb(II) to them. Acta Nat Univ Pekin 53(1):125–134

Joo JH, Hussein KA, Hassan SHA (2011) Biosorptive capacity of Cd(II) and Pb(II) by lyophilized cells of Pleurotus eryngii. Korean J Soil Sci Fert 44:615–624

Joutey NT, Savel H, Bahafid W, El Ghachtouli N (2015) Mechanism of hexava-lent chromum resistance and removal by microorganisms. Rev Environ Contam Toxicol 233:45–69

Kalac P, Svoboda L (2000) A review of trace element concentrations in edible mushrooms. Food Chem 69:273–281

Kamarudzaman AN, Tay CC, Amir A, Talib SA (2015) Biosorption of Mn(II) ions from aqueous solution by Pleurotus spent mushroom compost in a fixed-bed column. Proc Soc Behav Sci 195:2709–2716

Kapoor A, Viraraghavan T (1998) Biosorption of heavy metals on Aspergillus niger effect of pretreatment. Bioresour Technol 63:109–113

Khitous M, Moussous S, Selatnia A, Kherat M (2015) Biosorption of Cd(II) by Pleurotus mutilus biomass in fixed-bed column: experimental and breakthrough curve analysis. Desalination Water Treat 57(35):16559–16570. doi:10.1080/19443994.2015.1081625

Kim HY, Yoon DH, Lee WH, Han SK, Shrestha B, Kim CH, Lim MH, Chang W, Lim S, Choi S, Song WO, Sung JM, Hwang KC, Kim TW (2007) Phellinus linteus inhibits inflammatory mediators by suppressing redox-based NF-jB and MAPKs activation in lipopolysaccharide-induced RAW 264.7 macrophage. J Ethnopharmacol 114:307–315

Klein JM, Anziliero S, Camassola M, Grisa AMC, Brandalise RN, Zeni M (2012) Evaluation of metal biosorption by the fungus Pleurotus sajor-caju on modified polyethylene films. J Bioremed Biodeg 3:152. doi:10.4172/2155-6199.1000152:5

Kocaoba S, Arısoy M (2011) The use of a white rot fungi (Pleurotus ostreatus) immobilized on Amberlite XAD-4 as a new biosorbent in trace metal determination. Bioresour Technol 102:8035–8039

Kulshreshtha S, Mathur N, Bhatnagar P (2014) Mushroom as a product and their role in mycoremediation. AMB Express 4:29. doi:10.1186/s13568-014-0029-8

Lamrood PY, Ralegankar SD (2013) Biosorption of Cu, Zn, Fe, Cd, Pb and Ni by non treated biomass of some edible mushrooms. Asian J Exp Biol 4(2):190–195

Liew HH, Tay CC, Yong SK, Surif S, Abdul Talib S (2010) Biosorption character-istics of lead [Pb(II)] by Pleurotus ostreatus biomass. In: Abstracts of the proceedings of international conference on science and social research (CSSR), Kuala Lumpur, 2010

Majeed A, Jilani MI, Nadeem R, Hanif MA, Ansari TM (2012) Novel studies for the development of hybrid biosorbent. Int J Chem Biochem Sci 2:78–82

Majeed A, Jilani MI, Nadeem R, Hanif MA, Ansari TM (2014) Adsorption of Pb(II) using novel Pleurotus sajor-caju and sunflower hybrid biosorbent. Environ Prot Eng 40(2):5–15

Mandal TK, Baldrian P, Gabriel J, Nerud F, Zadraz̆il F (1998) Effect of mercury on the growth of wood-rotting basidiomycetes Pleurotus ostreatus, Pycnoporus cinnabarinus and Serpula lacrymans. Chemosphere 36(3):435–440

Manzi P, Aguzzi A, Pizzoferrato L (2001) Nutritional value of mushrooms widely consumed in Italy. Food Chem 73:321–325

Mosa KA, Saadoun I, Kumar K, Helmy M, Dhankher OP (2016) Potential biotechnological strategies for the cleanup of heavy metals and metal-loids. Front Plant Sci 7:1–14. doi:10.3389/fpls.2016.00303

Nnorom IC, Jarzyńska G, Falandysz J, Drewnowska M, Okoye I, Oji-Nnorom CG (2012) Occurrence and accumulation of mercury in two species of wild grown Pleurotus mushrooms from southeastern Nigeria. Ecotoxicol Environ Safe 84:78–83

Nnorom IC, Jarzyńska G, Drewnowska M, Dryżałowska A, Kojta A, Pankavec S, Falandysz J (2013) Major and trace elements in sclerotium of Pleurotus tuber-regium (Ôsŭ) mushroom—dietary intake and risk in southeastern Nigeria. J Food Compos Anal 29(1):73–81

Ogbo EM, Okhuoya JA (2011) Bio-absorption of some heavy metals by Pleurotus tuber-regium Fr. Singer (an edible mushroom) from crude oil polluted soils amended with fertilizers and cellulosic wastes. Int J Soil Sci 6:34–48

Oghenekaro AO, Okhuoya JA, Akpaja EO (2008) Growth of Pleurotus tuber-regium (Fr) Singer on some heavy metal-supplemented substrates. Afr J Microbiol Res 2:268–271

Osman MS, Bandyopadhyay M (1999) Bioseparation of lead ions from waste-water by using a fungus P. ostreatus. J Civil Eng 27:183–196

Oyetayo VO, Adebayo AO, Ibileye A (2012) Assessment of the biosorption potential of heavy metals by Pleurotus tuber-regium. Int J Adv Biol Res 2:293–297

Özdemir S, Okumuşa V, Kılınçb E, Bilgetekinc H, Dündara A, Ziyadanoğ̂ullarıb B (2012) Pleurotus eryngii immobilized Amberlite XAD-16 as a solid-phase biosorbent for preconcentrations of Cd^{2+} and Co^{2+} and their determi-nation by ICP-OES. Talanta 99:502–506

Palmieri G, Giardina P, Bianco C, Bianca F, Sannia G (2000) Copper induction of lactase isoenzymes in the ligninolytic fungus Pleurotus ostreatus. Environ Microbiol 66(3):920–924

Prakash V (2017) Mycoremediation of environmental pollutants. Int J Chem Tech Res 10(3):149–155

Prasad ASA, Varatharaju G, Anushri C, Dhivyasree S (2013) Biosorption of lead by Pleurotus florida and Trichoderma viride. Br Biotechnol J 3(1):66–78

Puentes-Cárdenas IJ, Pedroza-Rodríguez AM, Navarrete-López M, Villegas-Garrido TL, Cristiani-Urbina E (2012) Biosorption of trivalent chromium from aqueous solutions by Pleurotus ostreatus biomass. Environ Eng Manag J 11(10):1741–1752

Purkayastha RP, Mitra AK, Bhattacharyya B (1994) Uptake and toxicological effects of some heavy metals on Pleurotus sajor-caju (Fr.) Singer. Ecotoxi-col Environ Safe 27:7–13

Qazilbash AA (2004) Isolation and characterization of heavy metal tolerant biota from industrially polluted soils and their role in bioremediation. Biol Sci 41:210–256

Quarcoo A, Adotey G (2013) Determination of heavy metals in Pleurotus ostrea-tus (Oyster mushroom) and Termitomyces clypeatus (Termite mushroom) sold on selected markets in Accra, Ghana. Mycosphere 4(5):960–967

Radulescu C, Stihi C, Busuioc G, Gheboianu AI, Popescu IV (2010) Studies con-cerning heavy metals bioaccumulation of wild edible mushrooms from industrial area by using spectrometric techniques. Bull Environ Contam Toxicol 84:641–646

Raj DD, Mohan B, Vidya Shetty BM (2011) Mushrooms in the remediation of heavy metals from soil. Int J Environ Pollut Control Manag 3(1):89–101

Ruiz-Duenas FJ, Guille´n F, Camarero S, Pe´rez-Boada M, Martı´nez MJ, Martı´nez AT (1999) Regulation of peroxidase transcript levels in liquid cultures of the ligninolytic fungus Pleurotus eryngii. Appl Environ Micro-biol 65:4458–4463

Salman HA, Ibrahim MI, Tarek MM, Abbas HS (2014) Biosorption of heavy metals—a review. J Chem Sci Technol 3(4):74–102

Sarikurkcu C, Tepe B, Yamac M (2008) Evaluation of the antioxidant activity of four edible mushrooms from the Central Anatolia, Eskisehir—Turkey: Lactarius deterrimus, Suillus collitinus, Boletus edulis, Xeroco-mus chrysenteron. Bioresour Technol 99:6651–6655. doi:10.1016/j.biortech.2007.11.062

Singh J, Kant K, Sharma HB, Rana KS (2008) Bioaccumulation of cadmium in tissues of Cirrihna mrigala and Catla catla. Asian J Exp Sci 22:411–414

Soden DM, Dobson ADW (2001) Differential regulation of laccase gene expres-sion in Pleurotus sajor-caju. Microbiology 147:1755–1763

Suseem SR, Mary Saral A (2014) Biosorption of heavy metals using Pleurotus eous. J Chem Pharm Res 6(7):2163–2168

Synytsya A, Mickova K, Synytsya A, Jablonsky I, Spevacek J, Erban V (2009) Glucans from fruit bodies of cultivated mushrooms *Pleurotus ostreatus* and *Pleurotus eryngii*: structure and potential prebiotic activity. Carbohydr Polym 76:548–556. doi:10.1016/j.carbpol.2008.11.02

Tay CC, Redzwan G, Liew HH, Yong SK, Surif S, Abdul-Talib S Copper (II) (2010) Biosorption characteristic of *Pleurotus* spent mushroom compost. In: International conference on science and social research (CSSR 2010), Kuala Lumpur, Malaysia, Dec 5–7, 2010

Tay CC, Liew HH, Yong SK, Surif S, Abdul-Talib S (2009) Biosorption of lead(II) from aqueous solutions by *Pleurotus* as a toxicity biosorbent. In: Environmental science and technology conference (ESTEC2009), Kuala Terengganu Malaysia, Dec 7–8, 2009

Tay CC, Liew HH, Yin C-Y, Abdul-Talib S, Surif S, Abdullah A, Yong SK (2011) Biosorption of cadmium ions using *Pleurotus ostreatus*: growth kinetics, isotherm study and biosorption mechanism. Kor J Chem Eng 28(3):825–830

Tay CC, Redzwan G, Liew HH, Yong SK, Surif S, Abdul-Talib S (2012) Fundamental behavior for biosorption of divalence cations by *Pleurotus* mushroom spent-substrate. Malays J Sci 31:40–44

Tay CC, Liew HH, Abdul-Talib S, Redzwan G (2016) Bi-metal biosorption using *Pleurotus ostreatus* spent mushroom substrate (PSMS) as a biosorbent: isotherm, kinetic, thermodynamic studies and mechanism. Desalination Water Treat 57(20). http://www.tandfonline.com/action/showCitFormats?. doi: http://dx.doi.org/10.1080/19443994.2015.1027957

Velásquez L, Dussan J (2009) Biosorption and bioaccumulation of heavy metals on dead and living biomass of *Bacillus sphaericus*. J Hazard Mater 167:713–716. doi:10.1016/j.jhazmat.2009.01.044

Vimala R, Das N (2011) Mechanism of Cd(II) adsorption by macrofungus *Pleurotus platypus*. J Environ Sci 23:288–293

Vimala R, Charumathi D, Nilanjana Das (2011) Packed bed column studies on Cd(II) removal from industrial wastewater by macrofungs *Pleurotus platypus*. Desalination 275:291–296

Wu M, Xu Y, Ding W, Li Y, Xu H (2016) Mycoremediation of manganese and phenanthrene by *Pleurotus eryngii* mycelium enhanced by tween 80 and saponin. Appl Microbiol Biotechnol 100:7249–7261

Xiangliang P, Jianlong W, Daoyong Z (2005) Biosorption of Pb(II) by *Pleurotus ostreatus* immobilized in calcium alginate gel. Process Bio Chem 40:2799–2803

Xiangliang P, Jianlong W, Daoyong Z (2009) Biosorption of Co(II) by immobilised *Pleurotus ostreatus*. Int J Environ Pollut 37:289–298

Xu F, Liu X, Chen Y, Zhang K, Xu H (2016) Self-assembly modified-mushroom nano composite for rapid removal of hexavalent chromium from aqueous solution with bubbling fluidized bed. Sci Rep 6. 26201. doi: 10.1038/srep26201. http://www.nature.com/articles/srep26201

Yalçinkaya Y, Arica MY, Soysal L, Bektaş S (2002) Cadmium and mercury uptake by immobilized *Pleurotus sapidus*. Turk J Chem 26(3):441–452

Yang T, Chen M-L, Wang J-H (2015) Genetic and chemical modification of cells for selective separation and analysis of heavy metals of biological or environmental significance. TrAC Trends Anal Chem 66:90–102

Yazdani M, Chee KY, Faridah A, Soon GT (2010) An in vitro study on the adsorption, absorption and uptake capacity of Zn by the bioremediator *Trichodermaatro viride*. Environ Asia 3:53–59

Zhu FK, Qu L, Fan WX, Qiao MY, Hao HL, Wang XJ (2010) Assessment of heavy metals in some wild edible mushrooms collected from Yunnan Province, China. Environ Monit Assess 30:61–62

20

Enhanced thermal conductivity of waste sawdust-based composite phase change materials with expanded graphite for thermal energy storage

Haiyue Yang[1†], Yazhou Wang[2†], Zhuangchao Liu[3], Daxin Liang[1], Feng Liu[1], Wenbo Zhang[1], Xin Di[1], Chengyu Wang[1*], Shih-Hsin Ho[4*] and Wei-Hsin Chen[5]

Abstract

Background: With the current rapid economic growth, demands for energy are progressively increasing. Energy shortages have attracted significant attention due to the shrinking availability of non-renewable resources. Therefore, thermal energy storage is one of the solutions that lead to saving of fossil fuels and make systems more cost-effective by the storage of wasted thermal energy. In particular, the application of phase change materials (PCMs) is considered as an effective and efficient approach to thermal energy storage because of the high latent heat storage capacity at small temperature intervals. Nevertheless, leakage problems and low thermal conductivity limit the practical applications of PCMs. Therefore, form-stable phase change materials with high thermal conductivity are urgently needed.

Results: A novel form-stable composite phase change material was prepared by incorporating PEG into waste sawdust with 5% EG. In the composites, PEG served as a phase change material, while waste sawdust acted as a carrier matrix. EG was added to help increase the thermal conductivity of the composites. The melting temperature of CPCMs-4 with 5% EG was found to be 58.6 °C with a phase change enthalpy of 145.3 kJ/kg, while the solidifying temperature was 48.5 °C with a phase change enthalpy of 131.4 kJ/kg. The thermal conductivity of CPCMs-4 with 5% EG increased by 23.8% compared with that of CPCMs-4. Moreover, no obvious changes in melting, solidifying temperature, or latent heat after 200 heating–cooling cycles were detected. The supercooling extent of CPCMs-4 with 5% EG decreased by 19.2% compared with PEG. The volume change properties and wettability properties of CPCMs-4 with 5% EG are suitable for thermal energy in terms of practical application.

Conclusions: The prepared composites have excellent thermal and form-stable properties and they can be recognized as potential candidates for thermal energy storage as form-stable composite phase change materials. Using simple impregnation techniques with waste sawdust as a supporting material, this study demonstrates an innovative technology for practically and markedly enhancing the adsorption capacity of phase change materials.

Keywords: Phase change material, Waste sawdust, Second law of thermodynamics, Enhanced thermal conductivity, Wetting properties, Volume expansion percentage

*Correspondence: wangcy@nefu.edu.cn; stephen6949@hit.edu.cn
†Haiyue Yang and Yazhou Wang contributed equally to this study and share first authorship
[1] Key Laboratory of Bio-based Material Science and Technology, Ministry of Education, Northeast Forestry University, Harbin 150040, People's Republic of China
[4] State Key Laboratory of Urban Water Resource and Environment, School of Municipal and Environmental Engineering, Harbin Institute of Technology, Harbin 150090, People's Republic of China
Full list of author information is available at the end of the article

Background

With the current rapid economic growth, demands for energy are progressively increasing. Energy shortages have attracted significant attention due to the shrinking availability of non-renewable resources. Therefore, thermal energy storage, is one of the solutions that leads to saving of fossil fuels and makes systems more cost-effective by the storage of wasted thermal energy (Mateo et al. 2014). In particular, the application of phase change materials (PCMs) is considered as an effective and efficient approach to thermal energy storage because of the high latent heat storage capacity at small temperature intervals (Zeng et al. 2013; Zhang et al. 2013; Tang et al. 2015; Cai et al. 2015). In comparison to inorganic PCMs, organic PCMs can overcome a few of the most fatal deficiencies, including phase separation and super cooling (Qian et al. 2015b). Moreover, organic PCMs also offer some significant advantages, such as good thermal reliability and self-nucleating behavior (Memon 2014). Polyethylene glycol (PEG) has been used widely due to its appropriate phase change temperature and large heat storage capacity (He et al. 2014). In addition, PEG exhibits satisfactory melting behavior, non-toxic and non-corrosive features, excellent chemical and thermal properties, bio-degradability, low vapor pressure, and low cost (Qian et al. 2013). Nevertheless, leakage problems in the melting state still limit the downstream applications of PEG (Qian et al. 2015b).

To solve the problem, some form-stable CPCMs have been fabricated by mixing PEG and supporting materials due to the excellent shape-stabilized properties during the phase transition process (Xu and Li 2014; Chen et al. 2015; Sarı 2015; Lafdi et al. 2008; Tang et al. 2016; Qian et al. 2015a). To achieve the aim of high thermal energy capacity, selecting suitable porous materials with large specific surface areas as the carrier matrices is vital. However, the rising cost of using porous carriers will significantly hinder commercial feasibility. Waste sawdust, a solid waste obtained from furniture manufacturing companies or the papermaking industry, has been widely used in dealing with environmental problems (Luo et al. 2013). Waste sawdust has also been recognized as an absorbent for the removal of dyes, toxic salts, heavy metals, and waste oils from water (Shukla et al. 2002). Because waste sawdust is mainly composed of cellulose (40–50%), lignin (25–40%), and hemicellulose (10–25%) (Li et al. 2014; Collinson and Thielemans 2010), it contains large amounts of hydroxyl, carboxylic, and phenolic groups, which can provide the capacity and structural support needed for PEG. Thus, we demonstrate waste sawdust as a low-cost carrier matrix with great stability for composite phase change materials. Recently, phase change materials interlinked with biomass materials have been

gaining significant attention; however, relevant reports are still rare. For example, Cao et al. (Babapoor et al. 2016) produced shape-stabilized phase change materials based on fatty acid eutectics and cellulose composites. Shin et al. (2015) investigated the thermal property and latent heat energy storage behavior of sodium acetate trihydrate composites containing expanded graphite and carboxymethyl cellulose for phase change materials. However, little is known regarding waste sawdust absorption of phase change materials. In addition, a thorough understanding of the interaction between waste sawdust and phase change materials is still lacking.

In the past few years, a significant number of studies on increasing thermal energy absorbing/releasing speed and improving the heat transfer efficiency of organic PCMs have been conducted (Khodadadi et al. 2013; Fang et al. 2010). Due to the porous structure and enhanced thermal conductivity of expanded graphite (EG), many researchers have employed two or more materials to prepare composite shape-stable phase change materials (e.g., PCMs, EG) (Zhao et al. 2011; Zhong et al. 2010; Sarı and Karaipekli 2007; Zhang et al. 2012; Wang et al. 2014), indicating that efficient enhanced thermal conductivity occurred in composite shape-stable phase change materials. Some reports have suggested that EG has great application potentials in the field of phase change materials (Huang et al. 2014; Tian et al. 2015).

In this study, properties of novel PEG/waste sawdust composites with enhanced thermal conductivity for thermal energy storage were investigated. In the composites, PEG served as a phase change material while waste sawdust acted as a carrier matrix. To the best of our knowledge, waste sawdust used as supporting materials are firstly introduced into the phase change materials. In addition, because of its superior performance, EG was added to help increase the thermal conductivity of the composites. On the basis of the thermal conductivity meter (TCM), the thermal conductivity of the optimal sample increased by 23.8% compared to when there was no expanded graphite in the composites with a mass ratio of PEG to waste sawdust of 4:1. All the results of this study indicated that a waste sawdust-based form-stable CPCMs could be a potential candidate for efficient thermal energy storage.

Methods

Materials

Analytical reagent PEG with an average molecular weight of 10,000 was offered by Tianjin Tianli Chemical Co., Ltd. (Tianjin, China). Absolute ethanol (Analytical reagent) was obtained from Tianjin Chemical Reagent Co., Ltd (Tianjin, China). Expanded graphite was purchased from the Qingdao Graphite Co. Ltd (Qingdao, China). Waste sawdust (poplar, grain diameter of 178–250 μm) was a lab-made material derived in Carpenter's laboratory at Northeast Forestry University (Heilongjiang province, China).

Preparation of the composite phase change materials

The composite phase change materials (CPCMs) were obtained by mixing the waste sawdust, PEG, and EG with a weight ratio of 4:16:1. Firstly, waste sawdust was ultrasonically cleaned with absolute ethanol and deionized water for 10 min, respectively. Then it was dried at 60 °C for 12 h in an oven. After that, waste sawdust, PEG, and EG were each dispersed into absolute ethanol with quick stirring for 3 h at 25 °C, respectively. And then, the above three kinds of ethanol mixture were mixed again under stirring for 1 h at 25 °C. Finally the samples kept static at 110 °C in a vacuum oven for 12 h until the absolute ethanol was evaporated completely. The composite phase change materials were thus obtained. The synthetic route for the PEG/waste sawdust with EG composites is shown in Fig. 1.

In order to study the optimal mass fraction of PEG in the composites, a range of the PEG/waste sawdust mixtures were determined (mass fractions of PEG: 20, 40, 60, 80, 81, 82, 83, 84, and 85%). The liquid leakage test method was carried out as follows: different samples with various mass fractions of PEG were put on pieces of weigh paper, which have been weighed and numbered, respectively. Each sample kept static at 70 °C for 30 min before at 80 °C for 30 min in an oven. Then, each piece of weigh paper was weighed again. Compared with first weight, it can be determined whether there is a leakage happening in the processing of phase change. The optimal mass fraction of PEG in the composites was found to be 80%. Four samples were obtained, which were labeled as CPCMs-1, CPCMs-2, CPCMs-3, and CPCMs-4, individually. CPCMs-4 with 1% EG, CPCMs-4 with 3% EG, and CPCMs-4 with 5% EG were also obtained, respectively. The mass fractions of the PEG, waste sawdust, and EG in these samples are presented in Table 1.

Characterization analysis

A Fourier transformation infrared spectroscope (FTIR, Thermo Fisher Scientific Nicolet 6700, USA) was employed to estimate the chemical structure from 400 to 4000 cm^{-1} with a resolution of 4 cm^{-1}. The crystal structure was analyzed with an X-ray diffractometer (XRD, D/max 2200VPC, Rigaku, Japan) from 5° to 50° at room temperature. The surface microstructure of samples were observed with a scanning electron microscopy (SEM, Hitachi TM3030). The thermal properties of CPCMs are measured with a differential scanning calorimeter (DSC, Q20, TA, USA) from 25 to 100 °C at 5 °C/min under nitrogen at a flow rate of 20 mL/min. Thermogravimetric analysis (TGA) was carried out on a STA

Table 1 The conditions of different composite phase change materials

Samples	PEG (%)	Waste sawdust (%)	EG (%)
CPCMs-1	20	80	–
CPCMs-2	40	60	–
CPCMs-3	60	40	–
CPCMs-4	80	20	–
CPCMs-4 + 1% EG	80	20	1
CPCMs-4 + 3% EG	80	20	3
CPCMs-4 + 5% EG	80	20	5

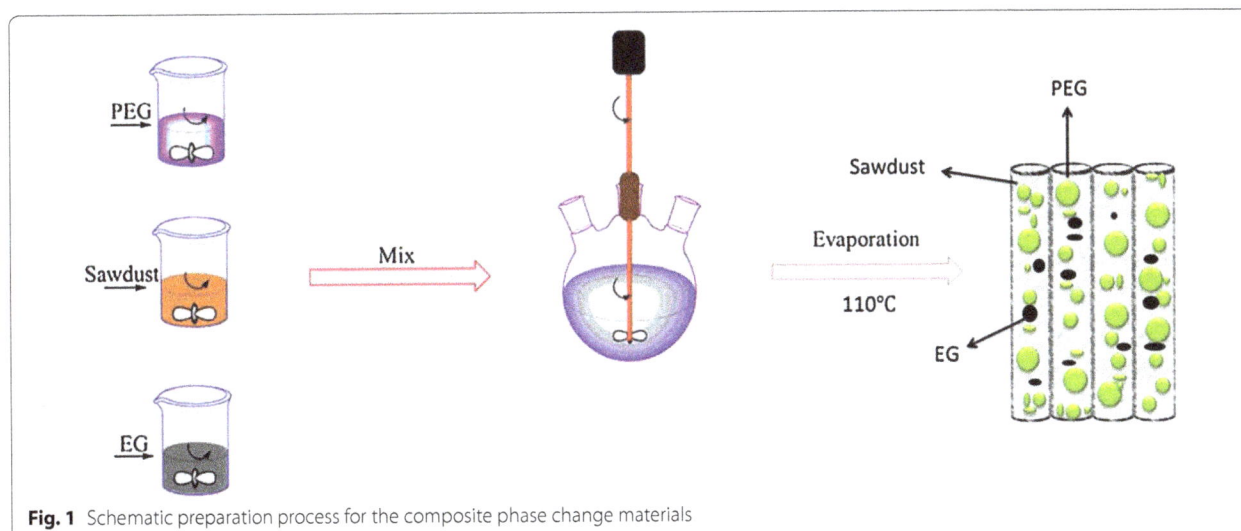

Fig. 1 Schematic preparation process for the composite phase change materials

6000-SQ8 thermal analyzer (0.1 mg, ± 1 °C) with heating rate of 10 °C/min under nitrogen atmosphere. The thermal conductivity of CPCMs was obtained by a thermal conductivity meter (TCM, TPS 2500s, Hot Disk, Sweden). The contact angle was collected using a video-based contact angle measuring device (OCA 20, Dataphysics, Germany).

Results and discussion
FTIR analysis
An FTIR analysis was used to estimate the chemical structures of the PEG, waste sawdust, and EG. The FTIR spectrums of (a) PEG, (b) CPCMs-4, (c) CPCMs-4 with 5% EG, and (d) waste sawdust are displayed in Fig. 2. In Fig. 2a, the two typical peaks of PEG were respectively 962 and 2906 cm^{-1}. The peak at 962 cm^{-1} is the stretching vibrations of C–H, and the peak at 2906 cm^{-1} belongs to –CH_2 stretching vibrations of PEG (Qian et al. 2015b). At wave numbers of 1109 and 3403 cm^{-1}, the stretching vibration peaks of C–O and –OH were discovered, respectively. Figure 2d is a typical spectrum of waste sawdust. Due to the existence of hydrogen and oxygen compounds in wood, such as cellulose, hemicellulose, and lignin, the peaks at around 3329 and 2280 cm^{-1} are, respectively, the stretching vibrations of –OH and –CH_2 in the waste sawdust. Figure 2c shows the characteristic absorption peaks at a wavenumber of 1636 cm^{-1}, corresponding to the stretching vibrations of C=O. In Fig. 2b, c, the stretching vibration peaks of –OH disappeared, suggesting that it may form a hydrogen bond caused by the –OH of PEG and the waste sawdust. Apart from some mild peak slips (e.g., –CH_2–OH), the main PEG and waste sawdust peaks appeared as expected. The results indicated that no significant new peaks were discovered, indicating that there are no chemical interactions

between PEG and waste sawdust, which is beneficial in terms of the form stability of the composites.

XRD analysis of the form-stable CPCMs
XRD was employed to investigate crystalline properties of the composites. Figure 3 displays the XRD patterns of (a) EG, (b) PEG, (c) CPCMs-4 with 5% EG, (d) CPCMs-4, and (e) waste sawdust. As shown in Fig. 3a, the steep peak at 26.5° was attributed to the crystal structure of EG. Two distinct steep peaks at about 19.3° and 23.2° were allocated to PEG crystal (Fig. 3b) while 22.2° was assigned to waste sawdust (curve e). Figure 3d displays the characteristic diffraction peaks of PEG and waste sawdust. In addition, in Fig. 3c, the PEG, waste sawdust, and EG-specific diffraction peaks appeared simultaneously, and no obvious peak changes were observed. These results suggest that the crystalline structures of the PEG, waste sawdust, and EG were not destroyed by the formation of the composite phase change material composed of PEG/waste sawdust and EG, which further suggests that no significant mutual effect occurred between the PEG, waste sawdust, and the EG.

Microstructure of the form-stable CPCMs
Figure 4 shows scanning electron microscope photographs of waste sawdust and CPCMs-1-4. Figure 5 presents the SEM photography of CPCMs-4 with EG. Figure 4c distinctly shows the microstructure of the waste sawdust. It reveals that a tubular structure emerged in the waste sawdust, along with numerous pores on the tube. Compared with Fig. 4c, d, f, shows that the PEG was absorbed into the pores and tubes, indicating that it was uniformly absorbed into the waste sawdust structure. In Fig. 4i, in addition, it can be seen that the waste sawdust

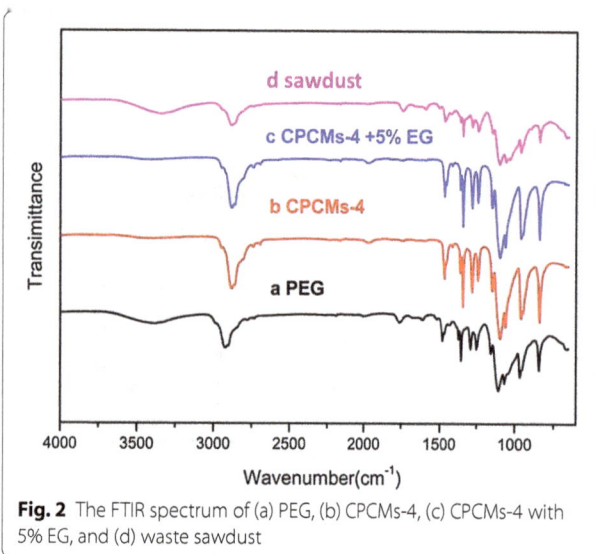

Fig. 2 The FTIR spectrum of (a) PEG, (b) CPCMs-4, (c) CPCMs-4 with 5% EG, and (d) waste sawdust

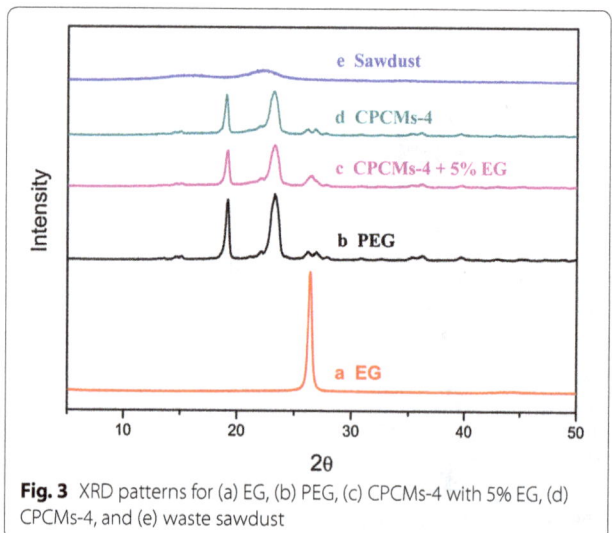

Fig. 3 XRD patterns for (a) EG, (b) PEG, (c) CPCMs-4 with 5% EG, (d) CPCMs-4, and (e) waste sawdust

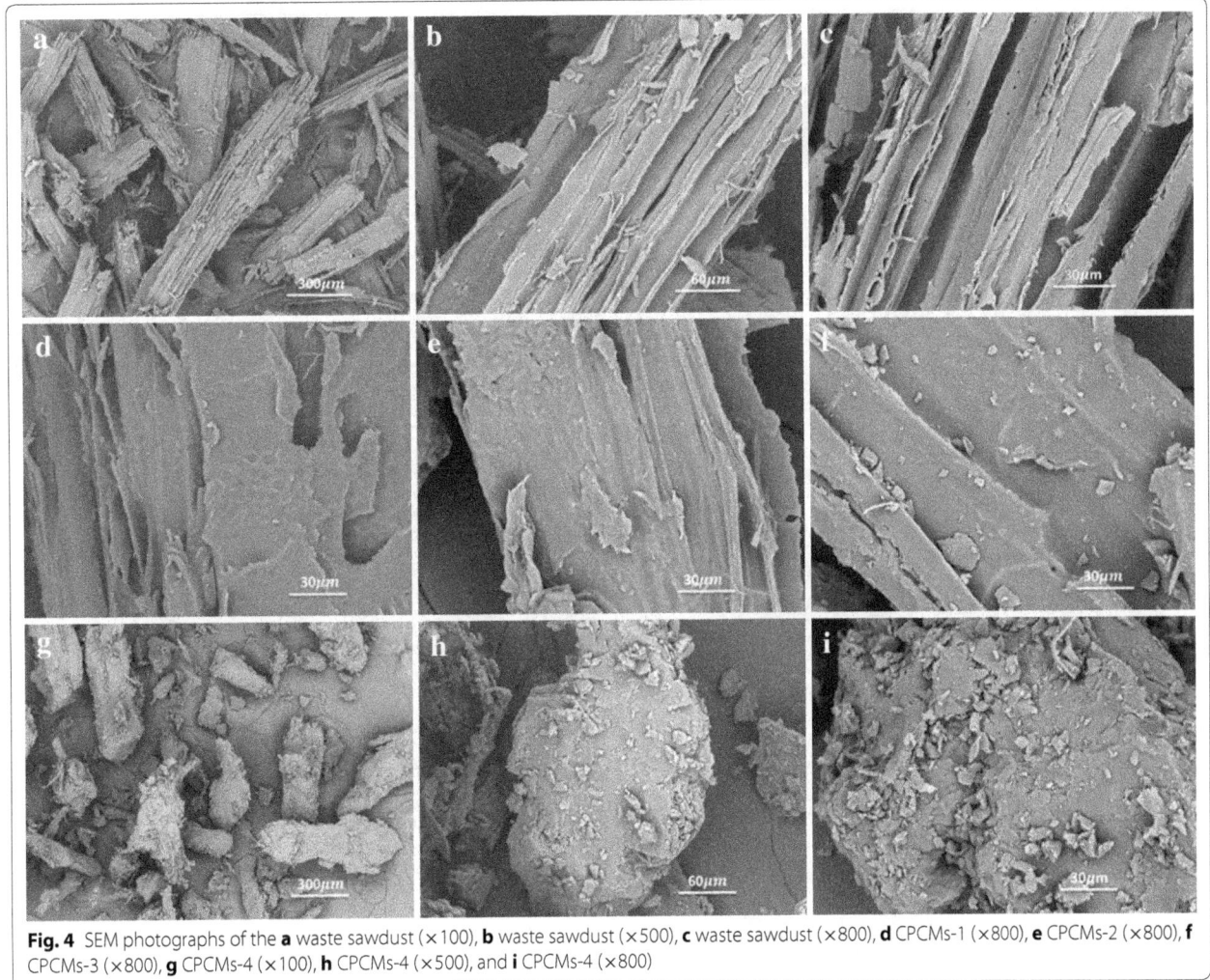

Fig. 4 SEM photographs of the **a** waste sawdust (×100), **b** waste sawdust (×500), **c** waste sawdust (×800), **d** CPCMs-1 (×800), **e** CPCMs-2 (×800), **f** CPCMs-3 (×800), **g** CPCMs-4 (×100), **h** CPCMs-4 (×500), and **i** CPCMs-4 (×800)

pores had absorbed much more PEG and that the tube structure and microstructure pores in the waste sawdust had disappeared. However, a comparison of Fig. 4g, h with Fig. 4a, b indicates no obvious changes in the fundamental shape, suggesting that the composites may remain form-stable during the phase change process.

Figure 5 displays the SEM images of CPCMs-4 with 1% EG, CPCMs-4 with 3% EG, and CPCMs-4 with 5% EG, demonstrating that EG disperses uniformly into CPCMs-4, which is in a good agreement with a previous study indicating that the addition of EG can help improve the thermal conductivity of PCMs and prevent liquids from leaking in composite phase change materials. Therefore, in this study, a novel form-stable composite phase change material with enhanced thermal conductivity was obtained.

Wetting properties of a composite phase change material

The absorption theory suggested that wettability properties of solid materials are especially significant in many fields (Wang et al. 2017). A low contact angle (< 90°) shows that the solid material is wettable. And the liquid will be absorbed into solid materials. The contact angle is the angle at which a liquid/vapor interface meets the solid surface. And it is determined by the interactions across the three interfaces (Fig. 6) (Zhenyu et al. 2016).

$$\cos\theta = (\gamma_{SV} - \gamma_{SL})/\gamma_{LV}, \tag{1}$$

where the parameters of γ_{SV}, γ_{SL}, and γ_{LV} are respectively the solid–vapor, solid–liquid, and liquid–vapor interfacial tensions, and where θ is the contact angle. In order to estimate the absorption performance indirectly, the wetting properties of different supporting materials were measured in this study. An ethanol solution of PEG could be quickly absorbed into waste sawdust within 2 s and the contact angle approaches 0°, which shows that waste sawdust has good wettability. The dynamic absorption process is shown in the Additional file 1: Video S1. Table 2 displays the maximum

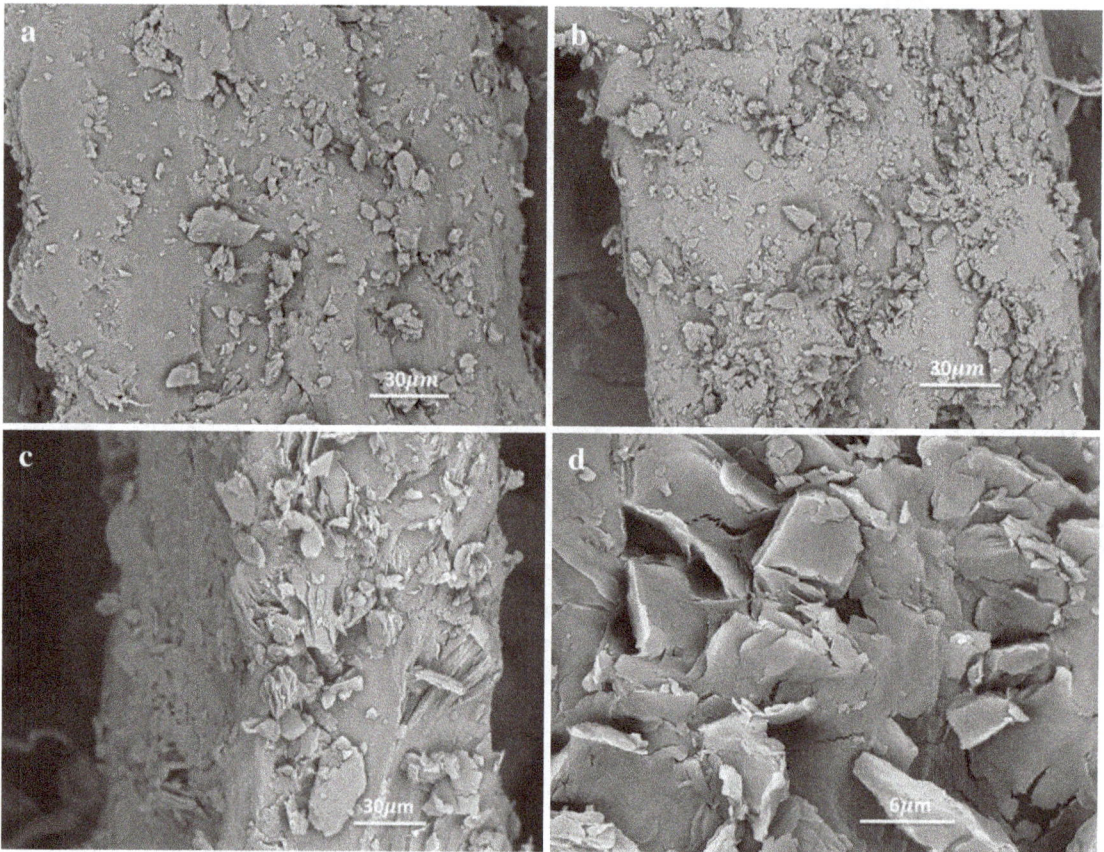

Fig. 5 SEM photographs of the **a** CPCMs-4 with 1% EG (×800), **b** CPCMs-4 with 3% EG (×800), **c** CPCMs-4 with 5% EG (×800), and **d** CPCMs-4 with 5% EG (×5.0k)

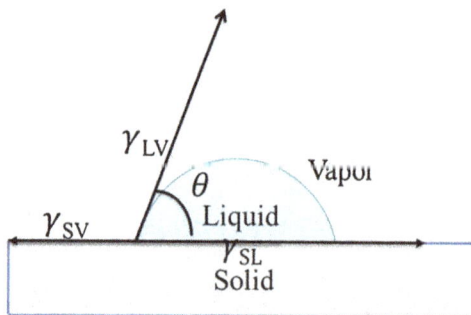

Schematic diagram of contact angle of waste sawdust -ethanol solution of PEG-vapor system

Fig. 6 A schematic diagram of the solid–liquid–vapor system contact angle

Table 2 Mass fraction of PEG and time required for the ethanol solution with PEG to diffuse into the supporting materials

Supporting material	Maximum mass fraction of PEG (wt%)	Time required for the ethanol solution with PEG to diffuse into the supporting material(s)
Hydroxyapatite	70 (Wang et al. 2017)	7
Calcium silicate	70 (Qian et al. 2015b)	10
Active carbon	70 (Feng et al. 2011)	11
Sawdust	80	2

mass fraction of PEG in the supporting materials and the time required for the ethanol solution with PEG to diffuse into the supporting materials. The time required for diffusing into waste sawdust was 2 s, which is much lower than that of hydroxyapatite (7 s), calcium silicate (10 s), active carbon (11 s), but the maximum mass fraction of PEG is larger than that of hydroxyapatite (70%), calcium silicate (70%), and active carbon (70%), indicating that waste sawdust has excellent wetting property as compared to hydroxyapatite, calcium silicate, and active carbon. The results confirm that waste sawdust has good wettability and that waste sawdust can be a potential supporting material for absorbing phase change materials.

Form-stable property

A drying oven was used to characterize the shape-stable properties of PEG (Fig. 7a), CPCMs-4 (Fig. 7b), and CPCMs-4 with 5% EG (Fig. 7c). Every given temperature was held for 30 min, and a digital camera was employed to record the results. As depicted in Fig. 7, at a temperature of 70 °C, the PEG melted quickly. On the contrary, the structures of both CPCMs-4 and CPCMs-4 with 5% EG remained constant. With an increase in the temperature to 80 °C, the PEG melted completely into liquid while CPCMs-4 and CPCMs-4 with 5% EG remained solid due to the presence of waste sawdust, which provided dramatic support to the composite structure. No liquid leakage was found on the CPCMs-4 and CPCMs-4 with 5% EG weigh paper, thus demonstrating that the masses and structure of the CPCMs-4 and CPCMs-4 samples with 5% EG remained constant both before and after the thermal test.

The results clearly indicate that the prepared composite phase change material was shape stable without any liquid leakage from the supporting materials, which infers a significant contribution to the field of thermal energy storage. In addition, in regard to the test of the CPCMs-4 heating–cooling cycles, Fig. 8 shows that there were no obvious changes in the melting and solidifying temperatures before and after 200 and 1000 heating–cooling cycles (denoted as CPCMs-4c, CPCMs-4c$_1$ after 200 and 1000 heating–cooling cycles, respectively). The melting temperature of CPCMs-4 before and after the 200 and 1000 heating–cooling cycles was, respectively, 58.1, 59.5, and 58.9 °C, while the solidifying temperatures of CPCMs-4 before and after the cycles were, respectively, 48.6, 48.7, and 48.4 °C. In addition, the latent heat required for melting for CPCMs-4 before and after the cycles was, respectively, 151.1, 143.0, and 126.4 kJ/kg. The latent heat required for solidifying of CPCMs-4 before and after the cycles was, respectively, 140.3, 133.7, and 117.5 kJ/kg. After the cycles, the melting enthalpy decreased by 5.4 and 16.4%, respectively, and the solidifying enthalpy decreased by 4.7 and 16.2%, respectively. This indicated that no obvious changes were found in terms of latent heat and phase change temperatures, thus verifying that CPCMs-4 has good thermal reliability.

Thermal properties of the composite phase change materials

The phase change latent and the temperature of CPCMs were obtained by a differential scanning calorimeter (DSC). The melting and solidifying DSC curves of (a) PEG, (b) CPCMs-1, (c) CPCMs-2, (d) CPCMs-3, and (e) CPCMs-4 are illustrated in Fig. 9. In addition, the melting and solidifying curves of (a) PEG, (b) CPCMs-4 with 5% EG, and (c) CPCMs-4 are depicted in Fig. 10. The thermal data of the CPCMs are shown in Table 3.

Fig. 7 Shape-stable images of **a** PEG, **b** CPCMs-4, and **c** CPCMs-4 with 5% EG

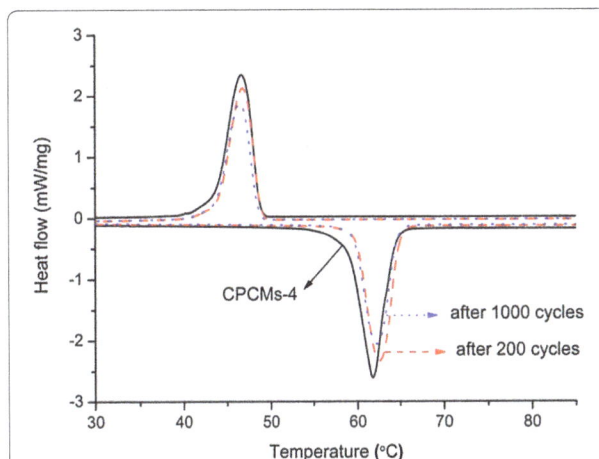

Fig. 8 The DSC curves of CPCMs-4 before and after experiencing either 200 or 1000 heating–cooling cycles

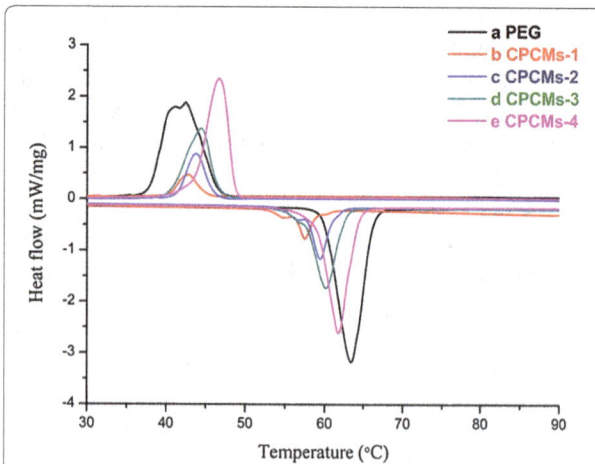

Fig. 9 The DSC curves of (a) PEG, (b) CPCMs-1, (c) CPCMs-2, (d) CPCMs-3, and (e) CPCMs-4

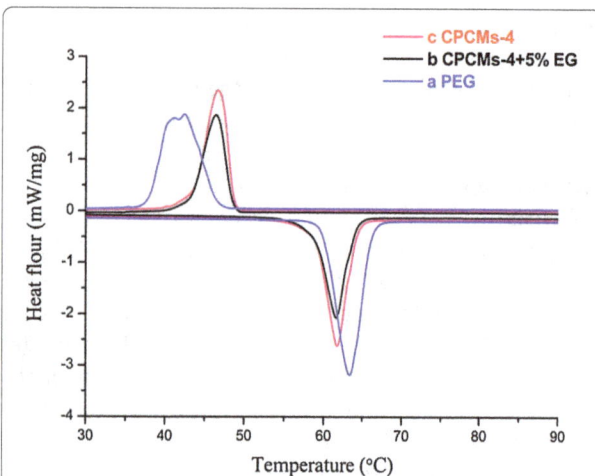

Fig. 10 The DSC curves of (a) PEG, (b) CPCMs-4 with 5% EG, and (c) CPCMs-4

The phase change latent is regarded as one of the most critical factors to assess the thermal energy storage capacity of CPCMs. As a matter of fact, the thermal energy capacity of CPCMs rests with the PEG content in the composites. A higher PEG content will significantly improve the phase change latent; however, the higher PEG content will also increase the possibility of liquid leakage. Taken together, both PEG content and form-stable properties are recognized as two critical factors, which can influence thermal energy storage capacity of CPCMs. In this study, the maximum mass fraction of PEG in the CPCMs was 80% without any liquid leakage during the phase transition (Fig. 7b). As shown in Table 3, the melting temperatures of PEG and CPCMs-1-4 were, respectively, 59.0, 56.8, 57.2, 57.3, and 58.1 °C while the melting enthalpy is 189.5, 37.8, 76.0, 111.2, and 151.1 kJ/kg. The solidifying temperatures of PEG and CPCMs-1-4 were, respectively, 46.5, 44.9, 45.9, 46.4, and 48.6 °C while the solidifying enthalpy is 172.5, 29.1, 69.9, 105.3, 140.3 kJ/kg. The melting temperatures of CPCMs-4 with 1% EG, CPCMs-4 with 3% EG, and CPCMs-4 with 5% EG were, respectively, 58.1, 58.0, and 58.6 °C. The solidifying temperatures were, respectively, 48.3, 48.0, and 48.5 °C. These results demonstrate that the phase change properties of CPCMs-4 with EG are very similar to those of the PEG mixture, suggesting that waste sawdust, EG, and PEG are compatible. The melting phase change enthalpies of CPCMs-4, CPCMs-4 with 1% EG, CPCMs-4 with 3% EG, and CPCMs-4 with 5% EG were, respectively, 151.1, 150.1, 148.6, and 145.3 kJ/kg, while their solidifying phase change enthalpies were 140.3, 138.3, 135.2, and 131.4 kJ/kg, respectively. Due to the low content of EG (i.e., 1–5%), the phase change enthalpy of the composites did not significantly decline, indicating that the heat storage capacity did not change significantly with a low mass fraction of EG. In composite phase change materials, the

Table 3 Thermal characteristics of CPCMs-1, CPCMs-2, CPCMs-3, CPCMs-4, CPCMs-4c, CPCMs-4c$_1$, CPCMs-4 with 1% EG, CPCMs-4 with 3% EG, CPCMs-4 with 5% EG and PEG

Samples	Mass fraction of PEG (wt%)	T_m (°C)	T_s (°C)	ΔH_m (kJ/kg)	ΔH_s (kJ/kg)
CPCMs-1	20.0	56.8 ± 0.1	44.9 ± 0.3	37.8 ± 1.7	29.1 ± 2.1
CPCMs-2	40.0	57.2 ± 0.1	45.9 ± 0.2	76.0 ± 2.4	69.9 ± 3.0
CPCMs-3	60.0	57.3 ± 0.1	46.4 ± 0.1	111.2 ± 2.4	105.3 ± 1.4
CPCMs-4	80.0	58.1 ± 0.1	48.6 ± 0.1	151.1 ± 1.8	140.3 ± 2.6
CPCMs-4c	–	59.5 ± 0.1	48.7 ± 0.2	143.0 ± 2.3	133.7 ± 2.7
CPCMs-4c$_1$	–	58.9 ± 0.1	48.4 ± 0.2	126.4 ± 2.5	117.5 ± 1.9
CPCMs-4 + 1% EG	79.2	58.1 ± 0.2	48.3 ± 0.1	150.1 ± 3.6	138.3 ± 2.0
CPCMs-4 + 3% EG	77.7	58.0 ± 0.3	48.0 ± 0.2	148.6 ± 2.4	135.2 ± 3.7
CPCMs-4 + 5% EG	76.2	58.6 ± 0.2	48.5 ± 0.2	145.3 ± 2.6	131.4 ± 2.5
PEG	100	59.0 ± 0.4	46.5 ± 0.2	189.5 ± 1.4	172.5 ± 1.9

phase change process comes from PEG, the phase change enthalpy of CPCMs can be obtained by the Eq. (2):

$$\Delta H_{CPCMs} = \Delta H_{PCM} \times \eta, \tag{2}$$

where the parameter of η is the mass fraction of PEG in the composites, and ΔH_{PCM} is the phase change enthalpy of the PCMs.

From CPCMs-1 to CPCMs-4, it displays that, with the increase of PEG, phase change temperatures rise. The reason would be explained by the second law of thermodynamics applying in the phase change process.

$$T = \frac{\Delta H}{\Delta S}, \tag{3}$$

where T is the phase change temperature of CPCMs, ΔH the phase change heat per unit mass, and ΔS is the entropy change during phase change transition. With the increasing mass fraction of PEG, ΔH improves while ΔS decreases. Therefore, the phase change temperature decreases. Compared with CPCMs-1-3, entropy change is the least and phase change enthalpy is the largest. That is the reason why DSC curves of CPCMs-4 is significantly different from CPCMs-1-3. According to the entropy formula, the entropy will be larger with mass fraction of PEG. Compared with PEG, waste sawdust has larger expansion ratio because of porous structure during the phase change transition. In consequence, the phase change temperature of CPCMs will become larger.

Supercooling as another important parameter in practical applications that should be taken into consideration. Lowering the extent of supercooling can assist the phase change transition process. In accordance with Table 3, the degree of supercooling can be obtained as the difference between the melting and solidifying temperature. Figure 11 shows the extent of supercooling for

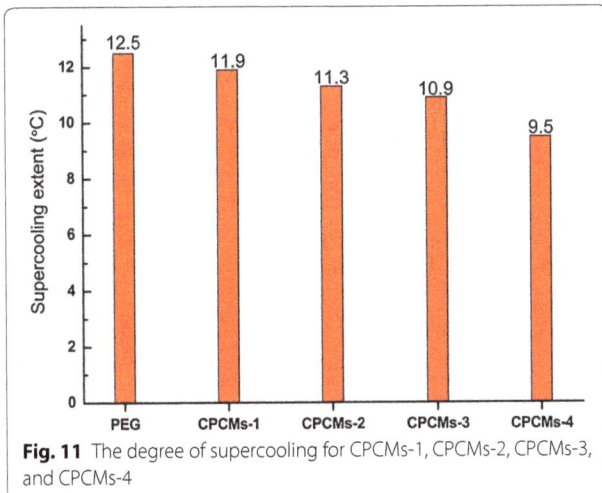

Fig. 11 The degree of supercooling for CPCMs-1, CPCMs-2, CPCMs-3, and CPCMs-4

PEG, CPCMs-1, CPCMs-2, CPCMs-3, and CPCMs-4. Compared with PEG, the degree of supercooling of the CPCMs gradually decreased with the increasing mass fraction of PEG. The degree of supercooling of CPCMs-1, CPCMs-2, CPCMs-3, CPCMs-4, CPCMs-4, CPCMs-4 with 1% EG, CPCMs-4 with 3% EG, and CPCMs-4 with 5% EG were, respectively, reduced by 4.8, 9.6, 12.8, 24.0, 21.6, 20.0, and 19.2% compared with the PEG. The results show that the degree of supercooling for the optimal sample can effectively reduce to 24.0% by adding 20% waste sawdust.

Volume change during the phase change transition process

Thermal expansion is the tendency of change in shape, area, and volume due to the change of temperature. The volume change of PCMs occur during the phase change transition process. Theoretically, a lower volume change of CPCMs is more suitable for thermal energy storage in practical applications because a large expansion percentage can cause damage to the surrounding environment, such as corrosion, volume deformation, and so on. In this study, the volume expansion ratio is defined as the following Eq. (4):

$$\gamma_V = (V_1 - V_0)/V_0, \tag{4}$$

where the parameters V_0 and V_1 are the volumes of the prepared sample at room temperature and at 70 °C maintained for 30 min, respectively. 0.6 g of waste sawdust, CPCMs-1, CPCMs-2, CPCMs-3, and CPCMs-4 and PEG were accurately weighed. Then, the samples were pressed into cylinders (the diameter was 13 mm) under a pressure of 15 Mpa. Table 4 shows the height, diameter, and volume ($V_0 = 0.25 \times \pi \times D_0^2 \times H_0$, $V_1 = 0.25 \times \pi \times D_1^2 \times H_1$) changes for the samples. The volume expansion ratios of waste sawdust, CPCMs-1, CPCMs-2, CPCMs-3, CPCMs-4, and PEG were, respectively, 6.50, 22.02, 23.59, 15.86, 14.22, and 25.64%. These results indicate that the volume expansion ratio of CPCMs-4 is much lower than that of PEG, which shows that CPCMs-4 have good thermal dimensional stability and it is promising for thermal energy storage in practical application.

Thermal conductivity of the form-stable composite phase change material

The thermal conductivity was estimated with a thermal conductivity meter (TCM) at the test temperature of 24 °C. As shown in Table 5, the thermal conductivity of CPCMs-4, CPCMs-4 with 1% EG, CPCMs-4 with 3% EG, and CPCMs-4 with 5% EG was, respectively, 0.1156 W/(m K), 0.1202 W/(m K), 0.1371 W/(m K), and

Table 4 The height, diameter, and volume change of tablet samples

Sample	D_0 (mm)	D_1 (mm)	H_0 (mm)	H_1 (mm)	V_0 (mm^3)	V_1 (mm^3)	(%)
CPCMs-1	13.00	13.12	3.94	4.72	522.70	637.79	22.02 ± 0.27
CPCMs-2	13.00	13.14	3.72	4.50	493.51	609.92	23.59 ± 0.26
CPCMs-3	13.00	13.08	3.60	4.12	469.63	553.33	15.86 ± 0.27
CPCMs-4	13.00	13.34	3.54	3.84	469.63	536.43	14.22 ± 0.30
PEG	13.00	–	3.6	–	477.60	600.02	25.64 ± 0.49

Table 5 Thermal conductivity of CPCMs-4, CPCMs-4 with 1% EG, CPCMs-4 with 3% EG, and CPCMs-4 with 5% EG

Samples	Thermal conductivity (W/(m))
CPCMs-4	0.1156 ± 0.0016
CPCMs-4 + 1%	0.1202 ± 0.0022
CPCMs-4 + 3%	0.1371 ± 0.0029
CPCMs-4 + 5%	0.1431 ± 0.0011

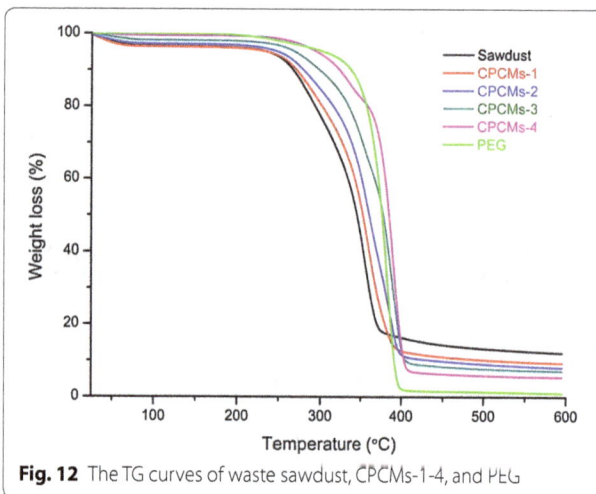

Fig. 12 The TG curves of waste sawdust, CPCMs-1-4, and PEG

0.1431 W/(m K). Compared with CPCMs-4, the thermal conductivity of CPCMs-4 increased with an increase in EG content, which is in a good agreement with the findings of a previous study (Tang et al. 2015). The greatest enhancement of 23.8% was obtained in CPCMs-4 with 5% EG, clearly indicating that adding small amounts of EG into the composites could significantly enhance the thermal conductivity. The increase in thermal conductivity is more likely due to the thermal conductive network formed by EG (Sobolciak et al. 2016).

Thermal stability

The thermal stability of the pure components (PEG and waste sawdust) and prepared composites (CPCMs-1-4) was evaluated using the TGA analysis, as shown in Fig. 12. In the waste sawdust curves, about 3% weight loss in the absorbed water, including the bound water and a small amount of free water, was observed. With increases in the mass fraction of PEG, the amount of absorbed water declined. The CPCMs-4 curve did not exhibit any absorbed water. Rapid weight loss for the waste sawdust, CPCMs-1-4, and PEG occurred at 356.1, 361.5, 384.4, 388.3, 389.6, and 383.0 °C, respectively. CPCMs were thermally stable when temperatures were below 150 °C. As a result, the CPCMs obtained in this study were quite stable at their working temperatures, which ensured their feasibility for practical application. The results clearly indicate that adding PEG into waste sawdust not only can improve the thermal stability of waste sawdust but can also enhance the humidity resistance of the waste sawdust, so its use can be applied in the real world.

Conclusions

In this study, a novel form-stable composite phase change material was prepared and comprehensively characterized. The melting temperature of CPCMs-4 with 5% EG was found to be 58.6 °C with a phase change enthalpy of 145.3 kJ/kg, while the solidifying temperature was 40.5 °C with a phase change enthalpy of 131.4 kJ/kg. The thermal conductivity of CPCMs-4 with 5% EG increased by 23.8% compared with that of CPCMs-4. Moreover, no obvious changes in melting, solidifying temperature, or latent heat after 200 heating–cooling cycles were detected. The supercooling extent of CPCMs-4 with 5% EG decreased by 19.2% compared with PEG. The volume change properties and wettability properties of CPCMs-4 with 5% EG are suitable for thermal energy in terms of practical application. All these results strongly suggest that the prepared composites have excellent thermal and form-stable properties and that they can be recognized as potential candidates for thermal energy storage as form-stable composite phase change materials. Using simple impregnation techniques with waste sawdust as a supporting material, this study demonstrates an innovative technology for practically and markedly enhancing the adsorption capacity of phase change materials. The superior performance and novel characterization methods were also discussed in detail.

In the future, different bio-based materials will be applied to the field of CPCMs due to their non-toxicity, recyclability, low cost, and environmental friendliness. And different properties of CPCMs, for example, fire retardation and intelligent response performance, will be studied in the future study trend.

Abbreviations
PEG: polyethylene glycol; EG: expanded graphite; CPCMs: composite phase change materials; TCM: thermal conductivity meter; FTIR: Fourier transformation infrared spectroscope; XRD: X-ray diffractometer; DSC: differential scanning calorimeter; CPCMs-1: 80% PEG + 20% waste sawdust composite phase change materials; CPCMs-2: 60% PEG + 40% waste sawdust composite phase change materials; CPCMs-3: 40% PEG + 60% waste sawdust composite phase change materials; CPCMs-4: 20% PEG + 80% waste sawdust composite phase change materials; SEM: scanning electron microscope; γ_{SV}, γ_{SL}, and γ_{LV}: the solid–vapor, solid–liquid, and liquid–vapor interfacial tensions, respectively; θ: contact angle; CPCMs-4c: after 200 heating–cooling cycles; CPCMs-4c$_1$: after 1000 heating–cooling cycles; η: the mass fraction of PEG in the composites; ΔH_{PCM}: the phase change enthalpy of the PCMs; T: phase change temperature of CPCMs; ΔH: the phase change heat per unit mass; ΔS: the entropy change during phase change transition.

Authors' contributions
HY and YW designed the study, performed experiments, analyzed data, and prepared the manuscript. ZL, DL, FL, WZ, and XD contributed to the discussion. CY, SHH, and WHC reviewed the results, helped in data analysis, and edited the manuscript. All authors read and approved the final manuscript.

Author details
[1] Key Laboratory of Bio-based Material Science and Technology, Ministry of Education, Northeast Forestry University, Harbin 150040, People's Republic of China. [2] School of Material Science and Engineering, Sun Yat-Sen University, Guangzhou 510275, People's Republic of China. [3] Guangdong Yihua Timber Industry Co., Ltd, Guangzhou 510000, People's Republic of China. [4] State Key Laboratory of Urban Water Resource and Environment, School of Municipal and Environmental Engineering, Harbin Institute of Technology, Harbin 150090, People's Republic of China. [5] Department of Aeronautics and Astronautics, National Cheng Kung University, Tainan 701, Taiwan.

Acknowledgements
This research was supported by the Fundamental Research Funds for Central Universities (2572015EB01) and the National Natural Science Foundation of China (31770605).

Competing interests
The authors declare that they have no competing interests.

References
Babapoor A, Karimi G, Khorram M (2016) Fabrication and characterization of nanofiber-nanoparticle-composites with phase change materials by electrospinning. Appl Therm Eng 99:1225–1235. https://doi.org/10.1016/j.

Cai Y, Sun G, Liu M, Zhang J, Wang Q, Wei Q (2015) Fabrication and characterization of capric–lauric–palmitic acid/electrospun SiO$_2$ nanofibers composite as form-stable phase change material for thermal energy storage/retrieval. Sol Energy 118:87–95. https://doi.org/10.1016/j.solener.2015.04.042

Chen C, Liu W, Wang Z, Peng K, Pan W, Xie Q (2015) Novel form stable phase change materials based on the composites of polyethylene glycol/polymeric solid-solid phase change material. Sol Energy Mater Sol Cells 134:80–88. https://doi.org/10.1016/j.solmat.2014.11.039

Collinson SR, Thielemans W (2010) The catalytic oxidation of biomass to new materials focusing on starch, cellulose and lignin. Coord Chem Rev 254(15–16):1854–1870. https://doi.org/10.1016/j.ccr.2010.04.007

Fang G, Li H, Chen Z, Liu X (2010) Preparation and characterization of flame retardant n-hexadecane/silicon dioxide composites as thermal energy storage materials. J Hazard Mater 181(1–3):1004–1009. https://doi.org/10.1016/j.jhazmat.2010.05.114

Feng L, Zheng J, Yang H, Guo Y, Li W, Li X (2011) Preparation and characterization of polyethylene glycol/active carbon composites as shape-stabilized phase change materials. Sol Energy Mater Sol Cells 95(2):644–650. https://doi.org/10.1016/j.solmat.2010.09.033

He L, Li J, Zhou C, Zhu H, Cao X, Tang B (2014) Phase change characteristics of shape-stabilized PEG/SiO2 composites using calcium chloride-assisted and temperature-assisted sol gel methods. Sol Energy 103:448–455. https://doi.org/10.1016/j.solener.2014.02.042

Huang Z, Gao X, Xu T, Fang Y, Zhang Z (2014) Thermal property measurement and heat storage analysis of LiNO$_3$/KCl—expanded graphite composite phase change material. Appl Energy 115:265–271. https://doi.org/10.1016/j.apenergy.2013.11.019

Khodadadi JM, Fan L, Babaei H (2013) Thermal conductivity enhancement of nanostructure-based colloidal suspensions utilized as phase change materials for thermal energy storage: a review. Renew Sustain Energy Rev 24:418–444. https://doi.org/10.1016/j.rser.2013.03.031

Lafdi K, Mesalhy O, Elgafy A (2008) Graphite foams infiltrated with phase change materials as alternative materials for space and terrestrial thermal energy storage applications. Carbon 46(1):159–168. https://doi.org/10.1016/j.carbon.2007.11.003

Li H, Lei Z, Liu C, Zhang Z, Lu B (2014) Photocatalytic degradation of lignin on synthesized Ag-AgCl/ZnO nanorods under solar light and preliminary trials for methane fermentation. Bioresour Technol 175C:494–501. https://doi.org/10.1016/j.biortech.2014.10.143

Luo H, Ning XA, Liang X, Feng Y, Liu J (2013) Effects of sawdust-CPAM on textile dyeing sludge dewaterability and filter cake properties. Bioresour Technol 139:330–336. https://doi.org/10.1016/j.biortech.2013.04.035

Mateo S, Gonzalez del Campo A, Canizares P, Lobato J, Rodrigo MA, Fernandez FJ (2014) Bioelectricity generation in a self-sustainable microbial solar cell. Bioresour Technol 159:451–454. https://doi.org/10.1016/j.biortech.2014.03.059

Memon SA (2014) Phase change materials integrated in building walls: a state of the art review. Renew Sustain Energy Rev 31:870–906. https://doi.org/10.1016/j.rser.2013.12.042

Qian Y, Wei P, Jiang P, Li Z, Yan Y, Liu J (2013) Preparation of a novel PEG composite with halogen-free flame retardant supporting matrix for thermal energy storage application. Appl Energy 106:321–327. https://doi.org/10.1016/j.apenergy.2012.12.070

Qian T, Li J, Ma H, Yang J (2015a) The preparation of a green shape-stabilized composite phase change material of polyethylene glycol/SiO$_2$ with enhanced thermal performance based on oil shale ash via temperature assisted sol–gel method. Sol Energy Mater Sol Cells 132:29–39. https://doi.org/10.1016/j.solmat.2014.08.017

Qian T, Li J, Min X, Deng Y, Guan W, Ma H (2015b) Polyethylene glycol/mesoporous calcium silicate shape-stabilized composite phase change material: preparation, characterization, and adjustable thermal property. Energy 82:333–340. https://doi.org/10.1016/j.energy.2015.01.043

Sarı A (2015) Fabrication and thermal characterization of kaolin-based composite phase change materials for latent heat storage in buildings. Energy Build 96:193–200. https://doi.org/10.1016/j.enbuild.2015.03.022

Sarı A, Karaipekli A (2007) Thermal conductivity and latent heat thermal energy storage characteristics of paraffin/expanded graphite composite as phase change material. Appl Therm Eng 27(8–9):1271–1277. https://doi.org/10.1016/j.applthermaleng.2006.11.004

Shin HK, Park M, Kim H-Y, Park S-J (2015) Thermal property and latent heat energy storage behavior of sodium acetate trihydrate composites containing expanded graphite and carboxymethyl cellulose for phase change materials. Appl Therm Eng 75:978–983. https://doi.org/10.1016/j.applthermaleng.2014.10.035

Shukla A, Zhang Y-H, Dubey P, Margrave JL, Shukla SS (2002) The role of sawdust in the removal of unwanted materials from water. J Hazard Mater 95(1):137–152

Sobolciak P, Karkri M, Al-Maadeed MA, Krupa I (2016) Thermal characterization of phase change materials based on linear low-density polyethylene, paraffin wax and expanded graphite. Renew Energy 88:372–382. https://doi.org/10.1016/j.renene.2015.11.056

Tang F, Su D, Tang Y, Fang G (2015) Synthesis and thermal properties of fatty acid eutectics and diatomite composites as shape-stabilized phase change materials with enhanced thermal conductivity. Sol Energy Mater Sol Cells 141:218–224. https://doi.org/10.1016/j.solmat.2015.05.045

Tang B, Wang L, Xu Y, Xiu J, Zhang S (2016) Hexadecanol/phase change polyurethane composite as form-stable phase change material for thermal energy storage. Sol Energy Mater Sol Cells 144:1–6. https://doi.org/10.1016/j.solmat.2015.08.012

Tian H, Wang W, Ding J, Wei X, Song M, Yang J (2015) Thermal conductivities and characteristics of ternary eutectic chloride/expanded graphite thermal energy storage composites. Appl Energy 148:87–92. https://doi.org/10.1016/j.apenergy.2015.03.020

Wang S, Qin P, Fang X, Zhang Z, Wang S, Liu X (2014) A novel sebacic acid/expanded graphite composite phase change material for solar thermal medium-temperature applications. Sol Energy 99:283–290. https://doi.org/10.1016/j.solener.2013.11.018

Wang Y, Liang D, Liu F, Zhang W, Di X, Wang C (2017) A polyethylene glycol/hydroxyapatite composite phase change material for thermal energy storage. Appl Therm Eng 113:1475–1482. https://doi.org/10.1016/j.applthermaleng.2016.11.159

Xu B, Li Z (2014) Paraffin/diatomite/multi-wall carbon nanotubes composite phase change material tailor-made for thermal energy storage cement-based composites. Energy 72:371–380. https://doi.org/10.1016/j.energy.2014.05.049

Zeng J-L, Zhu F-R, Yu S-B, Xiao Z-L, Yan W-P, Zheng S-H, Zhang L, Sun L-X, Cao Z (2013) Myristic acid/polyaniline composites as form stable phase change materials for thermal energy storage. Sol Energy Mater Sol Cells 114:136–140. https://doi.org/10.1016/j.solmat.2013.03.006

Zhang Z, Zhang N, Peng J, Fang X, Gao X, Fang Y (2012) Preparation and thermal energy storage properties of paraffin/expanded graphite composite phase change material. Appl Energy 91(1):426–431. https://doi.org/10.1016/j.apenergy.2011.10.014

Zhang Z, Shi G, Wang S, Fang X, Liu X (2013) Thermal energy storage cement mortar containing n-octadecane/expanded graphite composite phase change material. Renew Energy 50:670–675. https://doi.org/10.1016/j.renene.2012.08.024

Zhao J, Guo Y, Feng F, Tong Q, Qv W, Wang H (2011) Microstructure and thermal properties of a paraffin/expanded graphite phase-change composite for thermal storage. Renew Energy 36(5):1339–1342. https://doi.org/10.1016/j.renene.2010.11.028

Zhenyu S, Zhanqiang L, Hao S, Xianzhi Z (2016) Prediction of contact angle for hydrophobic surface fabricated with micro-machining based on minimum Gibbs free energy. Appl Surf Sci 364:597–603. https://doi.org/10.1016/j.apsusc.2015.12.199

Zhong Y, Li S, Wei X, Liu Z, Guo Q, Shi J, Liu L (2010) Heat transfer enhancement of paraffin wax using compressed expanded natural graphite for thermal energy storage. Carbon 48(1):300–304. https://doi.org/10.1016/j.carbon.2009.09.033

Permissions

The contributors of this book come from diverse backgrounds, making this book a truly international effort. This book will bring forth new frontiers with its revolutionizing research information and detailed analysis of the nascent developments around the world.

We would like to thank all the contributing authors for lending their expertise to make the book truly unique. They have played a crucial role in the development of this book. Without their invaluable contributions this book wouldn't have been possible. They have made vital efforts to compile up to date information on the varied aspects of this subject to make this book a valuable addition to the collection of many professionals and students.

This book was conceptualized with the vision of imparting up-to-date information and advanced data in this field. To ensure the same, a matchless editorial board was set up. Every individual on the board went through rigorous rounds of assessment to prove their worth. After which they invested a large part of their time researching and compiling the most relevant data for our readers.

The editorial board has been involved in producing this book since its inception. They have spent rigorous hours researching and exploring the diverse topics which have resulted in the successful publishing of this book. They have passed on their knowledge of decades through this book. To expedite this challenging task, the publisher supported the team at every step. A small team of assistant editors was also appointed to further simplify the editing procedure and attain best results for the readers.

Apart from the editorial board, the designing team has also invested a significant amount of their time in understanding the subject and creating the most relevant covers. They scrutinized every image to scout for the most suitable representation of the subject and create an appropriate cover for the book.

The publishing team has been an ardent support to the editorial, designing and production team. Their endless efforts to recruit the best for this project, has resulted in the accomplishment of this book. They are a veteran in the field of academics and their pool of knowledge is as vast as their experience in printing. Their expertise and guidance has proved useful at every step. Their uncompromising quality standards have made this book an exceptional effort. Their encouragement from time to time has been an inspiration for everyone.

The publisher and the editorial board hope that this book will prove to be a valuable piece of knowledge for researchers, students, practitioners and scholars across the globe.

List of Contributors

Abraham Figueiras Abdala and Eleazar Máximo Escamilla Silva
Departamento de Ingeniería Química, Instituto Tecnológico de Celaya, Av. Tecnológico y A.G. Cubas s/n, 38010 Celaya, Gto, Mexico

Alfonso Pérez Gallardo
Facultad de Química, Universidad Autónoma de Querétaro, Cerro de las Campanas s/n, 76010 Santiago de Querétaro, Querétaro de Arteaga, Mexico

Lorenzo Guevara Olvera
Departamento de ingeniería Bioquímica, Instituto Tecnológico de Celaya, Av. Tecnológico y A.G. Cubas s/n, 38010 Celaya, Gto, Mexico

Ning Li
Laboratory of Applied Biocatalysis, School of Food Sciences and Engineering, South China University of Technology, Guangzhou 510640, China

Min-Hua Zong, Wen-Yong Lou and Pei Xu
State Key Laboratory of Pulp and Paper Engineering, South China University of Technology, Guangzhou 510640, China

Gao-Wei Zheng
State Key Laboratory of Bioreactor Engineering, East China University of Science and Technology, Shanghai 200237, China

Daobing Yu, Yanke Shi and Xin Zhang
College of Forestry and Biotechnology, Zhejiang Agriculture and Forestry University, Lin'an 311300, Zhejiang, People's Republic of China

Qun Wang and Yuhua Zhao
College of Life Sciences, Zhejiang University, Hangzhou 310058, Zhejiang, People's Republic of China

Hongchang Yu and Baowu Zhu
School of Life and Environmental Sciences, Guilin University of Electronic Technology, Guilin 541004, People's Republic of China

Yulian Zhan
School of Life and Environmental Sciences, Guilin University of Electronic Technology, Guilin 541004, People's Republic of China
State Key Laboratory of Bioreactor Engineering, East China University of Science and Technology, Shanghai 200237, People's Republic of China

Ying Jiang, Yue-Peng Shang, Hao Li, Chao Zhang, Jiang Pan, Yun-Peng Bai and Chun-Xiu Li
State Key Laboratory of Bioreactor Engineering, East China University of Science and Technology, Shanghai 200237, People's Republic of China

Jian-He Xu
State Key Laboratory of Bioreactor Engineering, East China University of Science and Technology, Shanghai 200237, People's Republic of China
Shanghai Collaborative Innovation Center for Biomanufacturing Technology, East China University of Science and Technology, Shanghai 200237, People's Republic of China

Wen-Yong Lou, Yu-Han Cui and Pei Xu
Lab of Applied Biocatalysis, School of Food Science and Engineering, South China University of Technology, Guangzhou 510640, Guangdong, China

Jian Zhou
School of Chemistry and Chemical Engineering, South China University of Technology, Guangzhou 510640, Guangdong, China

Ping Wei and Min-Hua Zong
Lab of Applied Biocatalysis, School of Food Science and Engineering, South China University of Technology, Guangzhou 510640, Guangdong, China
School of Chemistry and Chemical Engineering, South China University of Technology, Guangzhou 510640, Guangdong, China

Ankita Juneja, Deepak Kumar and Vijay Singh
Ankita Juneja and Deepak Kumar are first authors Department of Agricultural and Biological Engineering, University of Illinois at Urbana-Champaign, Urbana, IL 61801, USA

Yu Wang, Jiao Liu, Qinggang Li, Zhidan Zhang, Ping Zheng and Jibin Sun
Key Laboratory of Systems Microbial Biotechnology, Chinese Academy of Sciences, Tianjin 300308, People's Republic of China
Tianjin Institute of Industrial Biotechnology, Chinese Academy of Sciences, Tianjin 300308, People's Republic of China

Fuping Lu
College of Biotechnology, Tianjin University of Science and Technology, Tianjin 300222, People's Republic of China

Xiaolu Wang
Key Laboratory of Systems Microbial Biotechnology, Chinese Academy of Sciences, Tianjin 300308, People's Republic of China
Tianjin Institute of Industrial Biotechnology, Chinese Academy of Sciences, Tianjin 300308, People's Republic of China
College of Biotechnology, Tianjin University of Science and Technology, Tianjin 300222, People's Republic of China

C. Su, M. Chakankar and H. Hocheng
Department of Power Mechanical Engineering, National Tsing Hua University, No. 101, Sec. 2, Kuang Fu Rd, 30013 Hsinchu, Taiwan ROC

U. Jadhav
Department of Microbiology, Savitribai Phule Pune University, Pune 411007, India

Yang Sun, Song Ding, He Huang and Yi Hu
School of Pharmaceutical Sciences, State Key Laboratory of Material-Oriented Chemical Engineering of Nanjing Tech University, Nanjing 211816, China

He Zhu, Chenba Zhu, Longyan Cheng and Zhanyou Chi
School of Life Science and Biotechnology, Dalian University of Technology, Dalian 116024, China

Kai Li and Chen-Guang Liu
State Key Laboratory of Microbial Metabolism, School of Life Sciences and Biotechnology, Shanghai Jiao Tong University, Shanghai 200240, China

Li-Yang Liu
Department of Wood Science, University of British Columbia, Vancouver V6T1Z4, Canada

School of Life Science and Biotechnology, Dalian University of Technology, Dalian, Liaoning 116023, China. 4 Department of Bioinformatics and Biotechnology, Government College University Faisalabad, Faisalabad 38000, Pakistan

Jin-Cheng Qin
State Key Laboratory of Microbial Metabolism, School of Life Sciences and Biotechnology, Shanghai Jiao Tong University, Shanghai 200240, China
School of Life Science and Biotechnology, Dalian University of Technology, Dalian, Liaoning 116023, China

Muhammad Aamer Mehmood
State Key Laboratory of Microbial Metabolism, School of Life Sciences and Biotechnology, Shanghai Jiao Tong University, Shanghai 200240, China
Department of Bioinformatics and Biotechnology, Government College University Faisalabad, Faisalabad 38000, Pakistan

Songsong Wei, Xingxing Jian and Jun Chen
State Key Laboratory of Bioreactor Engineering, East China University of Science and Technology, 130 Meilong Road, Shanghai 200237, People's Republic of China

Cheng Zhang
State Key Laboratory of Bioreactor Engineering, East China University of Science and Technology, 130 Meilong Road, Shanghai 200237, People's Republic of China
Science for Life Laboratory, KTH-Royal Institute of Technology, Stockholm SE-171 21, Sweden

Qiang Hua
State Key Laboratory of Bioreactor Engineering, East China University of Science and Technology, 130 Meilong Road, Shanghai 200237, People's Republic of China
Shanghai Collaborative Innovation Center for Biomanufacturing Technology, 130 Meilong Road, Shanghai 200237, China

Jhe-Wei Wu and I-Son Ng
Department of Chemical Engineering, National Cheng Kung University, Tainan 70101, Taiwan

Qiao Yu
East China University of Science and Technology, Shanghai, China

Yong Wang
Key Laboratory of Synthetic Biology, Institute of Plant Physiology and Ecology, Shanghai Institutes for Biological Sciences, Chinese Academy of Sciences, Shanghai, China

Shengyun Zhao
Fujian Key Laboratory of Eco-Industrial Green Technology, Wuyi University, Wuyishan, China

Yuhong Ren
State Key Laboratory of Bioreactor Engineering, New World Institute of Biotechnology, East China University of Science and Technology, Shanghai, China

I-Son Ng
Department of Chemical Engineering, National Cheng Kung University, Tainan 70101, Taiwan

Chung-Chuan Hsueh and Bor-Yann Chen
Department of Chemical and Materials Engineering, National I-Lan University, I-Lan 26047, Taiwan

Lin-Hao Lai and Wei-Ming Gu
Department of Food Science, Foshan University (Northern Campus), Nanhai, Foshan 528231, China

Jun Xiong and Hang Yin
Laboratory of Applied Biocatalysis, School of Food Science and Engineering, South China University of Technology, No.381 Wushan Road, Guangzhou 510640, China

Shi-Lin Cao
Department of Food Science, Foshan University (Northern Campus), Nanhai, Foshan 528231, China. Laboratory of Applied Biocatalysis, School of Food Science and Engineering, South China University of Technology, No.381 Wushan Road, Guangzhou 510640, China

Xue-Hui Li and Jian Zhou
School of Chemistry and Chemical Engineering, South China University of Technology, No. 381 Wushan Road, Guangzhou 510640, China

Min-Hua Zong
Laboratory of Applied Biocatalysis, School of Food Science and Engineering, South China University of Technology, No.381 Wushan Road, Guangzhou 510640, China

School of Chemistry and Chemical Engineering, South China University of Technology, No. 381 Wushan Road, Guangzhou 510640, China

Wen-Yong Lou and Pei Xu
Laboratory of Applied Biocatalysis, School of Food Science and Engineering, South China University of Technology, No.381 Wushan Road, Guangzhou 510640, China
Guangdong Province Key Laboratory for Green Processing of Natural Products and Product Safety, South China University of Technology, No. 381 Wushan Road, Guangzhou 510640, China

Hong Xu
Laboratory of Applied Biocatalysis, School of Food Science and Engineering, South China University of Technology, No.381 Wushan Road, Guangzhou 510640, China
School of Chemistry and Chemical Engineering, South China University of Technology, No. 381 Wushan Road, Guangzhou 510640, China
Guangdong Province Key Laboratory for Green Processing of Natural Products and Product Safety, South China University of Technology, No. 381 Wushan Road, Guangzhou 510640, China

Yong-Zheng Ma
School of Marine Science and Technology, Tianjin University, 92 Weijin Road, Tianjin 300072, China

Yusuke Asai, Atsushi Kouzuma and Kazuya Watanabe
School of Life Science, Tokyo University of Pharmacy and Life Sciences, Tokyo 192-0392, Japan

Morio Miyahara
School of Life Science, Tokyo University of Pharmacy and Life Sciences, Tokyo 192-0392, Japan. Meidensha Corporation, Shinagawa, Tokyo 141-8616, Japan

Sarita Sachdeva
Department of Biotechnology, Manav Rachna International University, Sector 43, Faridabad 121004, India

Meena Kapahi
Department of Biotechnology, Manav Rachna International University, Sector 43, Faridabad 121004, India
Manav Rachna University, Sector 43, Faridabad 121004, India

Haiyue Yang, Daxin Liang, Feng Liu, Wenbo Zhang, Xin Di and Chengyu Wang
Key Laboratory of Bio-based Material Science and Technology, Ministry of Education, Northeast Forestry University, Harbin 150040, People's Republic of China

Yazhou Wang
School of Material Science and Engineering, Sun Yat-Sen University, Guangzhou 510275, People's Republic of China

Zhuangchao Liu
Guangdong Yihua Timber Industry Co., Ltd, Guangzhou 510000, People's Republic of China

Shih-Hsin Ho
State Key Laboratory of Urban Water Resource and Environment, School of Municipal and Environmental Engineering, Harbin Institute of Technology, Harbin 150090, People's Republic of China

Wei-Hsin Chen
Department of Aeronautics and Astronautics, National Cheng Kung University, Tainan 701, Taiwan

Index

www.ingramcontent.com/pod-product-compliance
Lightning Source LLC
Chambersburg PA
CBHW082024190326
41458CB00010B/3269